Terotechnology XII

12th International Conference on Terotechnology,
20-21 October 2021, Kielce, Poland

Editor
Norbert Radek

Peer review statement

All papers published in this volume of "Materials Research Proceedings" have been peer reviewed. The process of peer review was initiated and overseen by the above proceedings editors. All reviews were conducted by expert referees in accordance to Materials Research Forum LLC high standards.

Copyright © 2022 by authors

Published under License by **Materials Research Forum LLC**
Millersville, PA 17551, USA

Published as part of the proceedings series
Materials Research Proceedings
Volume 24 (2022)

ISSN 2474-3941 (Print)
ISSN 2474-395X (Online)

ISBN 978-1-64490-204-2 (Print)
ISBN 978-1-64490-205-9 (eBook)

This book contains information obtained from authentic and highly regarded sources. Reasonable efforts have been made to publish reliable data and information, but the author and publisher cannot assume responsibility for the validity of all materials or the consequences of their use. The authors and publishers have attempted to trace the copyright holders of all material reproduced in this publication and apologize to copyright holders if permission to publish in this form has not been obtained. If any copyright material has not been acknowledged please write and let us know so we may rectify in any future reprint.

Distributed worldwide by

Materials Research Forum LLC
105 Springdale Lane
Millersville, PA 17551
USA
http://www.mrforum.com

Manufactured in the United States of America
10 9 8 7 6 5 4 3 2 1

Table of Contents

Preface

Terotechnology is the technology of installation, commissioning, maintenance, replacement, and removal of plant machinery and equipment, feedback on operation and design, and related subjects and practices.

It has been twenty years since the first Conference on Terotechnology was held. The conference is still a venture of Kielce University of Technology and the Polish Maintenance Society. The first conference was born in 2001, and subsequent meetings were held from 2005 to 2009 in the annual cycle. In 2011, a biannual conference cycle was established.

Unfortunately, due to the COVID-19 pandemic, we were forced to cancel the traditional form of the meeting. However, not wanting to deprive the participants of the opportunity to inform their colleagues about the work results, we decided to enable the publication of conference articles.

The 12th International Conference on Terotechnology was held virtually at the Kielce University of Technology, Kielce, POLAND, on 20-21 September 2021, situated halfway between Kraków and Warsaw. We hope that in the future, the pandemic will only be a historical footnote and that we will be able to return to the formula of a face-to-face meeting.

The selected papers from the 12th Conference have been reviewed by at least two experts and prepared for this volume of *Materials Research Proceedings*. The articles were chosen based on their quality and relevance to the conference. The volume represents the recent advances in materials, technologies, and methods.

The present Conference coincides with the celebration of the 25th anniversary of the Centre for Laser Technologies of Metals of Kielce University of Technology, established in 1996.

The Organizing Committee is grateful to the deans and workers of the Centre for Laser Technologies of Metals of the Kielce University of Technology, Department of Applied Computer Science of the Cracow University of Technology, who helped with the organization of the 12th Conference and the edition of this volume. The Committee would also like to express their gratitude to all who have updated and reviewed the papers submitted to the conference, the scientific secretary for her editing work, and Materials Research Forum LLC for producing the volume.

Conference Organizers

Kielce University of Technology,
Polish Maintenance Society

Scientific Committee

Organizing Committee

Bogdan Antoszewski - chairman
Mieczysław Scendo - co-chairman
Augustin Sladek - co-chairman
Ivo Hlavaty - co-chairman
Norbert Radek - scientific secretary
Dariusz Gontarski - organizing secretary
Jacek Pietraszek
Renata Dwornicka
Agnieszka Szczotok
Artur Kalinowski
Piotr Sęk
Piotr Kurp
Izabela Pliszka
Szymon Tofil
Hubert Danielewski
Marek Michalski
Ewa Ziach

Terotechnology XII
Materials Research Proceedings **24** (2022) 1-8

Materials Research Forum LLC
https://doi.org/10.21741/9781644902059-1

The Influence of Oil Contamination on Flow Control Valve Operation

DOMAGALA Mariusz[1,a *], MOMENI Hassan[2,b]
and FABIS-DOMAGALA Joanna[1,c]

[1] Cracow University of Technology, Faculty of Mechanical Engineering, Al. Jana Pawla II 37
31-841 Cracow, Poland

[2] Department of Mechanical and Marine Engineering, Western Norway University of Applied
Sciences, N5020 Bergen, Norway

[a]mariusz.domagala@pk.edu.pl, [b] Hassan.Momeni@hvl.no, [c] joanna.fabis-domagala@pk.edu

Keywords: Flow Control Valve, Solid Particle Simulation, CFD Simulation

Abstract. This study presents CFD simulations of the flow of contaminated oil inside a control valve to investigate the influence of solid contaminants on the valve operational features. The Euler-Lagrange approach has been used to simulate the flow of oil contaminated with solid particles. The CFD simulations allowed determining the effect of solid contamination on the value of hydrodynamic force and a pressure drop for different contamination levels and valve opening.

Introduction

Fluid power systems are widely used in industrial applications due to their high power density, dynamic load dissipation possibilities and flexibility unattainable for other systems. They use pressure energy created by pumps, which further transfers accumulated energy to receivers. Fluid power systems have been used for decades and use hydro-mechanical components, which are often complemented with electronic control systems recently, making them still competitive among other drive systems. The complexity level of fluid power systems depends on realized tasks and may include from a few to hundreds of components. Despite the undoubted progress in fluid power systems reliability, some problems are still not solved and are objects of many recent studies. One of the main problems of these systems is the contamination of working fluid, which is recognized as a leading cause of failures [1, 2]. The recent efforts in preventing hydraulic systems against the negative influence of contaminants is mainly focused on monitoring contamination levels [3, 4]. However, the practice shows that fully prevention of hydraulic fluid against contamination is nearly impossible even by using sophisticated filtration systems. Contaminants might have different natures and sources. They can occur due to wear or ingression (solid particles) or effects of chemical agents (air and water). One of the primary contamination sources is the one generated by the hydraulic system during normal operation due to wear, erosion or corrosion. The component recognized as the primary source of solid (metal) contaminants are pumps, where relative motion and high structural load may intensify factors responsible for wear, erosion, cavitation or fatigue. In particular, the vane pumps are found to be the significant sources of oil contamination, and industrial standards use them as a reference component for testing oil contamination levels [5-7].

During operation, the solid contaminants gather momentum from working fluid and may affect surfaces, causing their erosion, what may lead to failure, malfunction or reduce the lifetime of critical components of the system. Such components are valves, which malfunction or failures may lead to severe damages and catastrophic events. The wear caused by solid contaminants on valves can be evaluated by experimental tests or numerical simulations in which computational fluid dynamics (CFD) tools are implemented [8, 9]. As the contamination of working fluid is inevitable, the industrial standards [10] defines the cleanliness class of fluid as a rate of particle size in specified oil capacity.

The failure or wear caused by solid contaminants is a long term process, and symptoms can be relatively easy to identify. However, their influence on the operational characteristics of crucial components of fluid power systems is unknown. This study attempts to evaluate oil contamination on flow control valve functional features achieved by implementing the Euler-Lagrange flow simulation approach.

The obtained results may be useful both for researchers using similar modeling methods [11-14] and industries with machines and devices equipped with valve hydraulics, such as biotechnological engineering [15], wastewater treatment [16, 17], and internal combustion engine accessories [18]. It should also be a guideline for the use of special coatings [19, 20], including modified ones [21, 22] in similar situations, to avoid disfunction. The schema of analysis itself can be inspiring both for similar modeling techniques [23-25] and for production management systems [26] and quality management systems [27-30].

Methodology

The flow simulation of fluid with solid particles has been performed by Euler-Lagrange multiphase flow in which hydraulic oil is considered as continuous Euler phase while particles as Lagrange phase. The CFD method employs RANS (Reynolds Averaged Navier-Stokes) equations which define scalars as mean values and fluctuations over this value. RANS defines fluid velocity as:

$$u_i = \overline{u}_i + u_i' \tag{1}$$

where: $\overline{u}_i$ – is the mean velocity, u_i' – is the fluctuating velocity, i – stands for velocity component.

Thus the RANS equations have the following form:

$$\frac{\partial \rho}{\partial t} + \frac{\partial \rho}{\partial x_i}(\rho u_i) = 0 \tag{2}$$

$$\frac{\partial \rho}{\partial t}(\rho u_i) + \frac{\partial \rho}{\partial x_j}(\rho u_i u_j) = -\frac{\partial p}{\partial x_i} + \frac{\partial}{\partial x_j}\left[\mu\left(\frac{\partial u_i}{\partial x_j} + \frac{\partial u_j}{\partial x_i} - \frac{2}{3}\delta_{ij}\frac{\partial u_i}{\partial x_i}\right)\right] + \frac{\partial}{\partial x_j}\left(-\rho\overline{u_i u_j}\right) \tag{3}$$

where: ρ – is the fluid density, u – is the fluid velocity, p – is pressure, μ – is the dynamic viscosity, δ – is the Kronecker function.

The term $-\rho\overline{u_i u_j}$ represents the effects of turbulence (Reynolds stress) and makes the set of Eq. (2) and Eq. (3) to be open. The RANS equations can be closed by the use of the Boussinesq hypothesis:

$$-\rho\overline{u_i u_j} = \mu_t\left(\frac{\partial u_i}{\partial x_j} + \frac{\partial u_j}{\partial x_i}\right) - \frac{2}{3}\delta_{ij}\left(\rho k + \mu_t\frac{\partial u_k}{\partial x_k}\right) \tag{4}$$

and employing the turbulence model.

Terotechnology XII   Materials Research Forum LLC
Materials Research Proceedings **24** (2022) 1-8   https://doi.org/10.21741/9781644902059-1

Solid particles are represented as discreet phases and are tracked into the fluid domain during fluid flow. The equation of motion for a single particle according to Basset, Boussinesq and Oseen is as follows:

$$m_p \frac{dU_p}{dt} = F_D + F_B + F_R + F_{VM} + F_P + F_{BA} \tag{5}$$

where: F_D – drag force, F_B – buoyancy force, F_R – Coriolis force, F_{VM} is inertia force of fluid occupied by a particle (Virtual Mass), F_P – pressure force, F_{BA} – Basset force

The following equation expresses the inertia force of fluid occupied by a particle (Virtual Mass):

$$F_{VM} = \frac{C_{VM}}{2} m_F \left(\frac{dU_F}{dt} - \frac{dU_P}{dt} \right) \tag{6}$$

Combining Eq. 5 and Eq. 6, we have:

$$\frac{dU_p}{dt} = \left(\frac{1}{m_p + \frac{C_{VM}}{2} m_F} \right) (F_D + F_B + F_R + F'_{VM} + F_P + F_{BA}) \tag{7}$$

where:

$$F'_{VM} = \frac{C_{VM}}{2} m_F (U_F \nabla U_F) \tag{8}$$

$m_P = \frac{\pi}{6} d_P^3 \rho_P$ is particle mass, $m_F = \frac{\pi}{6} d_P^3 \rho_F$ is fluid mass, d_p is the particle diameter, ρ_P, ρ_F is the density of particle and fluid, respectively.
Introducing R_{VM} as:

$$R_{VM} = \frac{m_P}{m_P + \frac{C_{VM}}{2} m_F} = \frac{\rho_P}{\rho_P + \frac{C_{VM}}{2} \rho_F} \tag{9}$$

we will get Eq. 5 in the following form:

$$\frac{dU_p}{dt} = \left(\frac{R_{VM}}{m_p} \right) (F_D + F_B + F_R + F'_{VM} + F_P + F_{BA}) \tag{10}$$

Case Study
The valve presented in Fig. 1 is a proportional solenoid controlled flow control valve whose primary purpose is to maintain a constant flow rate independently on a pressure difference between the valve supply and receiver lines.
The valve consists of two spools (2,3) assembled in the body (1) inside the sleeve (4). The value of flow rate is controlled by a solenoid, which directly acts on the spool (3). The position of the second spool (2) depends on the pressure at the entry flow ducts (valve supply line) and outlet flow ducts (receiver line) and is determined by the force balance on the spool (2). Hydrodynamic reactions are among those forces that play the leading role in valve proper operation.

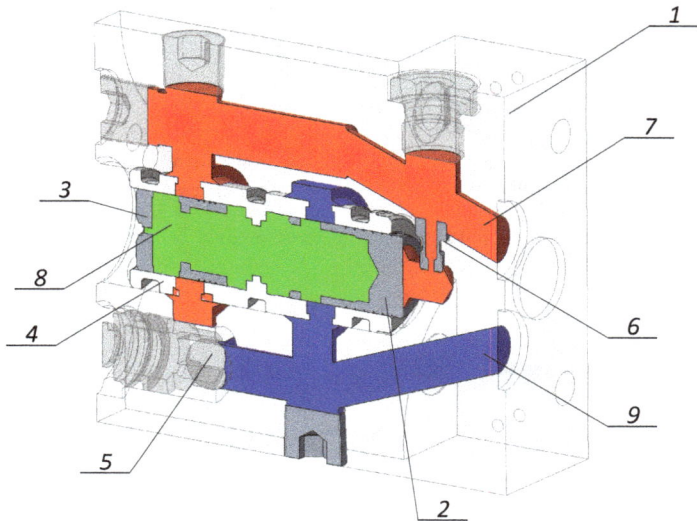

Fig. 1. *Flow control valve: 1 – valve body, 2,3 – spools, 4 – sleeve, 5 – non return valve, 6 – nozzle, 7 – entry flow ducts, 8 – flow duct inside spools, 9 – outlet flow duct.*

CFD Simulation

The main purpose of CFD simulation was to evaluate flow forces and pressure drops during the flow of contaminated oil inside the valve. The simulation was conducted with Euler-Lagrange approach in Ansys CFX code for both spools fixed position and steady state conditions. The following assumptions have been used:

Continous phase:

- fluid (hydraulic oil) has a constant properties: density 880 [kg/m^3], viscosity υ=40 [mm^2/s];
- flow is turbulent: the k-ϖ turbulence has been applied;
- heat transfer is neglected;
- cell type: hybrid, tetrahedral with prism.

Lagrangian phase:

- particles have spheres shape with a diameter of 1 [μm] and constant properties (steel), with 2.5% and 5% of fluid mass flow rate;
- interaction between particles and fluid is fully coupled;
- particles are uniformly injected over the valve inlet.

CFD model is presented in Fig. 2. The grid in gaps in which fluids flow has been refined to increase simulation accuracy. Due to the symmetry of flow ducts, half of the geometry was used, and symmetry boundary condition has been applied.

Terotechnology XII   Materials Research Forum LLC
Materials Research Proceedings **24** (2022) 1-8   https://doi.org/10.21741/9781644902059-1

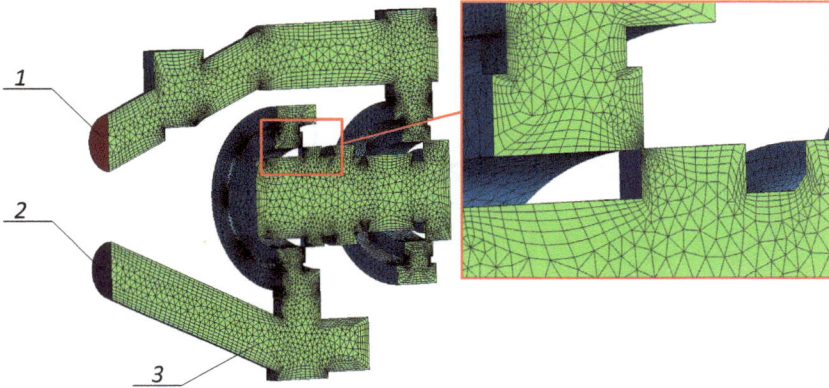

Fig. 2. *CFD model: 1 – inlet, 2 – outlet,3-symmetry plane.*

Numerical simulation has been performed for constant mass flow rate with a different contribution of solid contaminants. Two oil contamination levels have been checked: 4000 and 8000 particles per 1 [ml], which gives approximately 2.5% and 5% mass flow rates, respectively. It has to be mentioned that particle size used in the simulation, if far below the minimal value, is recognized as a contaminant by the relevant standard [10] and equals 4 [μm]. Results of CFD simulation have been presented in Fig. 3 as a fluid velocity on the symmetry plane and solid particle tracks and velocity.

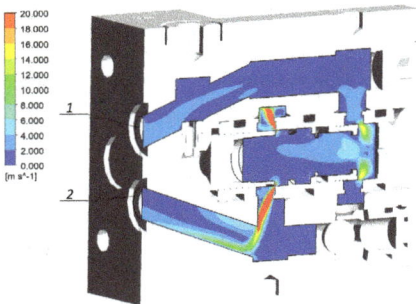

Fig. 3a. *Fluid velocity at normalized opening 0.1 in [m/s], 1 – inlet, 2 – outlet*

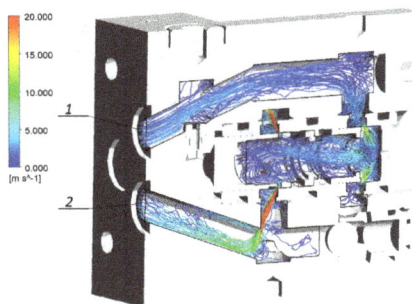

Fig. 3b. *Solid particle tracks at normalized opening 0.1 in [m/s], 1 – inlet, 2 – outlet*

The simulations have been performed for the constant flow rate (15 dm^3/min) and fixed position of spools (2,3). Figure 4 shows the value of the hydrodynamic force (flow force) on the spool (2) and total pressure drop for different spool (2) positions (valve opening).

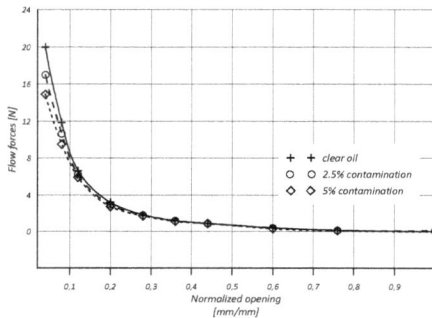

Fig. 4a. *Flow forces on the spool (3) for constant flow rate of 15 [dm³/min].*

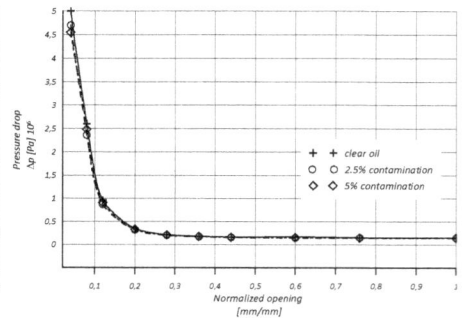

Fig. 4b. *Pressure drop for a constant flow rate of 15 [dm³/min].*

The above results indicate that solid contaminants affect the hydrodynamic force for small valve openings (up to 0.15) while differences are almost undistinguishable for larger openings. Less than minor pressure drop differences were recorded even for small valve openings.

Summary

Protection of hydraulic oil in fluid power systems against contamination is almost impossible in a real world application. Therefore, this study attempts to evaluate the influence of the level of solid contamination on the flow control valve feature, which is critical for valve operation. The proposed method uses computational fluid dynamics (CFD) tools and Euler-Lagrange approach to simulate contaminated fluid flow inside the flow control valve. The simulations have been conducted for two different contamination levels. The working fluid included metal particles. Obtained results show that the presented contamination levels (2.5% and 5% of fluid mass flow rate) affect flow forces while pressure drop changes are almost negligible. However, it has to be added that presented results correspond to the idealized situation in which metal particles have a uniform shape and size. In real world applications, the particle size and distribution may strongly depend on system complexity, implemented components or even working conditions. Despite this, the CFD tools seem to be effective for investigating solid particles' influence on fluid power components.

References

[1] S. Li, Z. Yang, H. Tian, C. Chen, Y. Zhu, S. Lu. Failure Analysis for Hydraulic System of Heavy-Duty Machine Tool with Incomplete Failure Data. Appl. Sci. 14 (2021) art. 1249. https://doi.org/10.3390/app11031249

[2] J. Fabis-Domagala, M. Domagala, H. Momeni. A Concept of Risk Prioritization in FMEA Analysis for Fluid Power Systems. Energies 14 (2021) art. 6482. https://doi.org/10.3390/en14206482

[3] L. Zeng, Z. Yu, H. Zhang, X. Zhang, H. Chen. A high sensitive multi-parameter micro sensor for the detection of multi-contamination in hydraulic oil, Sensors and Actuators A: Physical 282 (2018) 197-205. https://doi.org/10.1016/j.sna.2018.09.023.

[4] F. Ng, J.A. Harding, J. Glass. Improving hydraulic excavator performance through in line hydraulic oil contamination monitoring, Mech. Systems and Signal Process. 83 (2017) 176-193. https://doi.org/10.1016/j.ymssp.2016.06.006.

[5] ISO 20763:2004. Petroleum and Related Products—Determination of Anti-Wear Properties of Hydraulic Fluids—Vane Pump Method; International Organization for Standardization: Geneva, Switzerland, 2004

[6] DIN 51389-1. Determination of Lubricants; Mechanical Testing of Hydraulic Fluids in the Vane-Cell-Pump; General Working Principles; DIN Deutsches Institut für Normung e. V.: Berlin, Germany, 1982

[7] ASTM D7043-17. Standard Test Method for Indicating Wear Characteristics of Non-Petroleum and Petroleum Hydraulic Fluids in a Constant Volume Vane Pump; ASTM International: West Conshohocken, PA, USA, 2017

[8] M. Domagala, H. Momeni, J. Fabis-Domagala, G. Filo, M. Krawczyk, J. Rajda. Simulation of Particle Erosion in a Hydraulic Valve, Materials Research Proceedings 5 (2018) 17-24. https://dx.doi.org/10.21741/9781945291814-4

[9] Y. Yaobao, Y. Jiayang, G. Shengrong. Numerical study of solid particle erosion in hydraulic spool valves, Wear 392-393 (2017) 147-189. https://doi.org/10.1016/j.wear.2017.09.021

[10] ISO 4406: Hydraulic fluid power — Fluids — Method for coding the level of contamination by solid particles, International Organization for Standardization: Geneva, Switzerland, 2021

[11] M. Zmindak, L. Radziszewski, Z. Pelagic, M. Falat. FEM/BEM techniques for modelling of local fields in contact mechanics, Communications - Scientific Letters of the University of Zilina 17 (2015) 37-46.

[12] T. Lipiński. Corrosion resistance of 1.4362 steel in boiling 65% nitric acid, Manufacturing Technology 16 (2016) 1004-1009.

[13] Ł.J. Orman Ł.J., N. Radek, J. Pietraszek, M. Szczepaniak. Analysis of enhanced pool boiling heat transfer on laser-textured surfaces. Energies 13 (2020) art. 2700. https://doi.org/10.3390/en13112700

[14] E. Lisowski, J. Rajda, G. Filo, P. Lempa. Flow Analysis of a 2URED6C Cartridge Valve, Lecture Notes in Mechanical Engineering 24 (2021) 40-49. https://doi.org/10.1007/978-3-030-59509-8_4

[15] E. Skrzypczak-Pietraszek. Phytochemistry and biotechnology approaches of the genus Exacum. In: The Gentianaceae - Volume 2: Biotechnology and Applications, 2015, 383-401. https://doi.org/10.1007/978-3-642-54102-5_16

[16] M. Zenkiewicz, T. Zuk, J. Pietraszek, P. Rytlewski, K. Moraczewski, M. Stepczyńska. Electrostatic separation of binary mixtures of some biodegradable polymers and poly(vinyl chloride) or poly(ethylene terephthalate), Polimery/Polymers 61 (2016) 835-843. https://doi.org/10.14314/polimery.2016.835

[17] E. Radzyminska-Lenarcik, R. Ulewicz, M. Ulewicz. Zinc recovery from model and waste solutions using polymer inclusion membranes (PIMs) with 1-octyl-4-methylimidazole, Desalination and Water Treatment 102 (2018) 211-219. https://doi.org/10.5004/dwt.2018.21826

[18] M. Kekez, L. Radziszewski, A. Sapietova. Fuel type recognition by classifiers developed with computational intelligence methods using combustion pressure data and the crankshaft angle at which heat release reaches its maximum, Procedia Engineering 136 (2016) 353-358. https://doi.org/10.1016/j.proeng.2016.01.222

[19] A. Dudek, R. Wlodarczyk. Structure and properties of bioceramics layers used for implant coatings, Solid State Phenom. 165 (2010) 31-36. https://doi.org/10.4028/www.scientific.net/SSP.165.31

[20] N. Radek, J. Pietraszek, A. Gadek-Moszczak, Ł.J. Orman, A. Szczotok. The morphology and mechanical properties of ESD coatings before and after laser beam machining, Materials 13 (2020) art. 2331. https://doi.org/10.3390/ma13102331

[21] N. Radek, J. Pietraszek, A. Goroshko. The impact of laser welding parameters on the mechanical properties of the weld, AIP Conf. Proc. 2017 (2018) art.20025. https://doi.org/10.1063/1.5056288

[22] N. Radek, J. Konstanty, J. Pietraszek, Ł.J. Orman, M. Szczepaniak, D. Przestacki. The effect of laser beam processing on the properties of WC-Co coatings deposited on steel. Materials 14 (2021) art. 538. https://doi.org/10.3390/ma14030538

[23] J. Pietraszek, A. Gadek-Moszczak, N. Radek. The estimation of accuracy for the neural network approximation in the case of sintered metal properties. Studies in Computational Intelligence 513 (2014) 125-134. https://doi.org/10.1007/978-3-319-01787-7_12

[24] J. Pietraszek, E. Skrzypczak-Pietraszek. The uncertainty and robustness of the principal component analysis as a tool for the dimensionality reduction. Solid State Phenom. 235 (2015) 1-8. https://doi.org/10.4028/www.scientific.net/SSP.235.1

[25] J. Pietraszek, R. Dwornicka, A. Szczotok. The bootstrap approach to the statistical significance of parameters in the fixed effects model. ECCOMAS 2016 – Proc. 7th European Congress on Computational Methods in Applied Sciences and Engineering 3, 6061-6068. https://doi.org/10.7712/100016.2240.9206

[26] A. Maszke, R. Dwornicka, R. Ulewicz. Problems in the implementation of the lean concept at a steel works – Case study, MATEC Web of Conf. 183 (2018) art.01014. https://doi.org/10.1051/matecconf/201818301014

[27] T. Styrylska, J. Pietraszek. Numerical modeling of non-steady-state temperature-fields with supplementary data. Zeitschrift fur Angewandte Mathematik und Mechanik 72 (1992) T537-T539.

[28] J. Pietraszek. Response surface methodology at irregular grids based on Voronoi scheme with neural network approximator. 6th Int. Conf. on Neural Networks and Soft Computing JUN 11-15, 2002, Springer, 250-255. https://doi.org/10.1007/978-3-7908-1902-1_35

[29] J. Pietraszek, N. Radek, A.V. Goroshko. Challenges for the DOE methodology related to the introduction of Industry 4.0. Production Engineering Archives 26 (2020) 190-194. https://doi.org/10.30657/pea.2020.26.33

[30] D. Siwiec, R. Dwornicka, A. Pacana. Improving the non-destructive test by initiating the quality management techniques on an example of the turbine nozzle outlet, Materials Research Proceedings 17 (2020) 16-22. https://doi.org/10.21741/9781644901038-3

Terotechnology XII
Materials Research Proceedings 24 (2022) 9-14

Materials Research Forum LLC
https://doi.org/10.21741/9781644902059-2

Boiling Heat Transfer Performance of Pin-Fins During Boiling of Water and Ethanol

DZIEDZIC Joanna[1,a] and CEDRO Martyna[2,b] *

[1]Jan Kochanowski University of Kielce, ul. Uniwersytecka 15, 25-406 Kielce, Poland

[2]Kielce University of Technology, al. Tysiaclecia P.P.7, 25-314 Kielce, Poland

[a]s145704@ujk.edu.pl, [b]mcedro@tu.kielce.pl

Keywords: Boiling, Thermal Performance, Water, Ethanol

Abstract. In the article, heat transfer during boiling of two liquids: water and ethanol was described and test results were shown. Two samples of pin-fins were researched – of different heights of the microfins (namely 0.3 and 0.6 mm) made of pure copper. The higher fins performed better than the lower for both the liquids used in the experiments. It must be caused by larger surfaces and better heat exchange conditions during this phase change process. The results can be used in the industry for the production of heat exchangers.

Introduction

Pin–fins provide surface extension and can be thought of as successful heat exchangers that can exchange big amounts of heat during phase-change and not phase-change processes. In the case of boiling any structures that change the morphology of the heaters might cause the increase of heat fluxes as presented by Piasecka et al. [1, 2] regarding e.g. laser texturing used in the flow boiling mode. Kaniowski and Pastuszko [3] presented data on water boiling on the surfaces with microchannels (that created long fins 0.2 - 0.5 mm deep). It was presented that the maximal heat flux was even almost 2.5 times higher in comparison to the sample without such structure. On the other hand, the average diameter of the departing vapor bubbles rose and the departing frequency fell with the increasing heat flux values. Another paper by these authors [4] refers to the application of FC-72 as the boiling agent on such microstructures. Also in this case an improvement over the surface without any modifications was very clear. The heat transfer coefficient rose by over 500% and the values were comparable to those obtained in the case of using nanotubes. The authors also concluded at the end of the paper that the boiling process depends on surface wettability, its roughness, and contact angle. Orman et al. [5] investigated laser treatment for the development of longitudal microfins on horizontal copper substrates. The height of the fins was 0.25 mm and 0.55 mm, their widths 0.5 mm and 1.1 mm, and the groove's widths 0.60 mm and 1.15 mm. The results indicate significant possibilities of increasing the heat fluxes during boiling of water and ethyl alcohol if such microfins are used. In particular, a shift to the lower values of temperature differences was seen. Moreover, a modified correlation has been proposed for such structures based on the model presented in the literature for meshed heaters. Radek et al. [6-8] considered the laser treatment for the development of other types of surfaces with modified morphology for the improvement of tribological parameters.

The application of microstructures can be most favorable for heat pipes as their internal coatings. In this case, elevated heat fluxes can also be achieved as presented by Hrabovsky et al. [9] and Nemec [10]. Such structures might also find applications in ventilation systems and air handling units for heat recovery. Thus, higher thermal comfort for room users could be achieved as discussed by Kolková et al. [11] and Majewski et al. [12].

Terotechnology XII   Materials Research Forum LLC
Materials Research Proceedings **24** (2022) 9-14   https://doi.org/10.21741/9781644902059-2

The present manuscript discusses the potential of pin–fins for heat transfer during the boiling of water and ethanol. The obtained data can also be used for the design applications of such heaters in the industry.

Samples and Experiment
The experiments were done on micropins whose height was 0.3 mm and 0.6 mm, while other parameters (copper as the material, distance between the fins of 0.4 mm) were kept the same for both the specimens. The samples were made with micromachining. Thus, the spaces between the fins had increased roughness, which was not measured due to the fact that the roughness height was still much lower than the microfins. Fig. 1 presents the example photo of the specimen of the pin's height of 0.3 mm.

Fig. 1. *Sample with pin – fins (height 0.3 mm).*

Boiling heat transfer performance was determined with the sample acting like a heater located in the pool of water and ethanol. During the testing temperatures were recorded and, as a consequence, heat flux values could be calculated. The consecutive values of the heat fluxes were determined for the rising heat flux.

Results and Discussion
The tests were done under atmospheric pressure with vapor being returned to the vessel. The data obtained during the boiling of water have been presented in the form of boiling curves. Fig. 2 shows the test results for distilled water for both the specimens.

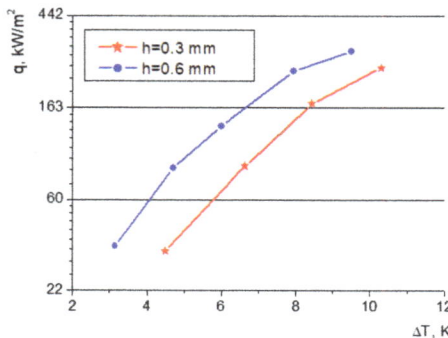

Fig. 2. *Boiling performance of the pin – fins: distilled water*

Terotechnology XII
Materials Research Proceedings **24** (2022) 9-14

Materials Research Forum LLC
https://doi.org/10.21741/9781644902059-2

It is very clear from the figure above that the higher microstructural sample performed better during the whole experiment and exchanged more heat flux (q) at the same temperature differences (ΔT). The curve for the micropin of 0.6 mm height was always higher than the curve for the 0.3 mm height. But at big temperatures, these differences were not so large. Maybe if more heat was provided to the samples from the electric heater those two curves could eventually meet, if film boiling conditions did not start first. Otherwise, the experimental procedure would need to be quickly stopped so that the experimental stand would not be burnt and destroyed (film boiling leads to very high temperatures under the specimens due to the problem of heat removal via the vapor blanket that acts as an insulator).

The same phenomenon of both these curves meeting at the end of the experimental procedure could be observed when the boiling agent was ethyl alcohol (which happened after water was removed from the vessel and the alcohol was provided there). The data for this liquid have been shown in Fig. 3 below.

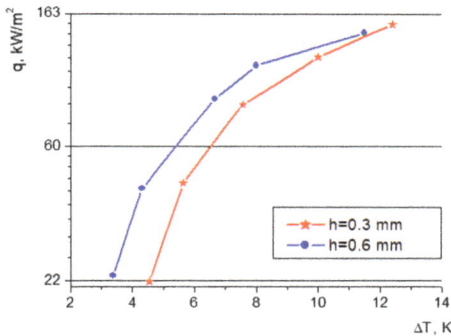

Fig. 3. *Boiling performance of the pin – fins: ethanol*

Here the results for the sample of 0.6 mm height were very close to the curve of 0.3 mm height at the largest heat fluxes (although the heat fluxes for water were higher than for ethanol). The fact that the longer microfins are better can be easily explained just by the bigger surface area that can exchange more heat and provide better cooling as a consequence. But when the heat flux increases and more vapor bubbles are made on the surface, the process undergoes a transformation into a different mode of heat transfer.

The improvement of heat transfer with the longer fins has the same character for both the liquids used in the experiments. Fig. 4 presents the enhancement ratio denoted as (k), which is a heat flux value from longer fins divided by a heat flux value from shorter fins. The changes for water and ethanol are of a similar type as evidenced by the figure below.

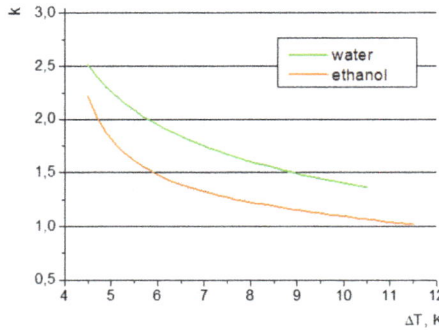

Fig. 4. *Enhancement ratio (k) for both boiling liquids.*

The performance of the samples can be determined with some models. Their development is important for all engineering sciences and various techniques are used in terms of mathematical modeling [13-15]. In the present paper, Smirnov et al. [16] model will be first used, because it was made for the regular geometry microstructures and can be applied here. The comparison of the test results generated with the above-mentioned model and the experimental data for the pin–fins has been presented in Fig. 5 and Fig. 6. It needs to be mentioned that the model required porosity, height, and other parameters and due to the fact that it was developed for the meshes, some modifications had to be made in order to generate the calculation results presented below.

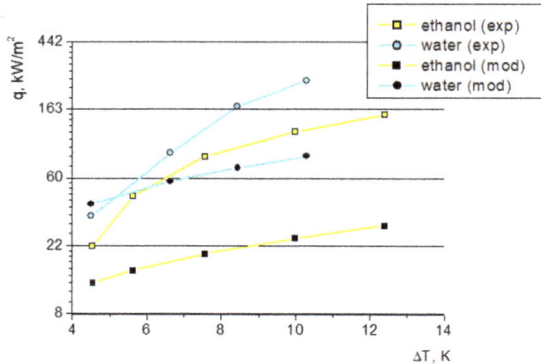

Fig. 5. *Comparison of the experimental results and the selected model (h = 0.3 mm).*

Terotechnology XII

Materials Research Forum LLC

Materials Research Proceedings **24** (2022) 9-14

https://doi.org/10.21741/9781644902059-2

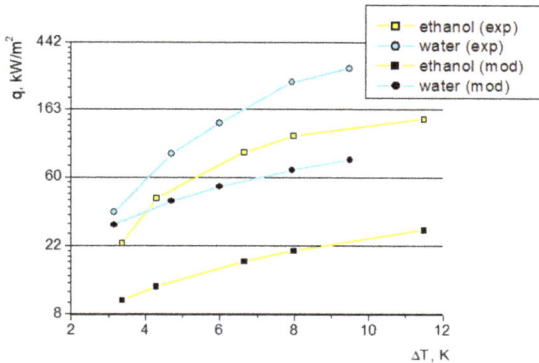

Fig. 6. Comparison of the experimental results and the selected model (h = 0.6 mm).

In the case of both kinds of microstructures, the model was only successful for water and it was only in the low range of temperature differences (up to 3 K). For larger ones (in the case of water) and generally for ethanol, the differences were quite large. Maybe a new model can be developed based on the existing one and modified – that will be able to provide better results for the pool boiling conditions of pin–fins boiling.

Summary and Conclusions

The paper presented the test results of boiling water and ethanol under atmospheric pressure conditions. Two specimens of different heights of the microfins were tested under the pool boiling conditions of heat transfer mode. It was clearly observed that higher pin–fins dissipated more heat than lower ones. The easiest explanation might be that the surface area is larger for convection (boiling) heat transfer mode. Thus, their performance will be better. Both the samples generate similar results at high-temperature differences, which can be explained by the dominant role of bubbles' movement and the creation of vapor film from the generated bubbles. In this case, the morphology of the microstructure might play a smaller role. In future research work, the experiments should be extended to other liquids used more commonly in the industry, for example, ammonia or commercially available fluids typically applied to refrigeration systems, where such structures can be used.

References

[1] M. Piasecka, K. Strąk. Influence of the surface enhancement on the heat transfer in a minichannel, Heat Transfer Engineering 40 (2019) 1162-1175. https://doi.org/10.1080/01457632.2018.1457264

[2] M. Piasecka, K. Strąk, B. Grabas. Vibration-assisted laser surface texturing and electromachining for the intensification of boiling heat transfer in a minichannel, Archives of Metallurgy and Materials 62 (2017) 1983-1990. https://doi.org/10.1515/amm-2017-0296

[3] R. Kaniowski, R. Pastuszko. Pool boiling of water on surfaces with open microchannels, Energies 14 (2021) art. 3062. https://doi.org/10.3390/en14113062

[4] R. Kaniowski, R. Pastuszko. Boiling of FC-72 on surfaces with open copper microchannel, Energies 14 (2021) art. 7283. https://doi.org/10.3390/en14217283

[5] Ł.J. Orman, N. Radek, J. Pietraszek, M. Szczepaniak. Analysis of enhanced pool boiling heat transfer on laser – textured surfaces, Energies 13 (2020) art. 2700. https://doi.org/10.3390/en13112700

[6] N. Radek, J. Pietraszek, A. Gądek-Moszczak, Ł.J. Orman, A. Szczotok. The morphology and mechanical properties of ESD coatings before and after laser beam machining, Materials 13 (2020) art. 2331. https://doi.org/10.3390/ma13102331

[7] N. Radek, J. Konstanty, J. Pietraszek, Ł.J. Orman, M. Szczepaniak, D. Przestacki. The effect of laser beam processing on the properties of WC-Co coatings deposited on steel. Materials 14 (2021) art. 538. https://doi.org/10.3390/ma14030538

[8] A. Szczotok, J. Pietraszek, N. Radek. Metallographic Study and Repeatability Analysis of γ' Phase Precipitates in Cored, Thin-Walled Castings Made from IN713C Superalloy. Archives of Metallurgy and Materials 62 (2017) 595-601. https://doi.org/10.1515/amm-2017-0088

[9] P. Hrabovsky, P. Nemec, M. Malcho. Compare cooling effect of different working fluid in thermosyphon, EFM14 Int. Conf. Experimental Fluid Mechanics 2014, Český Krumlov, Czech Republic, EPJ Web of Conferences (2015) art. 02024. https://doi.org/10.1051/epjconf/20159202024

[10] P. Nemec. Visualization of various working fluids flow regimes in gravity heat pipe, EFM16 Int. Conf. on Experimental Fluid Mechanics 2016, Marienbad, Czech Republic, EPJ Web of Conferences (2017) art. 02079. https://doi.org/10.1051/epjconf/201714302079

[11] Z. Kolková, P. Hrabovský, Z Florková, R. Lenhard R, Analysis of ensuring thermal comfort using an intelligent control system, Int. Conf. The Application of Experimental and Numerical Methods in Fluid Mechanics and Energy 2020, Slovakia, MATEC Web of Conferences (2020) 328, art. 03017. https://doi.org//matecconf/202032803017

[12] G. Majewski, Ł.J. Orman, M. Telejko, N. Radek, J. Pietraszek, A. Dudek. Assessment of thermal comfort in the intelligent buildings in view of providing high quality indoor environment, Energies 13 (2020) art. 1973. https://doi.org/10.3390/en13081973

[13] M. Krechowicz. Comprehensive risk management in horizontal directional drilling projects, Journal of Construction Engineering and Management 146 (2020) art. 04020034. https://doi.org/10.1061/(ASCE)CO.1943-7862.0001809

[14] B. Szeląg, J. Drewnowski, G. Łagód, D. Majerek, E. Dacewicz, F. Fatone. Soft sensor application in identification of the activated sludge bulking considering the technological and economical aspects of smart systems functioning, Sensors 20 (2020) art. 1941. https://doi.org/10.3390/s20071941

[15] I. Nová, K. Fraňa, T. Lipiński. Monitoring of the Interaction of Aluminum Alloy and Sodium Chloride as the Basis for Ecological Production of Expanded Aluminum. Physics of Metals and Metallography 122 (2021) 1288-1300. https://doi.org/10.1134/S0031918X20140124

[16] G. Smirnov, B.A. Afanasiev. Investigation of vaporization in screen wick – capillary structures, Advances in Heat Pipe Technology, Proc. 4th Int. Heat Pipe Conf. (1981) 405-413.

Terotechnology XII
Materials Research Proceedings **24** (2022) 15-20

Materials Research Forum LLC
https://doi.org/10.21741/9781644902059-3

Performance of Laser Treated Heaters in Pool Boiling

ORMAN Łukasz J.[1,a] *, RADEK Norbert[1,b] and DUDEK Agata[2,c]

[1]Kielce University of Technology, al. 1000-lecia P.P.7, 25-314 Kielce, Poland

[2]Czestochowa University of Technology, Al. Armii Krajowej 19, 42-200 Częstochowa, Poland

[a]orman@tu.kielce.pl, [b]norrad@tu.kielce.pl, [c]agata.dudek@pcz.pl

Keywords: Boiling, Heat Exchangers, Laser Treatment

Abstract. The article is focused on the heat transfer phenomenon during the boiling of specimens produced with the laser beam. The study discusses the enhancement of heat flux values, which is possible thanks to the use of surface extension with the laser treatment process. The visualization of the boiling process with the high-speed camera enables us to better understand the phenomenon of boiling. The paper also discusses the application of a selected model to determine heat flux values of the laser-treated sample during the process of distilled water and ethyl alcohol boiling under atmospheric pressure.

Introduction

Treatment of various materials with the laser beam alters the morphology of the surfaces, on which it is applied. The introduction of this technology into the heat transfer area has led to the discovery of significant possibilities that this technique has for improving heat transfer during phase change processes.

In fact, any change in the surface morphology might have a larger or smaller influence on the heat transfer phenomenon during pool boiling as well as flow boiling (where significant enhancement is possible as pointed out by Piasecka et al. [1, 2], who used laser texturing as a tool for surface modifications). In the case of pool boiling Kaniowski and Pastuszko [3] investigated the boiling of water on samples made in the form of microchannels, whose depth was from 0.2 to 0.5 mm. The maximal value of the heat flux was even two and a half times larger than in the case of the sample without such grooves. The authors also measured the average diameters of the departing vapor bubbles. These values increased and the departing frequency decreased as the heat flux rose. In their next paper [4] the authors tested FC – 72 as the boiling fluid. In this case, they also observed that there was an improvement caused by the modification of the surface. The values of the heat transfer coefficient increased by over five times and were considered to be similar to those generated with the nanotubes. The final remarks in the paper state that the boiling phenomenon is influenced by specimens' roughness, contact angle as well as wettability. In the paper [5] laser treatment was applied to produce samples of longitudinal fins on copper circular samples. The depth of the grooves was 0.25 mm and 0.55 mm, while their width was 0.60 mm and 1.15 mm. Such specimens were tested in water and ethanol and proved to be very efficient in dissipating heat. In the paper, a modification of the model was developed for these kinds of microstructures. It needs to be mentioned that the improvement of heat flux values can also be additionally obtained if nanofluids are used [6, 7]. In the case of using both nanofluids and laser treatment of samples, a significant enhancement might be obtained. Although in this case tests should also cover the long-term impact of the nanoparticles on the microstructure on the bottom of the grooves, which offers additional nucleation sites for further boiling enhancement.

Generally, many types of microstructures are produced and used for heat exchanger design, both on the laboratory scale as well as commercially. Some are part of heat pipes, which are very efficient and widely used devices for heat transfer in various applications as pointed out by Hrabovsky et al. [8] and Nemec [9]. They might also be applied in HVAC systems. A paper by Zender-Świercz [10] is focused on the issues of air and its quality considering the outdoor air properties.

The laser treatment is characterized by the production of grooves of regular and repeatable microgeometry. At their bottom, the morphology is altered, which might lead to increased heat fluxes, especially in the range of low-temperature differences. The current paper analyses the phenomenon of pool boiling on such microstructures during water and ethanol nucleate boiling conditions under ambient pressure.

Material and Method

Laser treatment of copper samples of 3 cm diameter enables the generation of various geometrical shapes and sizes of the microstructure. The precise design of samples enables a generation of specimens of optimal dimensions. Fig. 1 presents example samples produced with the SPI G3.1 SP20P pulsed fiber laser with an impulse frequency of 60 Hz and scanning velocity of 200 mm/s. The laser pulse during the fabrication of the specimens lasted 60 ns and the focal spot size was 35 μm.

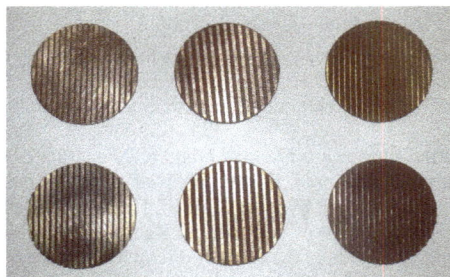

Fig. 1. *Examples of laser-treated samples of various groove depths and widths.*

The experimental determination of boiling of the laser-treated samples was done on the stand equipped with an electric heater, whose aim was to provide heat so that boiling could be sustained. The heat flux was changed using an autotransformer to generate high temperatures for nucleate boiling tests. Distilled water and ethyl alcohol were boiled and their vapor was recovered with the cooling coils. Before the actual tests both the liquid had to be degassed so that dissolved gases would not influence the experiment.

Results and Discussion

The performance of microstructure-covered samples can be done by determining how much heat is dissipated. However, it is also important to know how the physical process occurs. In order to do this, a high-speed camera can be applied. This device enables one to take a sequence of photos at extremely short time intervals. Fig. 2 presents example photos of the boiling phenomenon of distilled water at high heat flux values.

Terotechnology XII   Materials Research Forum LLC
Materials Research Proceedings **24** (2022) 15-20   https://doi.org/10.21741/9781644902059-3

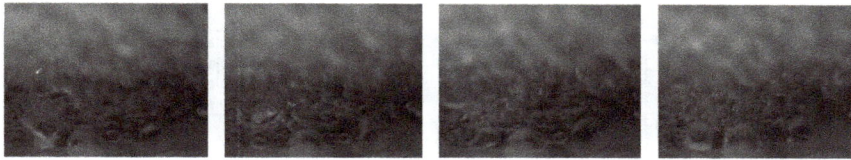

Fig. 2. *Pictures of boiling of water with the high speed camera – time between each frame: 0.02 s.*

As can be seen, the whole sample participates in the process of bubble generation and release. The convective forces seem to be quite strong at this stage of the developed nucleate boiling mode.

The study of the performance of the laser-treated samples indicates that some are more efficient than others. Fig. 3. presents the ratio of the heat flux dissipated by sample 1 (of 0.55 mm groove depth and its width of 1.15 mm) to the heat flux values dissipated by sample 2 (of 0.25 mm groove depth and its width of 0.60 mm) – based on data presented by the authors in [5].

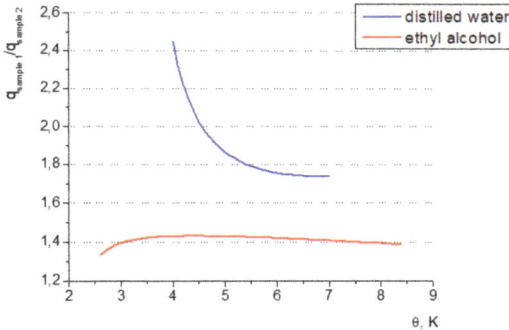

Fig. 3. *Enhancement ratio for two samples at water and ethanol boiling.*

Here we can see that sample 1, whose grooves are bigger and their width larger than sample 2, performs better during the boiling of both liquids. In the case of ethyl alcohol this enhancement is stable over various temperature differences and amounts to about 1.4. A different situation occurs for water. Here, sample number 1 outperformed number 2 to a larger degree, especially in the area of low-temperature differences. The different character of those changes for both the analyzed liquids might result from their different wetting characteristics. Surface tension for ethanol is smaller, while for water it is much larger.

The design of heat exchangers requires that their performance should be quite precisely determined with models or correlations. Different techniques can be applied in terms of mathematical modeling as the discussed example in [11 - 13], but their aim is to properly determine a certain quality. In the present paper, Nishikava et al. [14] model will be used. The experimental data taken from the paper by the authors [5] regarding samples 1 and 2 (mentioned earlier) have been compared with the calculation results according to the above-mentioned model and presented in Fig. 4 for distilled water and Fig. 5 for ethyl alcohol.

Materials Research Forum LLC

https://doi.org/10.21741/9781644902059-3

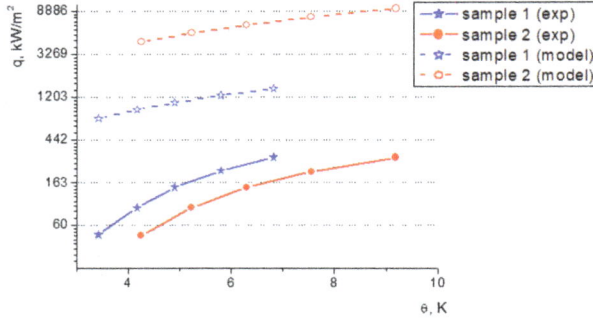

Fig. 4. *Comparison of the experimental and model calculated data for distilled water.*

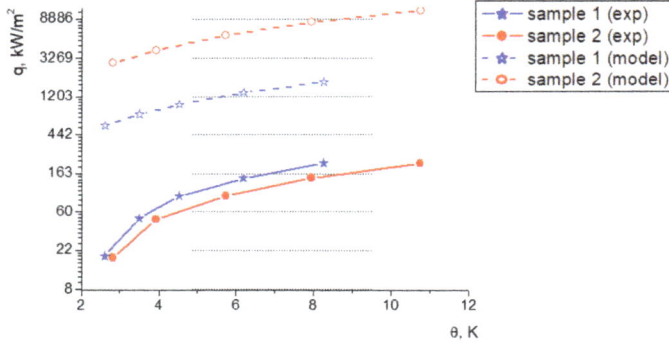

Fig. 5. *Comparison of the experimental and model calculated data for ethyl alcohol.*

The Nishikava et al. [14] model is quite simple among all the boiling models. It is the main assumption is that heat transfer within the microstructural element (in the analyses case longitudinal fins) occurs due to the conduction mode in the two-phase system composed of the boiling liquid and the solid material (in this case copper fins). The proposed correlation requires the calculation of the effective (substitute) thermal conductivity value as the product of the conductivity values of the liquid and solid phases as well as the volumetric porosity of the whole system. On the other hand, the heat flux values are calculated based on the already determined conductivity as well as the temperature difference between the heater and the saturation temperature. The height of the structure is also considered in the model.

The experimental and model data points differ significantly as can be seen in Fig. 4 and 5 for both the boiling liquids considered in the study. Undoubtedly, the simplicity of the model that was selected for calculations might be responsible for such results. The Nishikava et al. [14] model does not take the movement of vapor bubbles and the convection forces into account. It also needs to be noted that the structure generated by the laser beam is non-uniform and significant surface roughness is produced at the bottom of the grooves. Thus, additional nucleation sites are produced, which might lead to elevated heat flux and heat transfer coefficient values. The model adopted from literature does not take it into account either. Moreover, the laser-generated fins do not have the same width along with the height. It is reduced at the top and is larger at the bottom, which results from the laser treatment technique itself. While the calculations according to the considered model did not take this fact into account, the impact of this simplification might not be large enough to be visible in the graphs above.

Summary and Conclusions

Laser treatment is a modern and efficient method used to alter the morphology of various surfaces and can be effectively applied for boiling heat transfer enhancement. The various shapes and sizes that can be designed enable the production of highly efficient heaters that can considerably improve boiling conditions and lead to more heat being exchanged in such phase–change heat exchangers at the same temperature differences.

The differences between various samples with regard to heat flux values might be significant and depend on the temperature differences (as in the case of water), or be independent of it (as for ethyl alcohol). However, the proper design of such a heater can lead to higher values of heat transfer coefficient for such surfaces.

The comparison of the experimental data with the model calculations according to the calculation adopted from the literature has shown major differences. They might be related to the fact that the model is quite simple and does not consider the specific features of the laser-treated metal surfaces as those presented in the paper. However, a proper modification of the presented model, with the alternations mentioned earlier in the paper, could provide more accuracy for laser-treated heat exchanging surfaces [15-17]. It is a very interesting issue worth an in-depth study with the use of various computational and analytical tools [18-20].

References

[1] M. Piasecka, K. Strąk. Influence of the surface enhancement on the heat transfer in a minichannel, Heat Transfer Engineering 40 (2019) 1162-1175. https://doi.org/10.1080/01457632.2018.1457264

[2] M. Piasecka, K. Strąk, B. Grabas. Vibration-assisted laser surface texturing and electromachining for the intensification of boiling heat transfer in a minichannel, Archives of Metallurgy and Materials 62 (2017) 1983-1990. https://doi.org/10.1515/amm-2017-0296

[3] R. Kaniowski, R. Pastuszko. Pool boiling of water on surfaces with open microchannels, Energies 14 (2021) art. 3062. https://doi.org/10.3390/en14113062

[4] R. Kaniowski, R. Pastuszko. Boiling of FC-72 on surfaces with open copper microchannel, Energies 14 (2021) art. 7283. https://doi.org/10.3390/en14217283

[5] Ł.J. Orman, N. Radek, J. Pietraszek, M. Szczepaniak. Analysis of enhanced pool boiling heat transfer on laser – textured surfaces, Energies 13 (2020), art. 2700. https://doi.org/10.3390/en13112700

[6] S. Wciślik. Efficient Stabilization of Mono and Hybrid Nanofluids, Energies 13 (2020) art.3793. https://doi.org/10.3390/en13153793

[7] S. Wciślik. A simple economic and heat transfer analysis of the nanoparticles use, Chemical Papers 71 (2017) 2395-2401. https://doi.org/10.1007/s11696-017-0234-4

[8] P. Hrabovsky, P. Nemec, M. Malcho. Compare cooling effect of different working fluid in thermosyphon, EFM14 Int. Conf. on Experimental Fluid Mechanics 2014, Český Krumlov,

Czech Republic, EPJ Web of Conferences (2015), art. 02024.
https://doi.org/10.1051/epjconf/20159202024

[9] P. Nemec. Visualization of various working fluids flow regimes in gravity heat pipe, EFM16
Int. Conf. on Experimental Fluid Mechanics 2016, Marienbad, Czech Republic, EPJ Web of
Conferences (2017), art. 02079. https://doi.org/10.1051/epjconf/201714302079

[10] E. Zender-Świercz. Analysis of the impact of the parameters of outside air on the condition
of indoor air, Int. Journal of Environmental Science and Technology 14(8) (2017) 1583-1590.
https://doi.org/10.1007/s13762-017-1275-5

[11] M. Krechowicz. Comprehensive risk management in horizontal directional drilling projects,
Journal of Construction Engineering and Management 146 (2020) art. 04020034.
https://doi.org/10.1061/(ASCE)CO.1943-7862.0001809

[12] B. Szeląg, J. Drewnowski, G. Łagód, D. Majerek, E. Dacewicz, F. Fatone. Soft sensor
application in identification of the activated sludge bulking considering the technological and
economical aspects of smart systems functioning, Sensors 20 (2020) art. 1941.
https://doi.org/10.3390/s20071941

[13] B. Szeląg, K. Barbusiński, J. Studziński. Activated sludge process modelling using selected
machine learning techniques. Desalination and Water Treatment 17 (2018) 78–87.
https://doi.org/10.5004/dwt.2018.22095

[14] K. Nishikawa, T. Ito, K. Tanaka K. Enhanced heat transfer by nucleate boiling on a sintered
metal layer, Heat Transfer – Japanese Research 8 (1979) 65-81.

[15] N. Radek, J. Pietraszek, A. Gadek-Moszczak, Ł.J. Orman, A. Szczotok. The morphology
and mechanical properties of ESD coatings before and after laser beam machining, Materials 13
(2020) art. 2331. https://doi.org/10.3390/ma13102331

[16] I. Nová, K. Fraňa, T. Lipiński. Monitoring of the Interaction of Aluminum Alloy and
Sodium Chloride as the Basis for Ecological Production of Expanded Aluminum. Physics of
Metals and Metallography 122 (2021) 1288-1300. https://doi.org/10.1134/S0031918X20140124

[17] N. Radek, J. Konstanty, J. Pietraszek, Ł.J. Orman, M. Szczepaniak, D. Przestacki. The effect
of laser beam processing on the properties of WC-Co coatings deposited on steel. Materials 14
(2021) art. 538. https://doi.org/10.3390/ma14030538

[18] T. Styrylska, J. Pietraszek. Numerical modeling of non-steady-state temperature-fields with
supplementary data. Zeitschrift fur Angewandte Mathematik und Mechanik 72 (1992) T537-
T539.

[19] J. Pietraszek. Response surface methodology at irregular grids based on Voronoi scheme
with neural network approximator. 6th Int. Conf. on Neural Networks and Soft Computing JUN
11-15, 2002, Springer, 250-255. https://doi.org/10.1007/978-3-7908-1902-1_35

[20] J. Pietraszek, E. Skrzypczak-Pietraszek. The uncertainty and robustness of the principal
component analysis as a tool for the dimensionality reduction. Solid State Phenom. 235 (2015) 1-
8. https://doi.org/10.4028/www.scientific.net/SSP.235.1

Terotechnology XII
Materials Research Proceedings **24** (2022) 21-26

Materials Research Forum LLC
https://doi.org/10.21741/9781644902059-4

On the Method of Multi-Layer Laser Micro Machining

WITKOWSKI Grzegorz[1,a] *

[1]Kielce University of Technology, Faculty of Mechatronics and Mechanical Engineering, Al. 1000-lecia Państwa Polskiego 7, 25-314 Kielce, Poland

[a]gwitkowski@tu.kielce.pl

Keywords: Laser Devices, Surface Treatment, Micro Machining, Laser Ablation

Abstract. The presented method was developed for laser micromachining of material surfaces. The method allows to the creation of a spatial structure (2.5D) with the usage of a laser beam. The method presented in the article should be classified as removal techniques using the phenomenon of cold ablation. The method can be successfully used for pico- and femtosecond lasers equipped with a galvo scanning head. The advantage of the method is that it generates trajectories and modulates beam power based on the spatial geometry of the structure contained in the 3D data exchange file. The method uses proprietary solutions allowing for proper modulation of the laser beam power depending on the required geometry. The control application for the laser device and galvo head was developed based on the National Instruments LabView environment.

Introduction

Developing spatial micro textures with a controlled geometry is a technologically difficult task. This is mainly due to the texture dimension not exceeding one millimeter. At present, there is a narrow set of technologies capable of carrying out such a task, for example, classic milling, EDM machining, or laser processing [1]. Another issue during micro texturing is the need to maintain a proper surface condition near the machining area. The occurrence of cracks, melting or degradation of material near the micro-texture is unacceptable for almost all applications [2]. It is also necessary to control the occurrence of a heat-affected zone when using methods in which thermal exposure can occur. One of the supreme methods for manufacturing micro textures is laser processing. Laser forming a microstructure that is a vertical projection of a given shape onto a surface is a simple and well-known process. The task of creating spatial structures is definitely more difficult and complex.

Laser machining involves exposing the material to a concentrated, coherent, and focused beam of light [3]. The effects of the exposure depend on the length of the emitted light wave, the method of laser work, the pulse duration, and the frequency of pulses. The positioning accuracy of the laser beam is around 1-1.5 [μm] when using a galvanometric head. Depending on the beam mode and the length of the emitted beam, the trace of the beam can be circular with a diameter of 10 [μm]. For galvanometric heads, the beam feed rate can be close to 10 [m/s]. A significant difference between laser removal of material and the conventional methods used so far is the much better repeatability and speed of the process. As a result of the phenomenon of cold ablation in the process of laser machining of the surface, the influence of thermal exposures is reduced [4]. There is also no influence of mechanical forces in the process in question. Furthermore, it is possible to make a precise surface structure on thin-walled elements, as opposed to the precision milling process. Laser technology is also applicable in machining susceptible materials, including soft plastics. The so-far known and developed methods of making microstructures, with the use of laser technology, allow only perform cavities of the assumed flat geometry. There are only a few commercial full 3D laser machining methods.

Terotechnology XII   Materials Research Forum LLC
Materials Research Proceedings **24** (2022) 21-26   https://doi.org/10.21741/9781644902059-4

Outline of the Method

Despite the numerous advantages mentioned above, laser processing technology also has its limitations. According to the author, the most important limitation of the method is the possibility of making micro textures with a fixed outline in the cross-section. This is mainly due to the method of programming the laser equipment derived from the welding and cutting technology. Most commercial systems provide the possibility of programming the laser beam trajectory on a certain plane. The spatial effect of the structure is achieved through repeated execution of the indicated contours of the prepared program. It is an inefficient and tedious method. Laser beam guidance is usually provided by specialized galvanometric heads or, less frequently, by Cartesian robots [4]. The geometry of structures is the result of extruded cutting of certain flat shapes along the third dimension. A graphical interpretation of the limitations of the methods used is shown in Fig. 1.

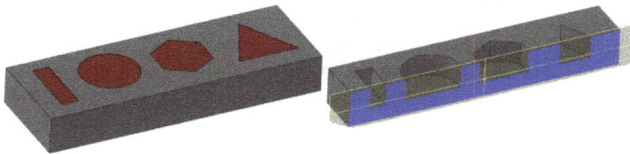

Fig. 1. An example of spatial textures that can be obtained using the classic laser micromachining method.

The presented method allows obtaining spatial micro textures using a laser device emitting a concentrated beam of optical radiation with a wavelength of 343 [nm] and a pulse duration of 6.2 [ps]. The use of such a laser allows for ablative material removal, significantly eliminating the area of the heat-affected zone. The main property of the described technology is a new way of planning the laser beam trajectory that is different from that used so far. The planned micro texture was assumed to be processed in layers from top to bottom. The 3D geometry of the planned micro-texture is transferred in the form of a three-dimensional data exchange file, e.g. STL. The input 3D file containing the spatial geometry was divided into layers of thickness corresponding to the depth of the cavity after a single laser scanning process. The concept of dividing into layers is shown in Fig. 2.

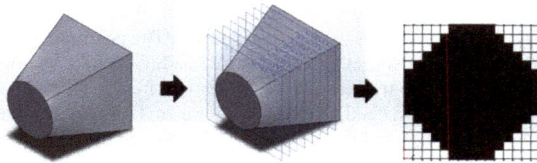

Fig. 2. Stages of geometry conversion for the developed technology.

The division of the STL file into layers was developed with the use of the LabView environment. The final quality of laser machining will depend on the number of intermediate layers created. Each of obtained layers was transformed into a raster image containing a set of pixels with coordinates in a flat arrangement. The size of a single pixel corresponds to the size of the laser dot and is controlled depending on the laser device and optics used. Each pixel created will become the point of exposure of the beam on the workpiece. Pixel coordinates in the created raster were transformed into a three-dimensional matrix containing also the layer number. It can be stated that the final result of the procedure is the development of a set of "voxels", i.e. the smallest indivisible three-dimensional elements corresponding to pixels in raster graphics. A single "voxel" has a volume equal to the volume of the cavity after a single exposure to the laser beam. Based on the developed matrix, flat trajectories were generated each for the layer.

Method Implementation – Test Stand

For the development of the control application responsible for geometry division, the NI LabView environment was used. Laboratory Virtual Instrument Engineering Workbench LabView is a system-design platform and development environment for a visual programming language from National Instruments [5]. LabView is commonly used for data acquisition, instrument control, and industrial automation in a variety of operating systems. The programming language used in LabView, sometimes called G, is based on data availability. The structure code and front panel of the developed application are shown in Fig. 3 and Fig. 4 respectively.

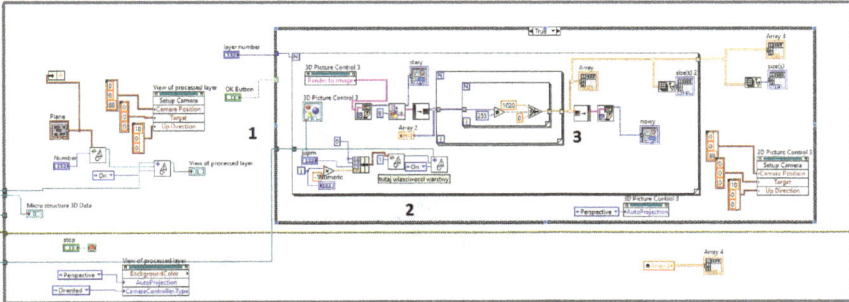

Fig.3. *LabView application source code.*

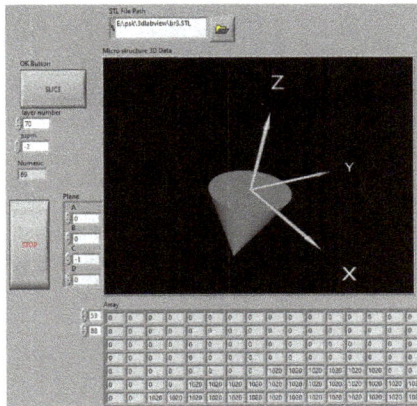

Fig.4. *The front panel of the application.*

Terotechnology XII
Materials Research Proceedings **24** (2022) 21-26

Materials Research Forum LLC
https://doi.org/10.21741/9781644902059-4

The presented code has been synthetically divided into three logical sections (Fig. 3). The first section (1) is responsible for uploading the proper STL file into the system. This section presents geometry in space also. The second section (2) is responsible for geometry slicing according to the input parameters such as the number and thickness of layers. The third section (3) contains procedures that create raster temporary layers and builds a three-dimensional power table for laser processing. In case the average RGB index of some pixel in the raster is not equal to 0 (not black), then at corresponding coordinates in power table system inputs max power of the laser. In other cases, the system enters the value 0. The obtained power matrix is transmitted successively to the laser device.

The presented laser station (Fig. 5) is part of the equipment of the Department of Terotechnology and Industrial Laser Systems of Kielce University of Technology. Experimental laser TruMicro series 5000 is a solid-state laser, based on mono crystal disk Nd:Yb pumped by the laser diodes, with the basic wavelength of 1060 [nm]. The laser generates a third harmonic with a wavelength of 343 [nm]. It is an ultrashort pulse laser with a pulse duration of about 6.2 [ps] and a nominal frequency of 400 kHz. The frequency can be divided by a natural number in the range of 1-10000. The single pulse energy is in the range of 0.5 - 12.6 µJ and the maximum pulse power is 2.032 MW. The average continuous power is 5 W.

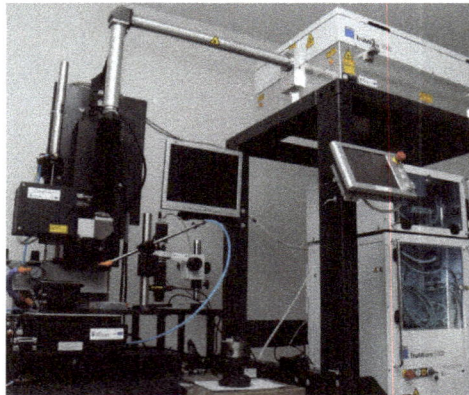

Fig.5. *Experimental Trumpf TruMicro laser station.*

The presented laser stand is used for drilling micro holes, surface texturing, and other laser micromachining technologies. The stand is equipped with a two-axis Scanlab galvo head enabling scanning at a speed of 3 [m/s]. The working area of the head is a square with a side length of 90 mm. The software provided with the laser has limited capabilities and is difficult to use, because of the experimental laser destination. The laser is fully integrated with the head using the control RTC4 PC interface board, with implemented real-time system.

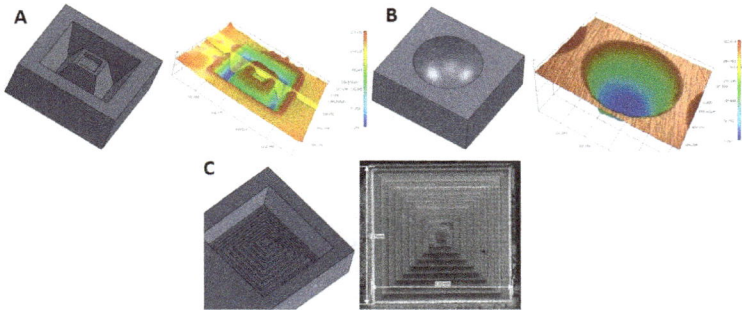

Fig.6. *Reference and microscopic image of obtained microstructure (A - truncated cone, B-sphere, C- step pyramid).*

Three preliminary experiments were conducted to verify the developed multi-layer laser machining method [6-8]. The samples of PMMA material were used in the research process. Machining parameters are listed in Table 1.

Table 1. *Laser machining parameters*

Pulse Energy [μJ]	Beam Velocity [mm/s]	Repetitions per layer	Pulse Duration [ps]
6	200	300	6.2

A CAD SolidWorks graphic program was used to create spatial geometries in STL files. The geometries in the form of a step pyramid, a sphere, and a truncated cone were made. The goal of the experiment was to obtain structures compatible with assumed and uploaded geometries. No specific depth of texture was assumed. Preliminary observations and microscopic photographs were made with the usage of a confocal microscope Hirox KH-8700. The comparison of the geometries and created structures uploaded is presented in Fig. 6 (A, B, and C respectively).

Summary
The results of the experiments confirm the effectiveness of the presented method for making micro textures with strictly assumed shapes and geometrical dimensions. With quite simple algebra manipulations it is possible to perform very complex micro textures (2.5D). The method can be easily used for most commercial, and industrial laser devices and their control systems. A hardware limitation of the presented method is the maximum depth of microstructure. It is caused by the length of the laser beam focus waist, which is different from the wavelength and focal length of the lens used. This problem can be easily solved by synchronization the head movement along the Z-axis during the machining process. A certain disadvantage of the solution is the inability to process the surface remaining in the shade. For this reason, this kind of machining is classified as 2.5D, not full 3D. The dimensional accuracy of the method requires proper resolution adjusting of the generated power and trajectory maps with the knowledge of the spot size of the beam. The obtained depth of microstructures depends on the number of programmed scan repetitions for each layer. The exact solution to these problems requires further development work.

Acknowledgment
The presented method was developed for the needs of the research project LIDER/30/0170/L-8/16/NCBIR/2017 financed by the National Centre for Research and Development.

References

[1] N. Radek, A. Szczotok, A. Gądek-Moszczak, R. Dwornicka, J. Bronček, J. Pietraszek. The impact of laser processing parameters on the properties of electro-spark deposited coatings. Archives of Metallurgy and Materials 63 (2018) 809-816. https://doi.org/10.24425/122407

[2] D. Bouilly, D. Perez, L.J. Lewis. Damage in materials following ablation by ultrashort laser pulses: A molecular-dynamics study, Phys. Rev. B 76 (2007) art. 184119. https://doi.org/10.1103/PhysRevB.76.184119

[3] N. Radek, K. Bartkowiak. Laser treatment of Cu-Mo electro-spark deposited coatings. Physics Procedia 12 (2011) 499-505. https://doi.org/10.1016/j.phpro.2011.03.061

[4] S. Tofil, G. Witkowski, K. Mulczyk. Control system of the ultrafast TruMicro experimental laser for surface microtreatment - Part II, Proc. 2018 19th Int. Carpathian Control Conf., ICCC 2018, Szilvasvarad, 2018, 528-531. https://doi.org/10.1109/CarpathianCC.2018.8399687

[5] R.W. Larsen. LabView for engineers, New Jersey, Pearson Education, 2010.

[6] J. Pietraszek, N. Radek, A.V. Goroshko. Challenges for the DOE methodology related to the introduction of Industry 4.0. Production Engineering Archives 26 (2020) 190-194. https://doi.org/10.30657/pea.2020.26.33

[7] Ł.J. Orman Ł.J., N. Radek, J. Pietraszek, M. Szczepaniak. Analysis of enhanced pool boiling heat transfer on laser-textured surfaces. Energies 13 (2020) art. 2700. https://doi.org/10.3390/en13112700

[8] N. Radek, J. Pietraszek, A. Gadek-Moszczak, Ł.J. Orman, A. Szczotok. The morphology and mechanical properties of ESD coatings before and after laser beam machining, Materials 13 (2020) art. 2331. https://doi.org/10.3390/ma13102331

Terotechnology XII
Materials Research Proceedings **24** (2022) 27-33

Materials Research Forum LLC
https://doi.org/10.21741/9781644902059-5

Surface Laser Micropatterning of Polyethylene Terephthalate (PET) to Increase the Shearing Strength of Adhesive Joints

TOFIL Szymon[1,a] *, MANOHARAN Manikandan[2,b] and
NATARAJAN Arivazhagan[2,c]

[1] Kielce University of Technology, Faculty of Mechatronics and Mechanical Engineering,

Av. 1000-lecia P.P. 7, 25-314 Kielce, Poland

[2] School of Mechanical Engineering, VIT University, Vellore – 632014, INDIA

[a]*tofil@tu.kielce.pl, [b]manikandan.manoharan@vit.ac.in, [c]narivazhagan@vit.ac.in

Keywords: UV Laser; Picosecond Laser; Adhesive Joints; Micromachining; Micropatterning; Polyethylene Terephthalate (PET)

Abstract. The method for increasing the shearing strength of adhesive joints in plastics was investigated. This method uses laser micropatterning of different construction materials to extend adhesive surfaces. In investigations, Polyethylene Terephthalate (PET) was used. TruMicro 5325c ultra-short pulse (picosecond pulse) laser and the SCANLAB GALVO scanning head applied in the research enable ablative removal of the material without the heat-affected zone (HAZ) in the rest of it. Ultra-short laser pulses (such as picosecond pulses) remove material without melting the rest of it. The presented method significantly increases the shearing strength of the formed joints in investigated materials made from plastics, which was proved in the results of laboratory tests. The laser device parameters used, which are described in this article, have not produced cracks in the microtreated materials. The research shows that bondings between elements with the appropriately machined microstructure are characterized by a severalfold increase in the strength of joints in relation to materials devoid of the microstructure. In addition, this study addresses practical solutions to the adhesive method used for joining of the polymers used for tests. This study could be helpful for each application where we have to connect these types of polymers.

Introduction

Over the last two decades, industrial technologies using laser machines were frequently addressed in many papers from different scientific fields [1-5]. The lasers have greatly facilitated the production in the automotive, aerospace, and electronics industry. Thanks to the rapid development of lasers, different modifications in the surface layers and in materials, which were previously considered difficult for machining for various reasons, are now possible. Currently, laser micro-treatment is an increasingly common method for modifying the surface properties of different construction materials [6-9]. The continuous improvement of the existing design solutions results in applying the latest achievements from materials engineering, especially in the above-mentioned industry branches. The use of lasers for this purpose seems to be justified. The impact of a laser beam on the surface of different materials was the subject of many scientific manuscripts presented in numerous publications [10,11]. In these publications was discussed the influence of laser parameters (such as light wavelength, frequency of impulses, duration of the impulse, etc.) on the surface of the material being machined. The use of different laser surface modification techniques of polymers was elaborated on in detail in papers [12,13].

Mechanical fasteners and welding belong to traditional joining techniques, with which many manufacturers feel comfortable. However, it has to be taken into consideration that these methods are not the most practical solutions to perform a modern assembly. The application of some mechanical fasteners can lead to an increase in production costs, and structure weight; and it can limit the material options to choose from or result in their fatigue, strain, or even tear. Strong adhesives or tapes can outperform mechanical fasteners in many structural applications ensuring a clean and durable structure. That is why a growing number of manufacturers from various industry branches are searching for industrial adhesives as an alternative to traditional joining techniques. Thanks to achieving high reliability of objects, more and more products are made as permanent assemblies, in which a bonding technology is widely used.

This manuscript is dedicated to the possibilities of using a laser micropatterning of a surface to obtain a stronger adhesive joint of the selected construction materials. In the tests, polyethylene terephthalate (PET) was used and joined with the Multibond 1101 epoxy glue. The surface of both sides of the material was micropatterned with a picosecond laser in order to obtain a better mechanical joint. The aim of this study was to offer an innovative joining solution that can reduce the power/pressure requirement for joining and provide a better joint. For this study, the effect of a pre-patterned surface of PET for joining will be demonstrated, and the strength tests of the joining process with and without micropatterning will be discussed. The research shows that bondings between elements with the appropriately machined microstructure are characterized by a severalfold increase in the strength of joints in relation to materials devoid of the microstructure.

This manuscript aims to present the impact of the type of laser-machined micro-texture on the material's surface on increasing the adhesive force in adhesive joints. In addition, this study addresses practical solutions to the adhesive method used for joining of the polymers used for tests. This study could be helpful for each application where we have to connect these types of polymers.

Materials and Methods

Polyethylene Terephthalate (PET), has special physical properties, and is ideal for the production of very precise parts and mechanical components, which are expected to be distinguished by resistance to very high loads and abrasion resistance. It is also used in electrical insulating components applied in electrical engineering, dimensionally stable and very precise mechanical parts (such as sleeves, guides, gear wheels, rolls, pump elements, etc.), heavily loaded bearings, sleeves, guides, and thrust washers. It is also widely applied in the automotive industry and the construction of yachts and ship vessels.

The MULTIBOND-1101 is a cold-hardened two-component epoxy adhesive with a fluid, semi-fluid, and gel-type consistency. It contains a colorless epoxy resin (component A) and yellow hardener (component B). MULTIBOND-1101 wets and even saturates porous materials and is easy to apply (mixing 1:1 by volume). The hardening process is homogeneous in the entire resin mass during the average time. It is applied for joining metals, stones, ceramics, and other hard materials except for polyolefins (PE, PP, PTFE). Typical applications include automotive elements and machine parts, elements of railway carriages, buses, and yachts. This adhesive has high resistance to water, oils, and petrol.

Micropatterning of tested materials was made on a laboratory stand equipped with a picosecond TRUMPF 5325c laser generating UV laser beam with a wavelength of 343 nm. The SNANLAB intelliSCAN 14 scanning head was used for positioning laser beams in the working area. Before laser micropatterning, the surface of the samples was cleaned with ethanol. The exemplary results of the conducted micro-treatment were shown in the images, which were taken with the use of the HIROX KH-8700 optical microscope (Fig. 1). After laser micropatterning samples were rinsed sequentially in an ultrasound bath with ethanol and deionized water. Next, an overlap joint between materials was formed using Multibond 1101 epoxy adhesive. The diagram of joining elements was presented and described below (Fig. 2). Subsequently, the INSTRON 4502 testing machine was used to perform strength tests, which enabled us to determine to what extent the micro-treatment of the selected geometric shape affected the increase in force needed to break the joint.

Results and Discussion

The main aspect of our work is to obtain a type of microstructure that improves the strength of the glued joints and at the same time doesn't destroy the surface properties of the material. The microstructure should change the properties of the material surface that will allow it to better penetrate by glue and increase the adhesive strength. Improper micropatterning may change the surface properties to hydrophobic, which will significantly impede the spread of the adhesive on the surface. The authors selected two common shapes parallel-line and circles which are presented in Fig. 1.

The effect of micro-treatment was obtained by ablative removal of material from the sample surface [10] by linear beam movement. This is one of the available options to perform microtexture measurements on this laboratory stand. Further research will be focused on other variants of shape, density, and depth of laser machined microtexture of other polymers. The parameters of the TruMicro 5325c laser machine, which was used to carry out microtexture measurements are following. Pulse energy – 12.6 μJ. Pulse repetition rate – 200 kHz. Laser scanning speed – 1000 mm/s. Type of shield gas – air.

According to the procedure described above, microstructures showing parallel lines perpendicular to the direction of the peeling force with a distance of approximately 50 μm and an average depth of 30 μm were performed on the surface of machined materials intended for bonding. They can be found in the following images. The microtexture that revealed circles covered 50% of the area intended for bonding. A diameter of a singular element was approximately 1 mm, and its average depth was 70 μm. The distribution and density of these elements were selected based on experimental tests conducted by the authors of this manuscript. Both parallel-line and circles textures were performed on both joined sample surfaces.

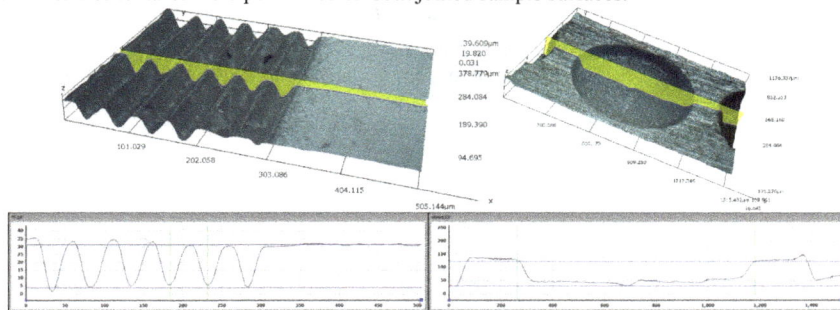

Fig. 1. *General view of the lateral profile of micropattern like parallel lines and circles made on the sample surface.*

Irrespective of the type of the applied adhesive, surface preparation is of key importance for ensuring a durable and stable adhesive joint due to the fact that the joint strength is determined to a large extent by the adhesion rate between the surface and the adhesive. Dirty surfaces inhibit adhesion and require cleaning to ensure an optimal joint. Some adhesives can bond by surface impurities, and others can require pre-cleaning before bonding. Each sample was cleaned with isopropanol to remove impurities before conducting the laser micro-treatment process. After that, the samples were purified with compressed air in order to remove any material residues. The samples prepared in this way were bonded in specially designed equipment, which enabled a constant and reproducible thickness of the 1 mm adhesive joint. This ensured a visual inspection of the bond. No additional pressure was necessary. Samples were left to fully dry for 24 hours. Then, the samples were placed in the INSTRON 4502 testing machine jaws and strength tests were performed.

During the tensile test, the sample was subjected to uniaxial strain. Measurements of forces were recorded. Such investigation is one of the fundamental sources that can provide information on the mechanical properties of plastics and adhesive joints. The values measured in these samples include strain (elongation) and strain force (fracture connection). Thus, tensile strength is the maximum stress that the material transfers during static tension. Tests were conducted for five samples with microstructure and five samples without microstructure.

Diagrams below show the tensile force of the formed joint. The plastic flow of the sample was not observed due to brittle cracking both in the applied binder and the examined material. The diagram of bonded elements was presented in Fig. 2, where 1 is texturing zone and 2 is the sample gripping section. Texture measurements were performed on the 12 mm x 12 mm surface on one side of bonded elements. Texturing zone was coated with adhesive, and an overlap joint was created, which was marked on a diagram. Samples for the static tensile test were prepared according to the requirements of PN-EN ISO 6892-1 standard.

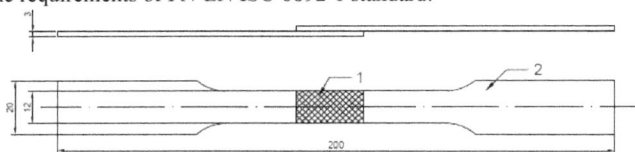

Fig. 2. Diagram of the lap joint.

Table 1 presents a comparison of the average results of the breaking strength of the joint made and the value of the increase in strength of the joint with the texture in relation to the joint without texture.

Diagrams in Figs. 3 and 4 illustrate the relationship between sample displacement and tensile force.

***Table 1.** Summary of the results of strength tests.*

	PET without micropattern	PET with micropattern like circle	PET with micropattern like perpendicular lines
Measurement 1 [N]	176.7	519.9	393.9
Measurement 2 [N]	188.2	532	386.4
Measurement 3 [N]	119.4	540.9	454.8
Measurement 4 [N]	216.6	555.8	347.4
Measurement 5 [N]	143	553	*X
Average [N]	168.78	540.32	395.62
Standard deviation [N]	38.18	14.89	44.40
Min [N]	119.4	519.9	347.4
Max [N]	216.6	555.8	454.8
The average increase in strength [%]	-	320.13	234.4

*X – sample accidentally destroyed during mounting in the testing machine holder

Materials Research Forum LLC

https://doi.org/10.21741/9781644902059-5

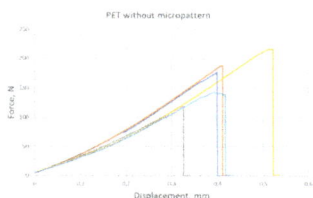

Fig. 3. PET samples joined without laser micro-treatment.

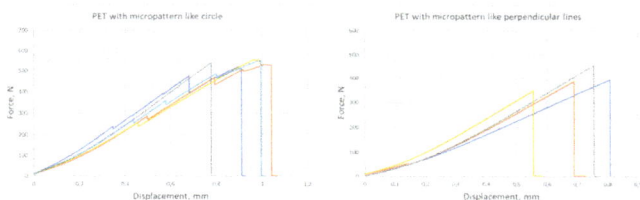

Fig. 4. PET – PET joints with laser micro-treatment.

Adhesively-bonded samples without micro-texturing were fractured with an average tensile force of 168.78 N. Specimens that were adhesively bonded with the structure in the form of circles were damaged with an average force of 540.32 N. The average force needed to destroy a sample, which is adhesively bonded with the structure in the shape of circles is higher by 320%. Specimens that were adhesively bonded with the structure in the form of parallel lines were damaged with an average force of 395.62 N. The average force needed to fracture a sample, which is adhesively bonded with the structure in the shape of circles is higher by 234%.

Most polymers are inherently either hydrophobic or only mildly hydrophilic, but they can be made at least temporarily hydrophilic by exposure to a laser beam. The laser treatment method is satisfactory to achieve superhydrophilic patterned surfaces. The implementation of the appropriate geometric shapes of the pattern increases the hydrophilic properties of the surface. The created micro-texture also increases the surface area of the bonded material, which has a considerable impact on the force needed to break the bonding. Unfortunately, improper fabrication of micro-textures affects changes in contact angle values and thus, changes in surface energy of the material. Due to the above, an additional series of tests, which would provide a more in-depth analysis of the obtained results, should be performed. Despite slight divergencies, the authors of the paper received satisfactory results in the perspective of conducting further research on the modification of shape and density parameters of the micro patterns. The authors of the paper noted that the appropriate modification of both the laser parameters and the texture itself will allow for obtaining even more durable connections. Both changing the operating parameters of the laser device and changing the properties of the pattern are of great importance for the properties of the resulting adhesive joint. These properties are closely correlated with each other. A slight change in one parameter can have a big impact on the end result of the process [13].

Conclusions
Ultrashort laser surface micropatterning is a very promising method for creating more advantageous hydrophilic conditions for better adhesives wetting. Many studies have shown their excellent properties for the ablative removal of thin layers of materials. Many methods have been developed for fabricating micro-pattern on the surface with the aim of industrial production. The studies that were conducted prove that the micro-texture type has a significant impact on creating a more durable adhesive bonding using laser micropatterning on the surface of bonded material. As regards PET material, the bonding between surfaces with the micro-texture exhibit much higher tear resistance during the static tensile test than the same bonding without surface patterning. Micropattern-like circles for PET are higher by 320% and it is a very promising result that gives hope of conducting future research on the modification of the shape and density of the microtexture distribution and the possibilities of its implementation on the processed material. Each of these factors contains one or more elements that must also be evaluated. Further strength tests of adhesive joints for other material pairs are also planned: metal - plastic; ceramics - plastic and metal - ceramics, with or without a textured surface, as well as with the use of another type of adhesive agent or without the use of additional adhesive.

Acknowledgments
The research reported herein was supported by a grant from the National Centre for Research and Development (NCBiR), project No. LIDER/30/0170/L-8/16/NCBR/2017.

References
[1] J. Pietraszek, N. Radek, A.V. Goroshko. Challenges for the DOE methodology related to the introduction of Industry 4.0. Production Engineering Archives 26 (2020) 190-194. https://doi.org/10.30657/pea.2020.26.33

[2] N. Radek, A. Szczotok, A. Gądek-Moszczak, R. Dwornicka, J. Bronček, J. Pietraszek. The impact of laser processing parameters on the properties of electro-spark deposited coatings. Archives of Metallurgy and Materials 63 (2018) 809-816. https://doi.org/10.24425/122407

[3] N. Radek, A. Sladek, J. Bronček, I. Bilska, A. Szczotok. Electrospark alloying of carbon steel with WC-Co-Al₂O₃: deposition technique and coating properties. Advanced Materials Research 874 (2014) 101-106. https://doi.org/10.4028/www.scientific.net/AMR.874.101

[4] A. Gądek-Moszczak, N. Radek, S. Wroński, J. Tarasiuk. Application the 3D image analysis techniques for assessment the quality of material surface layer before and after laser treatment. Advanced Materials Research 874 (2014) 133-138. https://doi.org/10.4028/www.scientific.net/AMR.874.133

[5] A.F. Lasagni, T. Roch, J. Berger, T. Kunze, V. Lang, E. Beyer. To use or not to use (direct laser interference patterning), that is the question. Proc. SPIE 9351 (2015) art. 935115. https://doi.org/10.1117/12.2081976

[6] B. Mao, A. Siddaiah, Y. Liao, P.L. Menezes. Laser surface texturing and related techniques for enhancing tribological performance of engineering materials: A review. J. Manuf. Process. 53 (2020) 153-173. https://doi.org/10.1016/j.jmapro.2020.02.009

[7] A. Garcia-Giron, J.M. Romano, A. Batal, B. Dashbozorg, H. Dong, E.M. Solanas, D.U. Angos, M. Walker, P. Penchev, S.S. Dimov. Durability and Wear Resistance of Laser-Textured

Terotechnology XII   Materials Research Forum LLC
Materials Research Proceedings **24** (2022) 27-33   https://doi.org/10.21741/9781644902059-5

Hardened Stainless Steel Surfaces with Hydrophobic Properties, Langmuir 35 (2019) 5353-5363. https://doi.org/10.1021/acs.langmuir.9b00398.

[8] J.M. Romano, A. Garcia-Giron, P. Penchev, S. Dimov. Triangular laser-induced submicron textures for functionalising stainless steel surfaces. Applied Surface Science 440 (2018) 162–169. https://doi.org/10.1016/j.apsusc.2018.01.086.

[9] S. Ravi-Kumar, B. Lies, X. Zhang, H. Lyu, H. Qin. Laser ablation of polymers – A review. Polym. Int. 68 (2019) 1391–1401. https://doi.org/10.1002/pi.5834

[10] G. Witkowski, S. Tofil, K. Mulczyk. Effect of laser beam trajectory on pocket geometry in laser micromachining. Open Engineering 10 (2020) 830-838. https://doi.org/10.1515/eng-2020-0093

[11] B. Antoszewski, P. Sęk. Influence of laser beam intensity on geometry parameters of a single surface texture element. Arch. Metall. Mater. 60 (2015) 2215–2219. https://doi.org/10.1515/amm-2015-0367

[12] J.M. Romano, M. Gulcur, A. Garcia-Giron, et al. Mechanical durability of hydrophobic surfaces fabricated by injection moulding of laser-induced textures. Applied Surface Science 476, (2019), 850-860. https://doi.org/10.1016/j.apsusc.2019.01.162.

[13] B. Antoszewski, Sz. Tofil, K. Mulczyk. The Efficiency of UV Picosecond Laser Processing in the Shaping of Surface Structures on Elastomers. Polymers, 12(9), (2020). https://doi.org/10.3390/polym12092041

Terotechnology XII
Materials Research Proceedings 24 (2022) 34-38

Materials Research Forum LLC
https://doi.org/10.21741/9781644902059-6

Laser Processing of WC-Co Coatings

RADEK Norbert[1,a *], PIETRASZEK Jacek[2,b], BRONČEK Jozef[3,c],
GONTARSKI Dariusz[1,d], SZCZOTOK Agnieszka[4,e], PARASKA Olga[5,f] and
MULCZYK Krystian[1,g]

[1]Kielce University of Technology, Faculty of Mechatronics and Mechanical Engineering,
Al. 1000-lecia Państwa Polskiego 7, 25-314 Kielce, Poland

[2]Cracow University of Technology, Faculty of Mechanical Engineering, Al. Jana Pawła II 37,
31-864 Cracow, Poland

[3]University of Zilina, Univerzitna 1, 01026 Zilina, Slovakia

[4]Silesian University of Technology, Faculty of Materials Engineering, str. Krasinskiego 8,
40-019 Katowice, Poland

[5]Khmelnitskiy National University, Department of Chemical Technology, st. Instytutska 11,
29016, Khmelnickiy, Ukraine

[a]norrad@tu.kielce.pl, [b]pmpietra@gmail.com, [c]jozef.broncek@fstroj.utc.sk, [d]gontar@tu.kielce.pl,
[e]agnieszka.szczotok@polsl.pl, [f]olgaparaska@gmail.com, [g]kmulczyk@tu.kielce.pl

Keywords: WC-Co Coating, Electro-Spark Deposition, Laser Beam Processing,
Properties

Abstract. The main objective of the present work was to determine the influence of laser beam processing (LBP) on the microstructure, microhardness, roughness, and corrosion resistance of coatings produced on C45 carbon steel by the electro-spark deposition (ESD) process. The studies were conducted using WC-Co electrodes produced by the Pulse Plasma Sintering method (PPS) of nanostructural powders. The coatings were deposited by means of the EIL-8A and they were laser treated with the Nd:YAG, BLS 720 model.

Introduction

Carbide coatings have numerous industrial applications. Characterized by high abrasion, sliding, and erosion resistance, they can be used as a substitute for hard chrome plating. Cemented carbides are cermets containing between 70 and 97 wt.% of refractory metal carbides (e.g. WC, TiC, TaC) and a metal binder which is most often cobalt, sometimes nicke,l and occasionally an iron-base alloy. At present, cemented carbides find numerous applications as wear parts as well as in all kinds of machining operations outperforming the conventional high-speed steel tools [1].

Cemented carbides are classified into categories depending on their chemical composition and WC grain/particle size. The latter parameter can vary widely, therefore cemented carbides are divided into four main groups [2]:

- coarse-grained - 3÷30 μm,
- standard - 1.5÷3 μm,
- fine-grained - 0.5÷1.5 μm,
- ultrafine-grained - < 0.5 μm.

ESD technology can be an alternative to other techniques for the production of engineering materials, e.g. sinters [3-5]

Terotechnology XII
Materials Research Proceedings **24** (2022) 34-38

Materials Research Forum LLC
https://doi.org/10.21741/9781644902059-6

We can produce carbide-based coatings using various technologies. Among the beam techniques, we observe a dynamic development of laser processing [6] and a well-established position among surface treatments using the ESD method [7].

To date, a number of alternative ESD techniques have been developed to suit various coating deposition conditions and meet surface topography requirements. ESD coatings may serve as both technological and functional surface layers and can be deposited using portable, manually operated equipment or fully automated systems.

ESD coatings have some disadvantages, such as high roughness and the presence of porosity and discontinuities but these can be eliminated by laser treatment. The laser beam is used for surface sealing, removing surface irregularities, homogenizing the chemical composition of the coating, and changing its phase composition [8]. Therefore laser-treated ESD coatings show lower porosity, better adhesion to the substrate, higher resistance to wear and seizure, higher fatigue strength due to the presence of compressive stress, and good resistance to corrosion. Analysis of properties of WC-Co coating systems requires many methods [9, 10].

This paper reports on the effects of laser treatment on microstructure, microhardness, roughness, and corrosion resistance of electro-spark WC-Co coatings.

Experimental Procedure

The coatings were deposited on the C45 grade plain-carbon steel by the ESD method. The electrodes, of composition 95% WC and 5% Co, were produced using the Pulse Plasma Sintering method.

The powders were mixed together in the right proportions and consolidated by means of a pressure-assisted, pulse-plasma sintering (PPS) method at the Faculty of Material Engineering, Warsaw University of Technology (Poland). The powder mix was held for 5 minutes at 1100°C and 50 MPa. PPS uses high-current pulses generated through continual discharging of a capacitor battery of 300 µF, thereby inducing several tens of kA current which flows through the consolidated powder within each millisecond pulse.

To deposit the WC-Co coatings the EIL-8A apparatus was operated at voltage, current, and capacitance of 230 V, 2.1 A, 300 µF, and 2 min/cm^2, respectively. The coatings were afterward treated with the Baasel Lasertechnik *720* Nd:YAG laser run in pulsed mode at a power level of 25 W, a spot size of 1 mm, and pulse duration of 0.5 ms and a pulse repetition frequency of 45 Hz. The sample movement rate and beam shift jump were set to 230 mm/min and 0.35 mm, respectively.

Results and Discussion

The morphology of WC-Co coatings was analyzed before and after laser treatment by means of the Quanta 3D FEG (SEM/FIB) scanning electron microscope, equipped with an integrated EDS/WDS/EBSD system (energy dispersion spectrometer EDS, wavelength dispersion spectrometer WDS and electron backscattered diffraction EBSD). The microstructures of ESD WC-Co coatings in both as-deposited and laser-treated conditions were observed by SEM.

Terotechnology XII

Materials Research Proceedings **24** (2022) 34-38

Materials Research Forum LLC

https://doi.org/10.21741/9781644902059-6

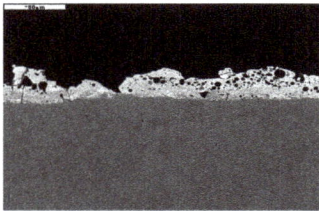

Fig. 1. *Microstructure of the WC-Co coating.*

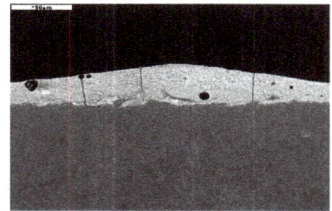

Fig. 2. *Microstructure of the WC-Co coating after laser treatment.*

A typical microstructure of the WC-Co coating is illustrated in Fig. 1. From the SEM analysis, it is evident that the as-deposited coating is porous and cracked, and has a thickness of between 25 and 35 μm. The heat-affected zone (HAZ) within the substrate ranges from 14 to 21 μm beneath the clearly seen coating-substrate interface. The ESD treatment homogenizes the chemical composition of the coatings and refines their microstructure. As seen in Fig. 2, the laser-modified outer layer is free from cracks and porosity The coating is 30-40 μm thick and perfectly adheres to the substrate, wherein the carbon-enriched HAZ extends from 25 down to 31 μm beneath the coating.

The roughness of the WC-Cu coatings was quantitatively assessed using the Talysurf CCI optical profiler and was measured in two perpendicular directions. The first measurement was made parallel to the electrode movement direction, while the second measurement was perpendicular to the scanning stitches. The average value of the Ra parameter for a given coating was calculated from these two measurements. Measurements of WC-Co coatings subjected to LBP were made in perpendicular and parallel directions to the path of a laser beam, and then the mean value of roughness was calculated. In most research studies, the measurements of surface roughness are measured along the path of the laser beam. The obtained results do not reflect the actual surface micro-geometry because the maximum height of irregularities occurs in the perpendicular direction.

WC-Co coatings were characterized by the value of the parameter Ra = 2.64÷3.16 μm, while after laser beam machining the arithmetic mean value of the profile ordinates was from 9.87÷10.57 μm. C45 steel substrates to which coatings were applied were characterized by Ra = 0.38÷0.42 μm. Selected profiles of the tested samples are presented in Fig. 3.

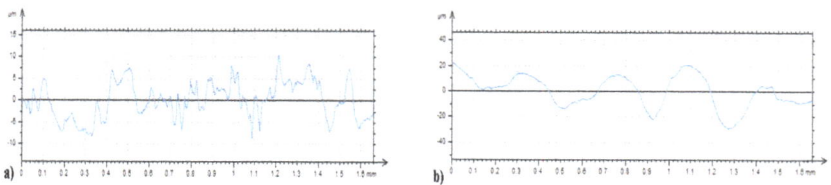

Fig. 3. *Examples of surface profiles of WC-Co coatings: a) before LBP, b) after LBP.*

Materials Research Forum LLC

https://doi.org/10.21741/9781644902059-6

Microhardness measurements were performed using the Vickers method. The microhardness was measured using the Microtech MX3 tester at a load of 0.4 N applied for 15 seconds. Indentations were made on perpendicular sections in three zones: in the coating (before and after LBP), in the heat-affected zone (HAZ) and in the substrate material. Each sample was subjected to 10 measurements. It was found that LBP caused a slight decrease in microhardness of the tested coatings. The microhardness of the WC-Co coating prior to LBP was ranging from 968 to 1065 HV0.4 and slightly decreased to between 937 and 995 HV0.4 after LBP. The C45 steel substrate of was not much affected by the laser treatment and its microhardness was ranging from between 461 and 528 HV0.4 in the HAZ to between 271 and 279 HV0.4 in the underlying substrate.

Corrosion resistance tests were carried out by the computerized Atlas'99 electrochemical analysis system using the potentiodynamic method. The cathodic and anodic polarization curves were acquired by polarizing the tested specimens at 0.2 mV/s (within the area of ± 200 mV from the corrosion potential) and 0.4 mV/s (within the area of higher potentials). Specimens with a 10 mm diameter separated area were polarized to 500 mV. In order to establish the corrosion potential the polarization curves were acquired 24 hours after exposure to the test solution (0.5M NaCl). All tests were carried out at $21\pm1°C$. The corrosion resistance results are shown in Fig. 4.

After laser treatment of C45 carbon steel a martensitic structure was obtained, which has a higher resistance to corrosion as compared to the ferritic-pearlitic structure. LBP improved the corrosion resistance of ESD coatings by about 21% due to the sealing effect. As a result of laser processing a decrease in the corrosion current density, from 11.6 to 9.2 $\mu A/cm^2$, and increase in the corrosion potential, from -595 mV to -585 mV, were observed.

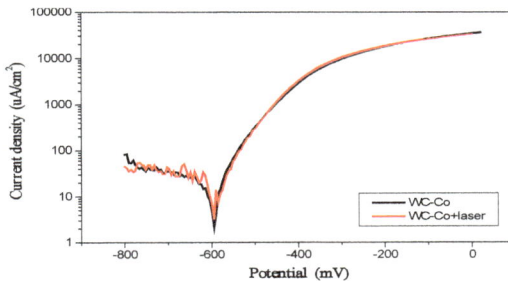

Fig. 4. Polarization curves of *WC-Co coatings in* as-deposited and laser treated *condition.*

Summary

1. The process of creating technological surface layers by the ESD method is associated with the transfer of mass and energy and the phenomenon of the formation of low-temperature plasma.
2. The laser beam can effectively modify ESD coatings and their functional properties.
3. Laser irradiation of coatings resulted in the healing of micro-cracks and pores.
4. The roughness of the ESD WC-Co coatings are more than tripled by the laser treatment.
5. The laser processed of electro-spark WC-Co coatings show improved resistance to corrosion (by ~21%) and slightly decreased microhardness (of ~6%).
6. The obtained results may of interests for industry branches using responsible parts of devices e.g. turbines [11], superheaters [12] or military equipment [13, 14].

References

[1] FANAR S.A. – production of cutting tools [in Polish], http://www.fanar.pl (23.01.2022).

[2] M. Wysiecki. Nowoczesne materiały narzędziowe, WNT, Warszawa, 1997.

[3] J. Borowiecka-Jamrozek. Engineering structure and properties of materials used as a matrix in diamond impregnated tools, Archives of Metallurgy and Materials 58 (2013) 5-8. https://doi.org/10.2478/v10172-012-0142-0

[4] J. Borowiecka-Jamrozek, J. Lachowski. Microstructure and mechanical properties of Fe-Cu-Ni sinters prepared by ball milling and hot pressing, Defect and Diffusion Forum 405 (2020) 379-384. https://doi.org/10.4028/www.scientific.net/DDF.405.379

[5] J. Borowiecka-Jamrozek. Processing and characterization of Fe-Cu-Ni sinters prepared by ball milling and hot pressing, Archives of Metallurgy and Materials 65 (2020) 1157-1161. https://doi.org/10.24425/amm.2020.133233

[6] P. Kurp, D. Soboń. The influence of laser padding parameters on the tribological properties of the Al_2O_3 coatings, METAL 2018: 27rd Int. Conf. Metallurgy and Materials, Ostrava, Tanger, 1157-1162.

[7] N. Radek, A. Sladek, J. Broncek, I. Bilska, A. Szczotok. Electrospark alloying of carbon steel with $WC\text{-}Co\text{-}Al_2O_3$: deposition technique and coating properties, Adv. Mater. Res. 874 (2014) 101-106. https://doi.org/10.4028/www.scientific.net/AMR.874.101

[8] N. Radek, A. Szczotok, A. Gądek-Moszczak, R. Dwornicka, J. Brončcek, J. Pietraszek.The impact of laser processing parameters on the properties of electro-spark deposited coatings, Archives of Metallurgy and Materials 63 (2018) 809-816. https://doi.org/10.24425/122407

[9] L. Radziszewski, M. Kekez. Application of a genetic-fuzzy system to diesel engine pressure modeling, The International Journal of Advanced Manufacturing Technology 46 (2010) 1-9. https:// doi.org/10.1007/s00170-009-2080-1

[10] M. Kekez, L. Radziszewski, A. Sapietova. Fuel type recognition by classifiers developed with computational intelligence methods using combustion pressure data and the crankshaft angle at which heat release reaches its maximum, Procedia Engineering 136 (2016) 353-358. https:// doi.org/ 10.1016/j.proeng.2016.01.222

[11]J. Pietraszek, A. Szczotok, N. Radek. The fixed-effects analysis of the relation between SDAS and carbides for the airfoil blade traces. Archives of Metallurgy and Materials 62 (2017) 235-239. https://doi.org/10.1515/amm-2017-0035"

[12]K. Trzewiczek, A. Szczotok, A. Gadek-Moszczak. Evaluation of the state for the material of the live steam superheater pipe coils of V degree. Advanced Materials Research 874 (2014) 35-42. https://doi.org/10.4028/www.scientific.net/AMR.874.35

[13]B. Szczodrowska, R. Mazurczuk. A review of modern materials used in military camouflage within the radar frequency range, Technical Transactions 118 (2021) art.e2021003. https://doi.org/10.37705/TechTrans/e2021003

[14]A. Kubecki, C. Śliwiński, J. Śliwiński, I. Lubach, L. Bogdan, W. Maliszewski. Assessment of the technical condition of mines with mechanical fuses, Technical Transactions 118 (2021) art. e2021025. https://doi.org/10.37705/TechTrans/e2021025

Terotechnology XII
Materials Research Proceedings **24** (2022) 39-45

Materials Research Forum LLC
https://doi.org/10.21741/9781644902059-7

Characterization and Properties of Diamond-Like Carbon Coatings

KOWALCZYK Joanna[1,a] *, KĘCZKOWSKA Justyna[1,b] and SUCHAŃSKA Małgorzata[1,c]

[1]Kielce University of Technology, al. Tysiaclecia P.P. 7, 25-314 Kielce, Poland

[a]jkowalczyk@tu.kielce.pl, [b]justynak@tu.kielce.pl, [c]suchanska@tu.kielce.pl

Keywords: Diamond-Like Carbon DLC, Tribology, Wear

Abstract. The paper presents the results of the research on the tribological properties of the DLC coating. The coating was applied using plasma-assisted chemical vapour deposition (PACVD). The obtained coating was determined to be a DLC coating using a Raman spectroscope. Friction tests were performed using a tribometer working in a ball-disc friction configuration at various loads applied. For the tests, a disc with a DLC coating was used as a sample, and a ball made of 100Cr6 steel was used as a counter-sample. The friction tests were carried out in the conditions of technically dry friction. Examination of the coating was done using a scanning microscope. Analysis of the geometric structure of the sample before and after the friction test was performed using a confocal microscope with an interferometric mode. The obtained test results indicated that the properties of DLC coatings are influenced by the deposition process conditions: the argon and methane flow ratio. Whereas the deposition time influences the tribological properties.

Introduction

DLC coatings have aroused great scientific and industrial interest since the beginning of the 70s due to their very good mechanical, physical [1], tribological and anti-corrosion properties [2]. The coatings have a very smooth surface [3–5], high hardness [1,3–6], low friction coefficient, high wear resistance [1,3,4,7,8], good thermal conductivity, high transparency [1,5], good chemical inertness [1,4–6] and biocompatibility [1,4,7,9]. In addition, the coatings can be applied for protective purposes, e.g. to protect other surfaces against corrosion [10,11].

Thanks to all these properties, DLC coatings have been widely used in the production of magnetic disks [5], cutting tools [3,5], mechanical parts, biomedical and optics devices etc. [5].

The DLC coating consists of hybridised bonds of sp1 type (acetylene-like), sp2 type (graphite) and sp3 type (diamond) [1,5,12]. The content of sp2 and sp3 bonds affects the optical, electrical and mechanical properties of the coating [1,5]. It is believed that sp2 and sp3 bonds also have an influence on some biomedical properties [1]. The relatively high content of sp3 hybridisation guarantees very good mechanical properties: Young's modulus of about 300 GPa, hardness above 17 GPa and the friction coefficient lower than 0.05. The physical and mechanical properties of the resulting (modified) surface depend on the type of substrate, method of deposition, layer thickness, applied precursors, etc. [4].

Currently, various methods of DLC coating deposition are used. Essentially, the methods can be divided into two groups: chemical vapour deposition (CVD) and physical vapour deposition (PVD) [3].

Raman spectroscopy is used to examine DLC coating. This method is widely used due to its comparability and non-destructive testing process [5].

Terotechnology XII   Materials Research Forum LLC
Materials Research Proceedings 24 (2022) 39-45   https://doi.org/10.21741/9781644902059-7

When analysing DLC coatings using Raman spectroscopy, a characteristic band is visible in the area of 1000-1700 cm^{-1}, which is band D and band G. The sp2 carbon atoms are centred in around 1355 cm^{-1} to 1550 cm^{-1} 1. The G band is called the graphite band and the D band is the "disorder" band from which useful information can be derived about DLC coating binding characterised by hardness and chemical composition. Raman spectra are used to determine sp2 bonds and the sp2/sp3 bonds ratio [6].

Ižák et al. [6], during their research, applied DLC coatings using the physical vapour deposition method with a pulsed arc system at various flow rates Ar and/or N2, various numbers of pulses and polarization voltage. The resulting coatings were tested with Raman spectroscopy. Changes in the number of pulses and the number of carbon atoms on the samples deposited in the vacuum chamber reflect the thickness and morphology of the DLC layers.

The paper describes a carbon coating tested with a Raman spectrometer to check if DLC was formed. Additionally, selected tribological properties of the deposited layer were examined.

Material and Method
The coating was deposited using the physical vapour deposition method (PVD) and plasma-assisted chemical vapour deposition method (PACVD). A Nanomaster NPE-4000 device was used for this purpose.

The sample was initially cleaned and then put into the chamber of the NPE-4000 device. A vacuum of approx. 3.5 · 10^{-5} Torr was generated for this test. Then the sample was cleaned in a vacuum using argon. The cleaning procedure was continued for 15 minutes. Chromium interlayer was deposited between the substrate and the DLC coating, to ensure adhesion of the coating to the substrate. The chromium deposition process took 30 minutes. The chromium interlayer was applied using the PVD method. A chrome target was used for the chromium application. DLC coating was applied using the PACVD method with the parameters given in Table 1.

Tabele 1. The coating application parameters

Cleaning				Interlayer				Coating			
Ar	CH₄	time	RF	Ar	CH₄	time	DC	Ar	CH₄	time	RF
30	0	15 min	200 W	30	0	30 min.	200 W	50	10	2 godz.	150 W

The resulting coating was tested to check if the obtained coating is a diamond-like coating. A Raman spectrometer was used for the test.

While using a scanning electron microscope Phenom XL equipped with an EDS microanalyser, the elements contained in the sample used were observed and identified

Tribological tests were conducted on a TRB³ working in the ball-on-disk pair in a reciprocating motion. The tests were conducted with the parameters summarized below:

- load P = 10 N i 1N,
- sliding speed v = 0.02 m/s,
- number of cycles = 10 000,
- moisture content 24 ± 2%,
- ambient temperature T_0 = 25 ± 2˚C,
- friction combination: steel ball (100Cr6, 6 mm) – discs with DLC coating.
- technically dry friction.

After the tribological tests the linear wear, friction coefficient and wear area were analysed. The wear area were observed using a confocal microscope with Leica DCM8 interferometric mode.

Terotechnology XII
Materials Research Proceedings **24** (2022) 39-45

Materials Research Forum LLC
https://doi.org/10.21741/9781644902059-7

Results and Discussion

The obtained coating was examined using a Raman spectrometer. Figure 1 shows the obtained examination results.

The obtained test results proved that the coating produced is a DLC coating. The spectrum obtained is typical for this kind of coating, namely the G and D bands are in the spectral range of 1200 to 1700 cm^{-1}. This is a characteristic feature of carbon structures containing graphite-like phases.

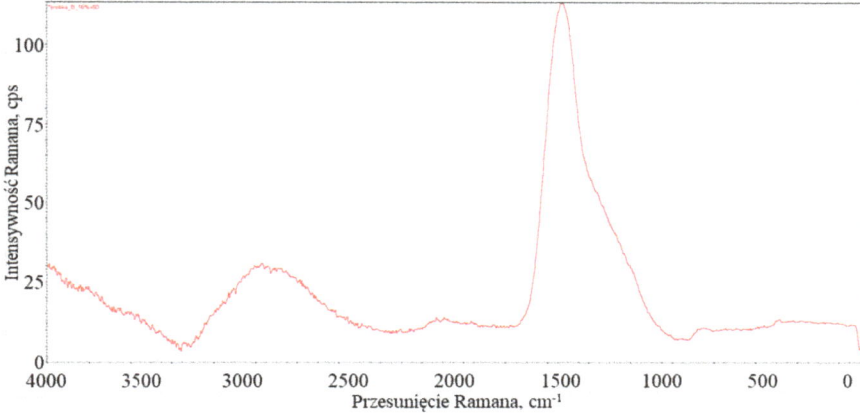

Fig. 1. Raman spectrum for the resulting coating.

Additionally, the deposited layer was examined using a scanning electron microscope equipped with an EDS analyser to check its chemical composition (Fig. 2).

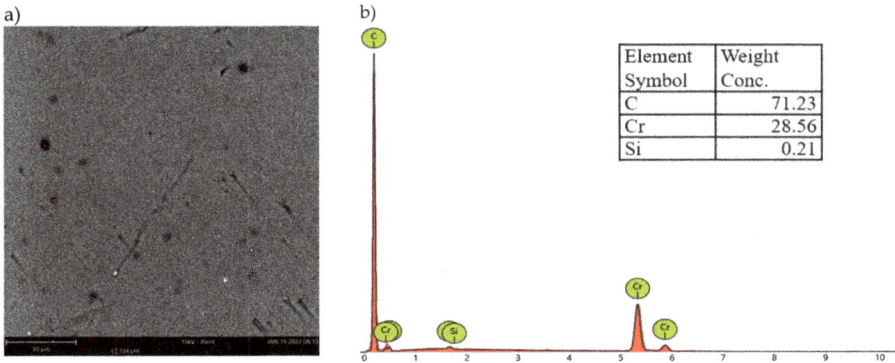

Element Symbol	Weight Conc.
C	71.23
Cr	28.56
Si	0.21

Fig. 2. *SEM: a) view of the coating and b) the X-ray spectrum together with the chemical composition in the micro-area.*

Terotechnology XII   Materials Research Forum LLC
Materials Research Proceedings **24** (2022) 39-45   https://doi.org/10.21741/9781644902059-7

The resulting coating is a DLC coating as a high carbon content was identified. Chromium was also observed as it was deposited in the interlayer. The coating obtained was as intended. Figure 3 shows isometric images and primary profiles of the disc before friction tests.

When analysing the maximum wear depth and areas of the discs, it was observed that the abrasion marks found on the disc with the a-C:H coating after technically dry friction were larger for the load of 10 N than for the load of 1 N. The maximum depth was more than half as large, and the worn area was 3.5 times larger. The abrasion mark on the disk was also wider for the higher load case.

Figure 4 shows isometric images and primary profiles of the discs after tribological tests.

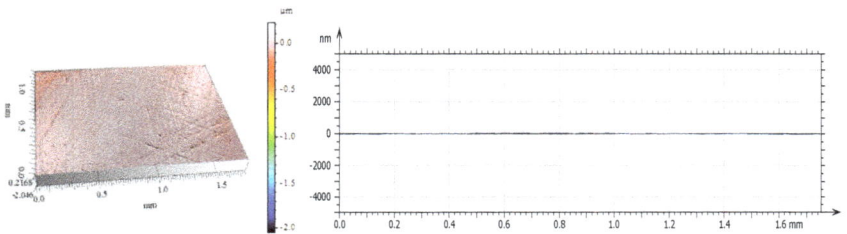

Fig. 3. *Surface textures before the tribological tests:*
isometric views and primary profiles the a-C:H coating.

1. 1 N

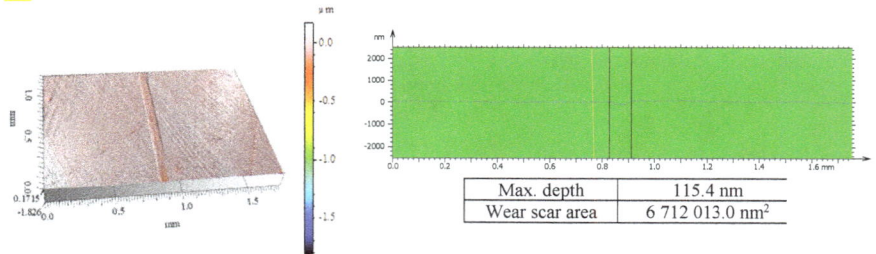

Max. depth	115.4 nm
Wear scar area	6 712 013.0 nm^2

2. 10 N

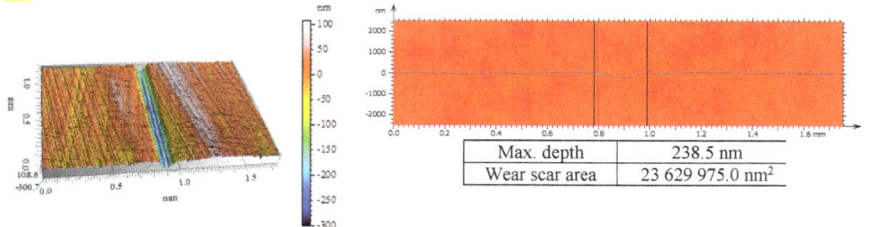

Max. depth	238.5 nm
Wear scar area	23 629 975.0 nm^2

Fig. 4. *Isometric views and primary profiles of the discs*
and balls after the tribological tests.

Table 2. *Surface texture parameters for the disc with coating DLC and before and after the tribological tests.*

Surface texture parameters	Before test	Past test	
		1 N	**10 N**
Sa [nm]	12.38	32.07	30.44
Sq [nm]	15.39	38.77	37.48
Sp [nm]	46.05	109.1	114.7
Sv [nm]	53.18	115.4	121.5
Sz [nm]	99.22	224.5	236.1
Ssk [-]	-0.30	0.65	0.39
Sku [-]	2.83	2.54	2.70

Table 2 presents the roughness of the discs before and after the tribological tests. When comparing the obtained parameters of the geometric structure of the disc surface, lower values of Sp, Sv, Sz, and Sku were found after technically dry friction for the lower load case (1 N). The value of the Sku parameter in all tested cases of the discs was close to 3, which is close to normal distribution. The distribution of unevenness on the measured surfaces was even [13]. Figure 5 shows results after the tribological tests.

a)   b)

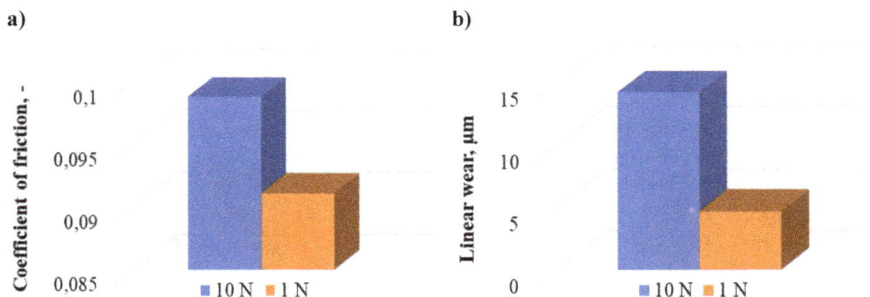

Fig. 5. *a) Coefficient of friction and b) linear wear.*

The wear process causes degradation of the material's strength properties. It occurs due to the interaction of load, such as pressure, thermal and chemical processes [14].

When analysing the results of tribological tests, it was found that lower values of the friction coefficient and linear wear were obtained during the test with a load of 1 N than the test with a load of 10 N. The values were lower by approx. 8% and 67%, respectively.

Summary

To sum up, the following conclusions were drawn from the research results obtained:
1. The tribological properties of the resulting coatings are influenced by the deposition conditions, such as the time and the substrate on which the coatings are deposited.
2. Raman spectroscopy is necessary to determine the type of coating formed. The results of these tests confirmed that diamond-like coatings were produced. A typical spectrum for DLC coatings was obtained. Bands G and D in the spectral range of 1200 to 1700 cm^{-1} were observed.

Terotechnology XII
Materials Research Proceedings **24** (2022) 39-45

Materials Research Forum LLC
https://doi.org/10.21741/9781644902059-7

3. Based on the analysis of the chemical composition performed with a scanning electron microscope equipped with EDS, it was determined that the coating consisted of carbon and chromium. Carbon came from the coating and chromium from the interlayer.
4. When analysing the results of tribological tests, it was found that higher loads translate into a higher friction coefficient and greater linear wear.

References

[1] P. Písařík, M. Jelínek, K.J. Smetana, B. Dvořánková, T. Kocourek, J. Zemek, D. Chvostová. Study of optical properties and biocompatibility of DLC films characterized by sp^3 bonds, Applied Physics A: Materials Science and Processing 112 (2013) 143–148. https://doi.org/10.1007/s00339-012-7216-8

[2] J. Kowalczyk, M. Madej, W. Dzięgielewski, A. Kulczycki, M. Żółty, D. Ozimina. Tribochemical Interactions between Graphene and ZDDP in Friction Tests for Uncoated and W-DLC-Coated HS6-5-2C Steel, Materials 14 (2021) art. 3529. https://doi.org/10.3390/ma14133529

[3] Y. Tokuta, M. Kawaguchi, A. Shimizu, S. Sasaki, Effects of Pre-heat Treatment on Tribological Properties of DLC Film, Tribology Letters 49 (2013) 341–349. https://doi.org/10.1007/s11249-012-0073-y

[4] K. Kyzioł, P. Jabłoński, W. Niemiec, J. Prażuch, D. Kottfer, A. Łętocha, Ł. Kaczmarek. Deposition, morphology and functional properties of layers based on DLC:Si and DLC:N on polyurethane, Applied Physics A: Materials Science and Processing 126 (2020) art. 751. https://doi.org/10.1007/s00339-020-03939-y

[5] H. Sheng, W. Xiong, S. Zheng, C. Chen, S. He, Q. Cheng. Evaluation of the sp^3/sp^2 ratio of DLC films by RF-PECVD and its quantitative relationship with optical band gap, Carbon Letters. 31 (2021) 929–939. https://doi.org/10.1007/s42823-020-00199-x

[6] T. Ižák, M. Marton, M. Vojs, R. Redhammer, M. Varga, M. Veselý. A Raman spectroscopy study on differently deposited DLC layers in pulse arc system, Chemical Papers 64 (2010) 46–50. https://doi.org/10.2478/s11696-009-0092-9.

[7] E.M. Cazalini, W. Miyakawa, G.R. Teodoro, A.S.S. Sobrinho, J.E. Matieli, M. Massi, C.Y. Koga-Ito. Antimicrobial and anti-biofilm properties of polypropylene meshes coated with metal-containing DLC thin films, Journal of Materials Science: Materials in Medicine 28 (2017) art.97. https://doi.org/10.1007/s10856-017-5910-y

[8] K. Milewski, J. Kudliński, M. Madej, D. Ozimina. The interaction between diamond like carbon (DLC) coatings and ionic liquids under boundary lubrication conditions, Metalurgija. 1–2 (2017) 55–58.

[9] M. Madej, D. Ozimina, K. Kurzydłowski, T. Płociński, P. Wieciński, M. Styp-Rekowski, M. Matuszewski. Properties of diamond-like carbon coatings deposited on CoCrMo alloys, Transactions of Famena 39 (2015) 79–88.

[10] N. Radek, A. Szczotok, A. Gądek-Moszczak, R. Dwornicka, J. Bronček, J. Pietraszek. The impact of laser processing parameters on the properties of electro-spark deposited coatings, Archives of Metallurgy and Materials 63 (2018) 809–816. https://doi.org/10.24425/122407

Terotechnology XII   Materials Research Forum LLC
Materials Research Proceedings **24** (2022) 39-45   https://doi.org/10.21741/9781644902059-7

[11] M. Madej, D. Ozimina, K. Marczewska-Boczkowska. Effect of tungsten on the durability of diamond-like carbon coatings in the chemical industry, Przemysł Chemiczny 93 (2014) 500–505.

[12] D. Ozimina, M. Madej, J. Kowalczyk. Determining the Tribological Properties of Diamond-Like Carbon Coatings Lubricated with Biodegradable Cutting Fluids, Archives of Metallurgy and Materials. 62 (2017) 2065–2072. https://doi.org/10.1515/amm-2017-0306

[13] M. Niemczewska-Wójcik, A. Mańkowska-Anopczyńska, W. Piekoszewski. The influence of the surface geometric structure of a titanium alloy on the tribological characteristics of a polymeric component, Tribologia 6 (2014) 97–112.

[14] M. Styp-Rekowski, E. Mańka, M. Matuszewski, M. Madej, D. Ozimina. Tribological problems in shaft hoist ropes wear process, Industrial Lubrication and Tribology 67 (2015) 47-51. https://doi.org/10.1108/ILT-01-2014-0004

Terotechnology XII
Materials Research Proceedings **24** (2022) 46-51

Materials Research Forum LLC
https://doi.org/10.21741/9781644902059-8

Selection of Process Parameters for Laser Texturing of DLC Coatings

KALINOWSKI Artur[1,a] *, RADEK Norbert[2,b], SĘK Piotr[2,c], PIETRASZEK Jacek[3,d] and ORMAN Łukasz[4,e]

[1]Kielce University of Technology, Faculty of Mechatronics and Mechanical Engineering, Al. 1000-lecia P.P. 7, 25-314 Kielce, Poland

[2]Kielce University of Technology, Faculty of Mechatronics and Mechanical Engineering, Al. 1000-lecia P.P. 7, 25-314 Kielce, Poland

[3]Cracov University of Technology, Faculty of Mechanical Engineering, Al. Jana Pawła II 37, 31-864 Kraków, Poland

[4]Kielce University of Technology, Faculty of Geomatics and Energy Engineering, Al. 1000-lecia P.P. 7, 25-314 Kielce, Poland

[a]akalinowski@tu.kielce.pl, [b]norrad@tu.kielce.pl, [c]psek@tu.kielce.pl, [d]jacek.pietraszek@pk.edu.pl, [e]orman@tu.kielce.pl

Keywords: Diamond-Like Carbon, Coatings, Laser Texturing

Abstract. This paper presents the results of the selection of parameters for the DLC coatings laser texturing process using a picosecond laser with a wavelength of 343 nm. It also shows the level of precision with which it is possible to generate a texture with this technology and examples of possible shapes created using TruTOPS PFO software with their selection for the best texture quality.

Introduction

Forming the micro-geometry of friction surfaces is a very common issue in modern machine constructions. Technological progress, including in the field of laser techniques, enables very high precision surface modification. Laser texturing allows for making micro-cavities on the surface of a given material in a quick and repeatable way.

Laser texturing is a type of subtractive micromachining process. The size of the material to be removed is in the range of micro and millimeters. The energy of the laser beam is used to remove material. Laser micromachining is a method commonly used when high and repeatable dimensional accuracy is required and for materials that are difficult to process by other methods. Laser texturing involves forming a desired geometric structure and/or distribution of physicochemical and mechanical properties on the surface of a processed material [1, 2].

The area of the beam impact on the material is determined by the size of the laser spot in focus. The time of the beam pulse is particularly important since different mechanisms of interaction with the material can be used depending on the intensity of the radiation and the exposure time. Pulses longer than 1 nanosecond are called long pulses. Pulses with durations between 1 picosecond and 1 nanosecond are called short pulses, and those with durations less than 1 picosecond are referred to as ultrashort pulses [3].

The crucial aspect is the proper selection of process parameters to obtain texture elements with desired and reproducible geometries and physicochemical properties. The size of the heat-affected zone, or lack of it, depends on the duration of the laser pulses. Treatment with long pulses leaves distinct remelting marks and heat-affected material alterations. Laser ablation induced by

Terotechnology XII   Materials Research Forum LLC
Materials Research Proceedings **24** (2022) 46-51   https://doi.org/10.21741/9781644902059-8

picosecond and femtosecond laser pulses is called cold ablation because no heat-affected zone, in its traditional sense, is observed in the material [4].

In recent years there has been great progress in the development of research and application topics related to carbon materials. It includes obtaining diamond-like coatings, DLC (Diamond-Like Carbon), and applying them by PVD and CVD methods [5-7]. DLC coatings can be used in a variety of tribological pairs [8].

The choice of using UV wavelength laser is because DLC coatings are mainly transparent to the visible and IR spectrum of light but can absorb UV radiation.

Experimental
The test specimens were 4H13 steel samples with antiwear a-C:H DLC coating containing vanadium and chromium interlayers. The coating was applied using PVD technology. The thickness of the coating was approximately 1.5 μm. DLC coatings are characterized by very good mechanical properties (including low coefficient of friction and high hardness). These properties make DLC coatings widely used for machine parts that are subjected to high loads and increased wear and subjected to increased wear. The composition of the steel used was as follows (wt.%): C: 0.36-0.45, Mn: 0.50-0.80, Cr: 12.0-14.0, Si: 0.60-0.80, Mo: 0.5-0.7, V: 0.2-0.3, Ni: 0.1-0.60, P: max 0.04, S: max 0.03, and the remainder is iron.

The process of PVD deposition of antiwear thin coatings is carried out at increased temperature. The coating was obtained by the following processes and temperatures:

- a-C:H by physical vapor deposition PVD sputtering at < 300°C;
- substrate material temperature 350°C.

Laser texturing of coatings is used to create a reservoir effect on the surface of interacting coatings. This phenomenon occurs when wear particles or a lubricant are stored in the cavities created by the texturing. This results in better distribution of the lubricant and also has a significant effect on reducing wear. The implementation of the experimental plan was divided into two stages. In the first stage, the selection of appropriate laser operating parameters was made in order to obtain a texture of optimal quality. The cavity quality, its width, and depth were taken into consideration. The depth was measured on a Talysurf CCI Lite optical profilometer. Six texture shapes were then designed using TruTops PFO software. A ring-shaped specimen of 4H13 steel with a DLC coating was sketched in the software along with different textures, and then its surface was divided into 6 regions to create texture on the surface of one specimen. The texture shapes were then examined for surface quality on a Hirox KH-8700 digital microscope.

Results and Discussion
Parameter selection began by determining the optimal laser power in terms of the highest fluence value and the quality of the groove produced on the surface of the specimen by laser. In order to find the optimal laser texturing conditions, an experiment was conducted. It consisted of evaluating the effect of the laser pulses on the coating surface while changing the scanning speed, with constant pulse frequency and laser power. A test was performed for 14 scanning speeds varied by 5 mm/s. The selection of the laser beam parameters was evaluated by observing the laser beam impact trace. Geometry evaluation was done using Mountains®8 software for surface topography. The evaluation took into account:

- groove depth,
- the presence and shape of the flotation,
- the shape of the groove.

Materials Research Forum LLC
https://doi.org/10.21741/9781644902059-8

The diameter of the laser spot in focus was determined to be 18.2 μm, and the area of interaction of the focused laser beam according to the specifications of the TruMicro 5000 laser was 259.2μm². Calculations from [4] were used to determine the other resulting parameters.

In order to select the optimal parameters for the laser texturing process, a preliminary test was carried out on one of the samples. The following laser parameters were used to perform the test:

- 70% laser power - adjustable power of 3.5 W;
- frequency 66.6 kHz;
- pulse divider - 2.

The examination of microstructure was performed using the Talysurf CCI optical profilometer for surface geometry structure investigation. The system enables the analysis of the geometric structure of the surface with a vertical resolution of up to 0.01 nm. Data analysis was performed using TalyMap Platinum 6.2.7200 software.

Considering the results from the optical measuring machine, sample No. 6 was selected at a beam speed of 110 mm/s with a maximum groove depth of 2.219 μm and a horizontal distance of 43.022 μm. Table 1 shows the parameter selection along with the calculated parameters.

Table 1. *Selecting the scanning speed of the TruMicro 5000 picosecond laser head*

No.	v [mm/s]	P [W]	f [kHz]	Pa [W]	Ie	I%	Io [μm]	In	F [J/cm²]	Maximal depth	Surface area [μm²]	Vertical distance [μm]
1	85	3.5	66.66	0.58	8.75	0.85	1.28	14	3.14	2.54	83.051	55.941
2	90	3.5	66.66	0.58	8.75	0.85	1.35	13	3.13	3.331	93.975	48.19
3	95	3.5	66.66	0.58	8.75	0.84	1.43	13	3.11	3.068	91.693	48.19
4	100	3.5	66.66	0.58	8.75	0.83	1.50	12	3.10	3.493	132.308	48.19
5	105	3.5	66.66	0.58	8.75	0.82	1.58	12	3.08	2.721	66.113	48.19
> >6	110	3.5	66.66	0.58	8.75	0.81	1.65	11	3.07	2.219	54.217	43.022
7	115	3.5	66.66	0.58	8.75	0.80	1.73	11	3.06	2.714	79.329	67.759
8	120	3.5	66.66	0.58	8.75	0.79	1.80	10	3.04	2.838	46.727	55.885
9	125	3.5	66.66	0.58	8.75	0.79	1.88	10	3.03	2.919	59.424	43.13
10	130	3.5	66.66	0.58	8.75	0.78	1.95	9	3.01	2.595	50.224	35.116
11	135	3.5	66.66	0.58	8.75	0.77	2.03	9	3.00	2.724	55.68	56.892
12	140	3.5	66.66	0.58	8.75	0.76	2.10	9	2.99	2.315	42.193	31.666
13	145	3.5	66.66	0.58	8.75	0.75	2.18	8	2.97	2.547	50.034	42.954
14	150	3.5	66.66	0.58	8.75	0.74	2.25	8	2.96	3.39	68.69	37.383

where:

- v – laser head scanning speed [mm/s];
- P – laser power [W];
- Pa – average laser radiation power [W];
- f – laser pulse frequency [kHz];
- Ie – single pulse energy [μJ];
- Io – the distance between pulse centers [μm];
- Ln – Number of pulses per workpiece point;
- F – fluence of laser radiation [J/cm²];

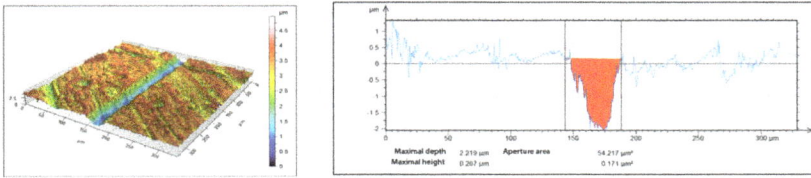

Fig. 1. *Surface structure after treatment with the selected laser*
parameters – 110 mm/s laser head scanning speed.

Fig. 1 shows the surface structure analysis after a single laser beam passage with the selected parameters.

Once the laser processing parameters were selected, the second stage of the experiment was started – texture shape selection. The texture shapes generated using TruTops PFO software were as follows:

- grit with 150 µm distance between beam passes;
- spiral with 100 µm distance between beam passes;
- lines with a distance between beam passes of 400 µm;
- circles 60 µm in diameter with a distance between centers of 100 µm;
- crescents - an arc with a radius of 60 µm and an angle of 180°;
- points with a diameter of 30 µm and a distance of 100 µm.

Fig. 2 shows the shapes of textures taken into consideration.

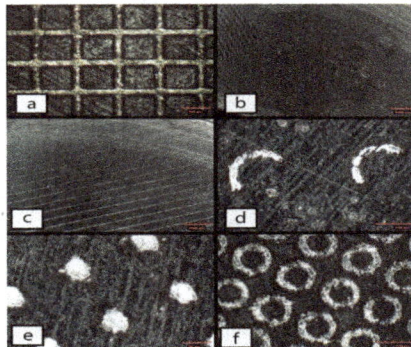

Fig. 2. *Different shapes of the texture with given magnification:*
a) Grid x2000 b) Spiral x35 c) Lines x35 d) Half circles x2000
e) Points x2000 f) Circles x2000.

The quality of the coating after texture generation did not deteriorate. The best texture quality (in terms of groove quality) is shown by the grid and spiral texture. The lines are clearly outlined, and their shape is regular. The point-shaped texture with a diameter of 30 µm shows some shape irregularity. The points do not have a uniform character. The circle-shaped texture also does not show the required repeatability of the cavity depth. The semi-circle-shaped texture like other non-orthogonal textures does not achieve the required coating depth uniformity. The areas where the texture was not made to the required depth are visible. Fig. 3 shows an example of the surface geometric texture results using a Hirox optical microscope. It can be seen that the groove depth is approximately 2 μm. This value matches the values achieved in the first stage of the experiment on the selection of laser texturing parameters.

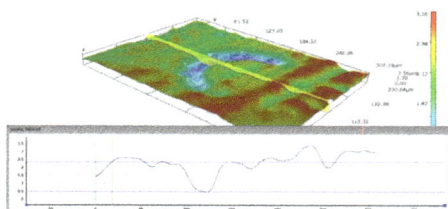

Fig. 3. *DLC coating surface microstructure after applying one type of texture.*

Summary

The aim of this work was to demonstrate the technological capabilities of texturing DLC coatings using a picosecond laser with a wavelength in the UV spectrum at which DLC coatings exhibit light absorption, making it possible to perform texturing on them. This laser is available at the Centre for Laser Technologies of Metals at the Kielce University of Technology. This paper presents the selection of parameters of the laser texturing process on the DLC coating applied on the substrate made of 4H13 steel. The parameters were chosen to produce grooves on the coating with a depth of about 2.13 μm. After selecting different texture shapes and examining them using microscopy, it was found that the best texture quality in terms of groove shape occurred for orthogonal shapes, i.e., shapes such as a grid or straight lines.

The presented results may be useful for people interested in significantly increasing the wear resistance of parts of machines and devices, e.g. biotechnological apparatus [9], chemical accessories [10-12], internal combustion engine accessories [13], or highly responsible elements of weapons [14, 15]. The analyzed process can also be an interesting experimental example for the analysis of adjustment calculus [16] and non-parametric methods [17-19].

References

[1] J. Pietraszek, N. Radek, A.V. Goroshko. Challenges for the DOE methodology related to the introduction of Industry 4.0, Production Engineering Archives 26 (2020) 190-194. https://doi.org/10.30657/pea.2020.26.33

[2] N. Radek, A. Szczotok, A. Gądek-Moszczak, R. Dwornicka, J. Bronček, J.. Pietraszek. The impact of laser processing parameters on the properties of electro-spark deposited coatings, Archives of Metallurgy and Materials 63 (2018) 809-816. https://doi.org/10.24425/122407

[3] W. Żórawski, R. Chatys, N. Radek, J. Borowiecka-Janrozek. Plasma sprayed composite coatings with reduced friction coefficient, Surface & Coatings Technology 202 (2008) 4578-4582. https://doi.org/10.1016/j.surfcoat.2008.04.026

[4] P. Sęk. Production and properties of textured surfaces, Ph.D. thesis, Kielce University of Technology, Kielce, 2014.

[5] M. Madej, D. Ozimina, K. Kurzydłowski, T. Płociński, P. Wieciński, M. Styp-Rekowski, M. Matuszewski. Properties of diamond-like carbon coatings deposited on CoCrMo alloys, Transactions of FAMENA 39/1 (2015) 79-88.

[6] M. Madej, K. Marczewska-Boczkowska, D. Ozimina. Effect of tungsten on the durability of diamond-like carbon coatings in the chemical industry, Przemysł Chemiczny 93 (2014) 500-505.

[7] K. Molewski, J. Kudliński, M. Madej, D. Ozimina. The interaction between diamond like carbon (DLC) coatings and ionic liquids under boundary lubrication conditions, Metalurgija 56 (2017) 55-58.

[8] M. Styp-Rekowski, E. Mańka, M. Matuszewski, M. Madej, D. Ozimina. Tribological problems in shaft hoist ropes wear process, Industrial Lubrication and Tribology 67 (2015) 157-164. https://doi.org/10.1108/ILT-01-2014-0004

[9] E. Skrzypczak-Pietraszek. Phytochemistry and biotechnology approaches of the genus Exacum. In: The Gentianaceae - Volume 2: Biotechnology and Applications, 2015, 383-401. https://doi.org/10.1007/978-3-642-54102-5_16

[10] M. Zenkiewicz, T. Zuk, J. Pietraszek, P. Rytlewski, K. Moraczewski, M. Stepczyńska. Electrostatic separation of binary mixtures of some biodegradable polymers and poly(vinyl chloride) or poly(ethylene terephthalate), Polimery/Polymers 61 (2016) 835-843. https://doi.org/10.14314/polimery.2016.835

[11] M. Dobrzański. The influence of water price and the number of residents on the economic efficiency of water recovery from grey water, Technical Transactions 118 (2021) art. e2021001. https://doi.org/10.37705/TechTrans/e2021001

[12] I. Nová, K. Fraňa, T. Lipiński. Monitoring of the Interaction of Aluminum Alloy and Sodium Chloride as the Basis for Ecological Production of Expanded Aluminum. Physics of Metals and Metallography 122 (2021) 1288-1300. https://doi.org/10.1134/S0031918X20140124

[13] M. Kekez, L. Radziszewski, A. Sapietova. Fuel type recognition by classifiers developed with computational intelligence methods using combustion pressure data and the crankshaft angle at which heat release reaches its maximum, Procedia Engineering 136 (2016) 353-358. https://doi.org/10.1016/j.proeng.2016.01.222

[14] B. Szczodrowska, R. Mazurczuk. A review of modern materials used in military camouflage within the radar frequency range, Technical Transactions 118 (2021) art.e2021003. https://doi.org/10.37705/TechTrans/e2021003

[15] A. Kubecki, C. Śliwiński, J. Śliwiński, I. Lubach, L. Bogdan, W. Maliszewski. Assessment of the technical condition of mines with mechanical fuses, Technical Transactions 118 (2021) art. e2021025. https://doi.org/10.37705/TechTrans/e2021025

[16] T. Styrylska, J. Pietraszek. Numerical modeling of non-steady-state temperature-fields with supplementary data. Zeitschrift fur Angewandte Mathematik und Mechanik 72 (1992) T537-T539.

[17] J. Pietraszek. Response surface methodology at irregular grids based on Voronoi scheme with neural network approximator. 6th Int. Conf. on Neural Networks and Soft Computing JUN 11-15, 2002, Springer, 250-255. https://doi.org/10.1007/978-3-7908-1902-1_35

[18] J. Pietraszek, E. Skrzypczak-Pietraszek. The uncertainty and robustness of the principal component analysis as a tool for the dimensionality reduction. Solid State Phenom. 235 (2015) 1-8. https://doi.org/10.4028/www.scientific.net/SSP.235.1

[19] J. Pietraszek, R. Dwornicka, A. Szczotok. The bootstrap approach to the statistical significance of parameters in the fixed effects model. ECCOMAS 2016 – Proc. 7th European Congress on Computational Methods in Applied Sciences and Engineering 3, 6061-6068. https://doi.org/10.7712/100016.2240.9206

Terotechnology XII

Materials Research Proceedings 24 (2022) 52-58

Materials Research Forum LLC

https://doi.org/10.21741/9781644902059-9

Properties of TiO₂ Coatings Obtained by Atomic Layer Deposition (ALD) on the Ti13Nb13Zr Titanium Alloy

PIOTROWSKA Katarzyna[1,a] *and MADEJ Monika[2,b]

[1] Kielce University of Technology, al. Tysiaclecia P.P. 7, 25-314 Kielce, Poland

[2] Kielce University of Technology, al. Tysiaclecia P.P. 7, 25-314 Kielce, Poland

[a]kpiotrowskapsk@gmail.com, [b]mmadej@tu.kielce.pl

Keywords: Ti13Nb13Zr, Coating, ALD and PVD Technique, Surface Technique, Properties

Abstract. The paper presents the results of research on the surface texture, adhesion, and the results of tribological tests of TiO₂ coatings obtained with the ALD technique. The geometric structure of the surface before and after the tribological tests was assessed using a confocal microscope with an interferometric mode. Model tribological tests were carried out for a reciprocating motion under the conditions of technically dry friction and friction lubricated with Ringer's solution. The tests showed that the TiO₂ coating has better tribological characteristics compared to Ti13Nb13Zr. The study of the geometric structure of the surface after tribological tests showed that the use of the lubricant resulted in a 3-times reduction in wear.

Introduction

Biomaterials consist of a number of materials with different compositions, structures, and properties. Regardless of whether they are made of a metal alloy, ceramic, or polymer, high biocompatibility, i.e. biological compatibility with a living organism, is required from them. After implantation, all biomaterials shouldn't cause allergic reactions and inflammation of the tissues surrounding the implant. They shouldn't also lead to implant rejection. Metallic materials account for the largest share of the global biomedical market. They are characterized by very good strength parameters, corrosion resistance, and well-mastered manufacturing techniques. Despite the continuous improvement of metallic materials, there is still the problem of their stabilization in the human body environment. Harmful products of their wear, including metal ions, entering the bloodstream may cause inflammatory processes and accumulate in tissues and organs [1].

Therefore, for many years, research has been conducted on surface modification of biomaterials to protect metal ions from entering the patient's body. The most commonly considered are various types of surface layers and coatings produced by physical and chemical methods [2-6].

The atomic thin film deposition (ALD) technique is used to produce nanoscale coatings. The system was invented by V. Aleskovsky and his associates in the 1960s. At the turn of the century, the technique was modified and applied mainly in electronics. The first publication using the ALD technique in corrosion protection was published in 1999 [7-9]. Then, there were further publications and patents on the application of the above technique in biomedical materials [10-12]. Previous research on thin coatings has focused on ceramic materials including Al₂O₃ aluminum oxide [13,14]. Although the coatings have very good properties, such as high hardness, and mechanical and thermal strength, they have some disadvantages which should be subject to extensive analysis. Aluminum oxide, Al₂O₃, has very excellent barrier properties, but can be chemically unstable and even dissolve in body fluid environments [15,16]. Therefore, the authors

Terotechnology XII   Materials Research Forum LLC
Materials Research Proceedings **24** (2022) 52-58   https://doi.org/10.21741/9781644902059-9

of this paper proposed the use of TiO_2 coating as a barrier/antiwear layer deposited on Ti13Nb13Zr titanium alloy.

Materials and Methods

The test substrate was titanium alloy Ti13Nb13Zr with the elemental composition shown in Table 1. The choice of material was dictated by its biocompatibility, mechanical properties, and high resistance to dissolution in body fluids [2]. In order to improve the functional and operational properties, a TiO_2 coating was applied to the Ti13Nb13Zr alloy using the atomic thin film chemical deposition (ALD) technique.

Table 1. *Chemical composition of the Ti13Nb13Zr, [wag. %]*

C	H	O	N	Fe	Nb	Zr	Ti
≤ 0.08	≤ 0.015	≤ 0.016	≤ 0.05	≤ 0.25	12.5–14.0	12.5–14.0	rest

The TiO_2 coating was deposited by means of the method of atomic layers deposition from a gaseous phase - ALD, using the Beneq system. Two precursors, Titanium Tetrachloride ($TiCl_4$) and water (H_2O) were used during the process, which was alternately introduced to the chamber. The process temperature was 130 °C, and 700 cycles have been applied. The assumed thickness of the deposited TiO_2 layer was 110 nm.

The surface texture before and after tribological tests was analyzed using a confocal microscope with Leica DCM8 interferometric mode. Axonometric images along with surface profiles on the cross-section are shown in Figure 1.

The tribological tests were carried out using an Anton Paar NTR[3] nanotribometer. Tests were carried out in a reciprocating motion under conditions of technically dry friction and friction in a Ringer solution environment. The applied loading force was 5 mN, the frequency was 1 Hz, cycle number was 10000. The counter-sample in the tested friction pairs was a ball 2 mm in diameter made of Al_2O_3. The lubricant with pH 5.5 and a volume of 1000 ml consisted of NaCl - 8.6 g, KCl 0.3 g, and $CaCl_2$ 0.48 g. The tests were repeated five times for each friction pair with the given parameters.

The adhesion of the coating to the Ti13Nb13Zr substrate was investigated using a scratch test. The test was carried out using an Anton Paar NHT[3] The test consisted in scratching the surface of the specimen with a Sphero-conical diamond indenter with a rounded radius of 2 μm at a load linearly increasing from 0.03 to 5 mN, a loading rate of 20 mN/min and a scratch length of 3 mm.

Results

The functional properties of a material determine the surface roughness. The geometric structure of the surface, especially of the mating elements, should be analyzed in 3D. Fig. 1 and Table 2 show axonometric (3D) images with surface profiles and amplitude parameters of the tested elements. The pink color indicates the average profile.

The analysis of the results of the geometric structure of the surface revealed that the values of all the analyzed amplitude parameters – Sp, Sv, Sz, Sa, Sq were lower for the TiO_2 coating. The positive value of the Ssk parameter for the reference sample informs about the presence of steep hills with sharp tops. The negative value of Ssk of the Ti13Nb13Zr TiO_2 sample proves its plateau surface formation – gentle slopes and rounded peaks.

The results of friction tests were summarized on a graph of the friction coefficient – μ (Fig. 2) of test elements depending on the types of materials, friction pairs and lubricants used.

Materials Research Forum LLC

https://doi.org/10.21741/9781644902059-9

The results of the friction tests carried out indicate that the TiO_2 coating has the best tribological characteristics. The observed friction coefficients were comparable for both materials studied. Tests conducted with a lubricant indicate that resistance to the motion was 4-fold lower compared to values obtained from technically dry friction.

Table. 2. *Parameters of the surface texture*

Parameter	Sp	Sv	Sz	Sa	Sq	Ssk	Sku
Unit	*[μm]*	*[μm]*	*[μm]*	*[μm]*	*[μm]*		
Ti13Nb13Zr	1.42	0.5	1.93	0.06	0.09	1.97	12.27
Ti13Nb13Zr TiO₂	1.24	0.4	1.65	0.05	0.06	-0.02	3.45

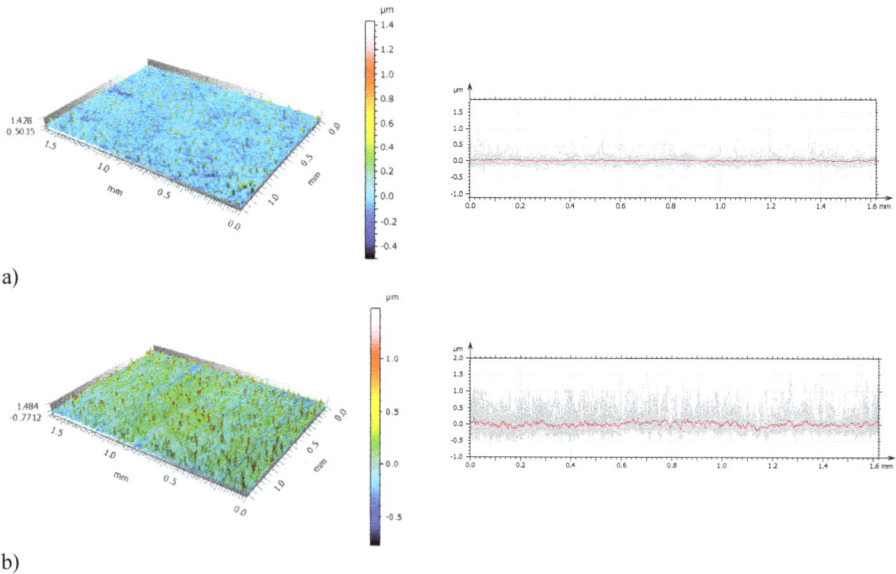

a)

b)

Fig. 1. *Axonometric images and surface profiles of the reference sample Ti13Nb13Zr (a) and TiO₂ coating (b).*

The measure of adhesion of the coating to the substrate is the critical force. This force first causes cracking of the coating L Ft 1 - L Ft 4, and then its destruction - L Ft 5. Fig. 3 shows the results of the scratch test. The critical force was assessed on the basis of the registered changes in the coefficient of friction and frictional force and the indenter penetration depth.

Materials Research Forum LLC
https://doi.org/10.21741/9781644902059-9

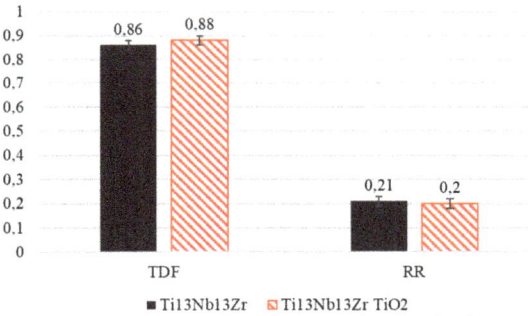

*Fig. 2. Average friction coefficient obtained under dry friction (TDF)
and with the use of Ringer's solution (RS) as a lubricant.*

Tests of adhesion of the TiO$_2$ coating to the substrate showed that the coating was completely delaminated as a result of the applied load. On the friction force and friction coefficient diagrams, characteristic changes proving that the deposited layer was breaking off were registered. These points are marked as L Ft 1 - L Ft 5. At the point L Ft 5, the coating has completely failed. The maximum indenter depth was approximately 160 nm.

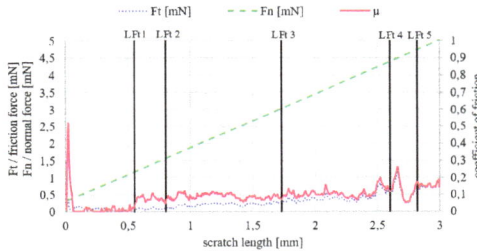

a)

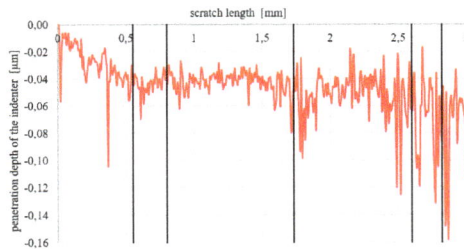

b)

*Fig. 3. Scratch test results of the TiO$_2$ coating a) graph of variation of friction force,
normal force and coefficient of friction, b) graph of penetration depth of the indenter.*

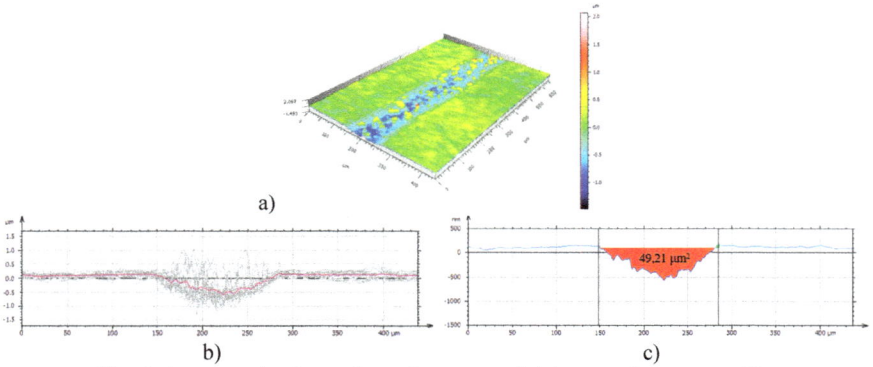

Fig. 4. *Axonometric views of sample wear track (a), mean the wear profile on transverse cross-section (b) volume of the wear track after the RS–lubricated sliding of Ti13Nb13Zr–Al₂O₃ friction pair.*

After tribological tests, the samples were subjected to wear trace measurements. The average depth and average area of wear of the samples after friction in Ringer's solution environment are shown in Fig. 4 and Fig. 5.

Fig. 4 and Fig. 5 show axonometric images of wear traces (a) wear profiles on a cross-section with the average profile marked (b), and the volume of the abrasion trace on the average profile (c) after friction with the use of Ringer's solution as a lubricant. It can be seen from the figure that the TiO_2 coating shows less wear compared to Ti13Nb13Zr. In the case of technically dry friction, the wear trace volume was 90.9 μm^2 for the titanium alloy and 49.5 μm^2 for the TiO_2 coating. The results of the geometric structure of the surface after tribological tests clearly indicate that the application of the coating favorably affects the functional properties of the Ti13Nb13Zr titanium alloy.

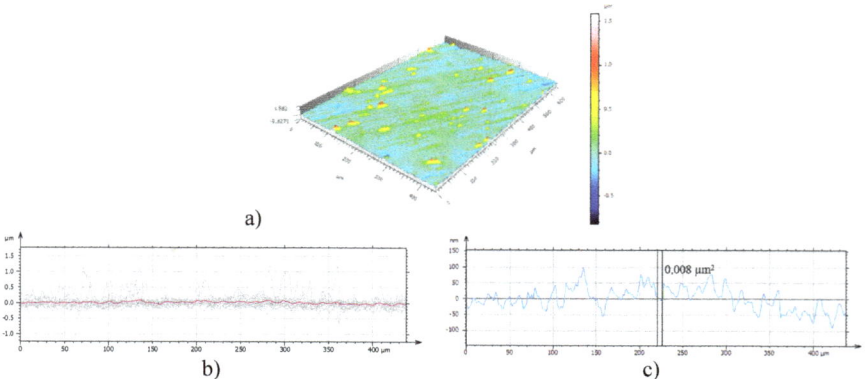

Fig. 5. *Axonometric views of sample wear track (a), mean the wear profile on transverse cross-section (b) volume of the wear track after the RS–lubricated sliding of Ti13Nb13Zr TiO₂–Al₂O₃ friction pair.*

Materials Research Forum LLC
https://doi.org/10.21741/9781644902059-9

Conclusions

The results of tribological tests indicated that both materials tested had similar coefficients of friction. In the case of technically dry friction, it amounted on average to 0.84, and in the case of friction in the environment of Ringer's solution to 0.2. Microscopic analysis of abrasion traces showed that the TiO_2 coating wore less, regardless of the environmental conditions of the tests conducted. Adhesion tests of the coating revealed that the deposited layer delaminated after reaching a critical force of about 4.4 mN. This is indicated by both the friction coefficient and force diagram as well as the indenter penetration change diagram. When the critical force of 4.5 mN was reached, the indenter was at a depth of about 120 nm and the coating thickness was 110 nm.

References

[1] D. Bociąga, K. Mitura. Biomedical effect of tissue contact with metallic material used for body piercing modified by DLC coatings, Diamond and Related Materials, 17 (2008) 1410-1415. https://doi.org/10.1016/j.diamond.2008.02.014

[2] K. Piotrowska, M. Madej, A. Granek. Assessment of Mechanical and Tribological Properties of Diamond-Like Carbon Coatings on the Ti13Nb13Zr Alloy, Open Engineering 10 (2020) 536–545.

[3] M. Madej. Tribological properties of diamond-like carbon coatings, Advanced Materials Research 847 (2014), 9-15. https://doi.org/10.4028/www.scientific.net/AMR.874.9

[4] K. Piotrowska, M. Madej, D. Ozimina. Assessment of the functional properties of 316L steel alloy subjected to ion implantation used in biotribological systems, Materials 14 (2021) art. 5525. https://doi.org/10.3390/ma14195525

[5] M. Vicen, J. Bronček, O. Bokůvka, R. Nikolić, N. Radek. Tribological Behaviour of the Sucaslide Diamond-like Carbon Coating, Transactions of Famena, XLV (2021) 31-40.

[6] N. Radek, A. Szczotok, A. Gądek-Moszczak, R. Dwornicka, J. Bronček, J. Pietraszek. The impact of laser processing parameters on the properties of electro-spark deposited coatings. Archives of Metallurgy and Materials 63 (2018) 809-816. https://doi.org/10.24425/122407

[7] A.A. Malygin, V.E. Drozd, A .A. Malkov, V.M. Smirnov. From V. B. Aleskovskii's "Framework" Hypothesis to the Method of Molecular Layering/Atomic Layer Deposition, Chem. Vap. Depos. 21 (2015) 216–240. https://doi.org/10.1002/cvde.201502013

[8] R. Matero, M. Ritala, M. Leskelä, T. Salo, J. Aromaa, O. Forsén. Atomic layer deposited thin films for corrosion protection, J. Phys. IV France 9 (1999) Pr8-493 - Pr8-499. https://doi.org/10.1051/jp4:1999862

[9] I. Spaji´c, P. Rodi˘c, G. Šekularac, M. Lekka, L. Fedrizzi, I. Milošev. The effect of surface preparation on the protective properties of Al_2O_3 and HfO_2 thin films deposited on cp-titanium by atomic layer deposition, Electrochimica Acta 366 (2021) art. 137431. https://doi.org/10.1016/j.electacta.2020.137431

[10] J.A. Carlisle, M.J. Pellin, J.W. Elam, J. Wang. Hermetic bio-inert Coatings for bio-implants Fabricated Using Atomic Layer Deposition, United States Patent No. 2006/0251875A1 (2006).

[11] D. Karunakaran, A. Jeyachandran. Analysis of Mechanical Properties of Al_2O_3 Coated Dental Implants, International Journal of Engineering Research & Technology 4 (2015) 373-375.

[12] P. Boryło, K. Lukaszkowicz, M. Szindler, J. Kubacki, K. Balin, M. Basiaga, J. Szewczenko. Structure and properties of Al_2O_3 thin films deposited by ALD process, Vacuum 131 (2016) 319-326. https://doi.org/10.1016/j.vacuum.2016.07.013

[13] M. K. Abbass, S.A. Ajeel, H.M. Wadullah. Biocompatibility, bioactivity and corrosion resistance of stainless steel 316L nanocoated with TiO_2 and Al_2O_3 by atomic layer deposition method, J. Phys. Conf. Ser. 1032 (2018) art. 136962. https://doi.org/10.1088/1742-6596/1032/1/012017

[14] D.S. Finch, T. Oreskovic, K. Ramadurai, C.F. Herrmann, S.M. George, R.L. Mahajan. Biocompatibility of atomic layer-deposited alumina thin films, J. Biomed. Mater. Res. A 87 (2008) 100-106. https://doi.org/10.1002/jbm.a.31732

[15] M. Ritala, H. Saloniemi, M. Leskelä, T. Prohaska, G. Friedbacher, M. Grasser-Bauer. Studies on the morphology of Al_2O_3 thin films grown by atomic layer epitaxy, Thin Solid Films 286 (1996) 54-58. https://doi.org/10.1016/S0040-6090(95)08524-6

[16] G.C. Correa, B. Bao, N.C. Strandwitz. Chemical stability of titania and alumina thin films formed by atomic layer deposition, ACS Appl. Mater. Interfaces 7 (2015) 14816-14821. https://doi.org/10.1021/acsami.5b03278

Terotechnology XII
Materials Research Proceedings **24** (2022) 59-65

Materials Research Forum LLC
https://doi.org/10.21741/9781644902059-10

Evaluation of the Influence of Laboratory Weathering of Paint Systems used in Rolling Stock on Selected Mechanical Properties

GARBACZ Marcin[1,a*]

[1]Railway Research Institute, Materials and Structure Laboratory, Chlopicki 50 Street, 04-275 Warsaw, Poland

[*a] mgarbacz@ikolej.pl

Keywords: Rolling Stock, Coatings, Accelerated Weathering, Xenon-Arc Radiation, Hardness, Resistance to Cupping Test, Adhesion

Abstract. This paper evaluates the results of selected mechanical properties of outdoor painting systems used in the railway industry and the influence of laboratory weathering (simulation of outdoor conditions) with sunlight, temperature, and moisture on the change of these properties. The impact of the above-mentioned weather parameters, particularly solar radiation, was briefly described on the degradation of coating systems of weathering on selected mechanical properties such as hardness, resistance to gradual deformation (cupping test), and adhesion (fracture strength) was evaluated. The tests were carried out at the Railway Institute, Materials and Construction Laboratory, using two weathering methods: ISO 16474-2 [1] and the in-house method described in DN 001/08/A2/16 paragraph 4.1.8 [2].

Introduction

The primary purpose of using organic coating systems in rail transport is to protect the metal from damaging factors such as rust, salt, and dirt and improve the vehicle's aesthetics. Multi-layer solvent-based products, increasingly replaced by water-based ones, are used to protect metal surfaces in rolling stock. This is dictated by the European Union regulations being established to reduce organic compound emissions caused by organic solvents [3].

The application of organic coatings in rolling stock can cover many areas, including car bodies, undercarriages, roofs, axles and wheels, and the interior of train equipment. Depending on the application site, different properties and requirements are expected from the paint system used. The most exposed to the negative effects of weathering are the external coating locations - mainly the car bodies. Such coatings are affected by several factors that cause their degradation over time, ranging from environmental pollution (chemical and biological), through aggressive media used in the form of cleaning agents (often incompetently and contrary to the recommendations of the agent manufacturers), to harmful solar radiation with high temperature and humidity, which are the leading cause of weakening the functional and decorative properties of the systems discussed. The radiation mainly affects the top layers of the applied protective system. Currently, polyurethane and acrylic products are used as the last layer of protection in rail transport. They are available in the form of a topcoat or base paint finish with a clear varnish, which provides the desired decorative properties, weather resistance, and protection against graffiti [4].

Degradation factor: UV radiation

The process of coatings degradation by solar radiation begins with photolysis. By absorbing ultraviolet (UV) radiation, the polymer enters an excitation state (to a higher energy state), which contributes to the breaking of covalent chemical bonds in its structure and the formation of reactive free radicals that initiate further chemical transformations. The ease with which the covalent bonds break depends on the energy of these bonds. The next step in the photochemical degradation of the polymer is self-oxidation, which occurs due to the reaction of free radicals with oxygen - with the simultaneous formation of peroxyl radicals. Further reactions follow the general scheme of free radical reactions. Peroxyl radicals react with the polymer molecules, attacking the hydrogen atoms in the polymer chains, resulting in the formation of hydroperoxides and other free radicals. The hydroperoxides (which are highly susceptible to photolysis) are broken down and the free radicals formed during the entire process of coating material destruction damage the polymer molecules. Free radicals can cause depolymerisation of macromolecular compounds, chain scission and oxidation of smaller molecules [5].

It should be emphasised that if a polymer does not absorb radiation, it does not degrade but may nevertheless be susceptible to attack by free radicals from the absorption and excitation of other film-forming materials (e.g. fillers). Aliphatic polymers are UV-resistant as they do not absorb ultraviolet radiation and are permeable. These polymers include acrylates, polyvinyl acetate and aliphatic polyurethanes. Aromatic organics such as epoxies and phenolic resins, on the other hand, have high UV absorption and are subject to photolysis and photo-oxidation. UV radiation can also "attack" the UV-sensitive layers beneath transparent topcoats. When using such transparent varnishes, care must be taken to ensure that light-sensitive primers and base surfaces are adequately protected from ultraviolet radiation [6].

Many methods can be used to protect paint systems against UV radiation. One of these, known as selective absorption, involves the appropriate modification of non-light-resistant top coats with pigments, other absorbers, or UV reflectors. Selective absorption of ultraviolet light by pigments protects the binder and increases the durability of the paint film. Materials such as zinc oxide, zinc sulfide, red iron oxide, carbon black and rutile titanium dioxide are effective absorbers of ultraviolet light and provide good protection for sensitive polymers. Other methods of radiation protection for coatings include:
- use of reflective pigments which act as a mirror and reflect light (e.g. aluminum flake pigments),
- use of luminescent pigments, which absorb the energy of near-ultraviolet radiation and re-emit it as a longer wavelength to produce colors with a much higher saturation or intensity in the red, orange, yellow and green wavelengths,
- the use of organic additives that selectively absorb ultraviolet light and are essentially similar to those used in suntan lotions. Like pigments, they absorb short-wave ultraviolet light and convert it into heat energy without degrading it when exposed to ultraviolet light [6].

Degradation factor: temperature and humidity
In addition to destructive radiation, a factor contributing to the destruction of the polymer is the effect of thermal radiation, known as thermal degradation. The temperature of the material is influenced by many factors: air temperature (depending on latitude, prevailing climatic conditions, and season), the dose of infrared radiation, airflow within the specimen (thermal exchange of the material with the environment) and the properties of the material itself (color and thermal conductivity coefficient). The absorption of radiation is closely correlated with the color of the material exposed. Lighter materials reflect more radiation than darker ones, which absorb it. Also, in this case, the surface and structure of the material (e.g. its polishing) are of great importance. The actual temperature of a material is often up to 30°C higher than the measured ambient temperature. It is important to realize that temperature significantly affects the properties of a material by increasing the rate and number of chemical and photochemical reactions in the material caused by UV radiation. It also increases the rate of diffusion of fillers (e.g. stabilizers, plasticizers)

as well as additional components in the form of impurities from the atmosphere. Rapid temperature changes (cooling and heating) result in shrinkage and stretching of the material, with consequent cracking and peeling. The degradation of the material can be further increased by water absorption and desorption (hydrolytic degradation), so these two elements must be considered as a parameter that together increases the degradation of the material. In addition, rainwater can provide pollutants from the atmosphere and thus accelerate material degradation [7].

Other degradation factors
In addition to the factors mentioned above, factors contributing to the degradation of paint systems are mechanical degradation (under mechanical stress), oxidative degradation (under the influence of oxygen), hydrolytic degradation under the influence of an agent other than water, such as acids and alkalis, and biological degradation. In nature, we most often deal with a synergic correlation of all factors [8].

The above-mentioned degradation processes occurring in the polymer deteriorate the coating's physical, chemical, and mechanical properties, which in particular leads to a decrease in corrosion resistance and susceptibility to stretching and bending, while abrasion and scratch resistance (hardness) brittleness increase. Exploitation factors affecting the paint coating mainly increase the roughness of its surface and consequently its decorative properties, causing it to deteriorate, e.g. by loss of gloss, color change, fading, or local discoloration. External factors, coming into contact with polymeric coatings, generate their physical degradation in the form of silver cracks (their name derives from the characteristic silvery reflections visible in the reflected light of a microscope), etching and craters, blistering, inter-layer delamination, loss of adhesion between the coating and the substrate. There is also an increase in the roughness of the surface and a decrease in the resistance to chemicals used for cleaning or the effectiveness of anti-graffiti protection due to an increase in the adhesion of the coating. Consequently, the coating material degradation leads to a reduction in its service life [5,8].

Research methodology and research material:
The performance of a paint system under natural conditions can and should be checked by conducting accelerated laboratory tests simulating sunlight and other laboratory weather conditions such as temperature and humidity. Accelerated laboratory tests should simulate and approximate the conditions under which the materials are used as closely as possible. A large body of scientific literature on organic coatings shows that the best correlation between laboratory test results and results obtained under natural conditions is obtained by combining different degradation mechanisms. When only the effects of radiation, temperature, and humidity (including rain) are considered to simulate these conditions in the laboratory, many test methods have been developed worldwide, including standardized ones. One of such methods, which is the most commonly used globally for testing paints and varnishes, is the series of standards PN-EN ISO 16474, which collects (contains) the basic and most important information on carrying out such tests in laboratories [7].

Weathering tests in IK were carried out according to the second part of the standard mentioned above, which assumes the use of apparatuses equipped with xenon arc lamps for this purpose. The total irradiance was 324 MJ/m^2, theoretically about 1.5 years under natural conditions for a temperate latitude climate [9].

In addition, this paper presents similar test results for simulated varying atmospheric conditions according to the procedure developed at IK in which cyclic irradiation is used with a table lamp with a filament (heating function) and an ultraviolet radiator of 400W with a reflector made of varnished aluminum sheet. The light used has no UV filter and emits a discrete spectrum (range from 238 nm). Short wavelengths transmit higher energy, which can be the initiator of photochemical reactions and thus lead to faster destruction of coatings. Weathering in the climatic chamber has a different degradation mechanism than that carried out in chambers with xenon lamps; namely, high positive and negative temperatures and their rapid changes are simulated in the climatic chamber. Under the influence of the temperature shock, coatings undergo shrinkage and stretching of the material, which consequently causes cracks and peeling and a decrease in selected physical properties of the coating, such as elasticity or adhesion.

The exact weathering methodologies and assessment methods after aging and information on the testing apparatus used are shown in Table 1.

The research material consisted of exterior painting systems commonly used on Poland's railway car bodies and roofs. The tested systems were composed of two-component and multi-component systems, depending on their application place, while the base was different grades of structural steel. The compositions of the systems numbered IV, VI, and XI, which include train roofs as their primary application place, consisted of epoxy primers and polyurethane and acrylic topcoats or their mixtures. The remaining systems consisted of epoxy primers, epoxy or polyester putties, polyurethane, acrylic, and mixed filling primers and finishing coats of the topcoat type (systems I, II, IV, VI, VII, XI) or basecoat + clearcoat type (systems III, V, VIII, IX, X), also based on polyurethane, acrylic resins and their mixtures. The average thickness of the tested multi-component systems was within the range of (200 ÷ 300) μm, except for system number III (a thick putty layer was used here, and the average thickness was about 700 μm). In contrast, the average thickness of the two-component systems was about 150 μm.

Table 1. Research methodology along with the apparatus used for this purpose.

N o.	Type of test	Reference document (method)	Apparatus	Remarks:
			Weathering methods:	
1	Resistance to UV radiation (standardized method)	ISO 16474-2 [1]	Light test chamber, Atlas GmbH Xenotest 440, XENOSENSIV RC 34 BST	methodology: PN-EN ISO 16474-2, point 7.3 table 3 method A; filters simulating daylight with BST surface temperature control (65°C) and radiation control in the wavelength range (300 ÷ 400) nm; irradiance: 60W/ m²; chamber temperature 38°C, relative humidity 50%, test periods: 102 min dry, 18 min rain; total irradiation dose: 324 MJ/m² (approx.. 1500h).
2	Resistance to UV radiation (in-house method)	DN 001/08/A2/16 [2]	Climatic chamber SECASI SI550C150F40H lamp Famed-1 type L8/59 (UV-C)	Cyclic exposure to varying weather conditions: 1) T: + 60°C, RH: 95%, time 12 h; 2) T: - 20°C, time 6 h; 3) T: + 60°C, RH: 65%, time 6 h; 4) UV irradiation for 4 h every 48 h (Famed-1 lamp type L8/59). Testing time: 1500 h.
			Assessment methods after aging:	
3	Determination of hardness	ISO 1522 [10]	TQC Sheen TI SP 0500	König's pendulum was used for the measurements; the measured hardness was related to the hardness against the glass (calibration)

| 4 | Determination of resistance to indentation | ISO 1520 [11] | Erichsen 202 EM | minimum depth of indentation to cause failure to be determined (steady rate 0,2 mm/s); to be assessed with the normal corrected vision |
| 5 | Determination of adhesion/cohesion (fracture strength - Cross-cut test or X-cut test) | ISO 2409 [12] ISO 16276-2 [13] | Hand-held multi-blade cutting tool by Erichsen (type B, C) or cutter with a rigid blade by Erichsen van Laar (with a V-shaped cutting edge) | method dependent on coating thickness, above 250 μm an X-shaped cut was used; Tesa tape used with adhesion strength according to the standard |

Evaluation of results and conclusions

Table 2. *Summary of results of adhesion, indentation, and hardness measurements before and after weathering test according to ISO 16474-2 [1] and DN 001/08/A2/16 [2].*

System* No.	Fracture strength – ISO 2409 [12] / 16276-2 [13], [degree of adhesion]. (P=95%, k=2, ± 1 s.p.)			Indentation – ISO 1520 [11], [mm]. (P=95%, k=2, ± 0,4 mm)			Hardness – ISO 1522 [10] [absolute value related to the hardness of the glass]. König (P=95%, k=2, ± 0,04)		
	unaged material	ISO 16474-2	DN 001/08/A2/16	unaged material	ISO 16474-2	DN 001/08/A2/16	unaged material	ISO 16474-2	DN 001/08/A2/16
I	1	2	2	11.4	3.7	5.3	0.33	0.44	0.43
II	1	2	2	9.0	1.1	1.0	0.35	0.45	0.48
III	1	1	1	0.8	0.4	0.7	0.33	0.58	0.72
IV	1	2	1	7.6	2.3	1.7	0.51	0.73	0.80
V	0	0	0	8.7	1.9	2.0	0.26	0.59	0.62
VI	0	0	1	10.0	7.8	7.9	0.41	0.48	0.55
VII	0	0	0	6.6	5.7	5.3	0.08	0.12	0.12
VIII	1	1	1	7.2	6.5	6.0	0.07	0.08	0.11
IX	1	2	-	-	-	-	0.17	0.35	-
X	1	1	-	-	-	-	0.41	0.53	-
XI	1	0	-	-	-	-	0.52	0.53	-

* bold indicates systems with topcoat: basecoat + clearcoat, all other cases topcoat.

The adhesion test carried out using the cross-cut method or X-cut method (depending on the thickness of the system) did not show any significant deterioration of this parameter after the aging tests. The most significant difference achieved in pre-aging and post-aging results for both aging methods was, in each case, only 1 degree of adhesion. In each case, only a decrease in adhesion was observed between the layers of the system (no delamination from the base).

In the case of tests on resistance to indentation, in most cases, relatively large drops of this parameter after aging were observed; however, some of the systems tested retained their properties at a satisfactory level (systems VI, VII, VIII). Also, in this case, similar results were obtained for both aging methods used, where the average drops in adhesion for all tested systems were 4.0 mm for the ISO 16474-2 method [1] and 3.9 mm for the method according to DN 001/08/A2/16 [2], respectively.

When evaluating hardness, an increase in this parameter was observed in every specimen tested. A more significant increase was observed in the case of aging according to an in-house aging method.

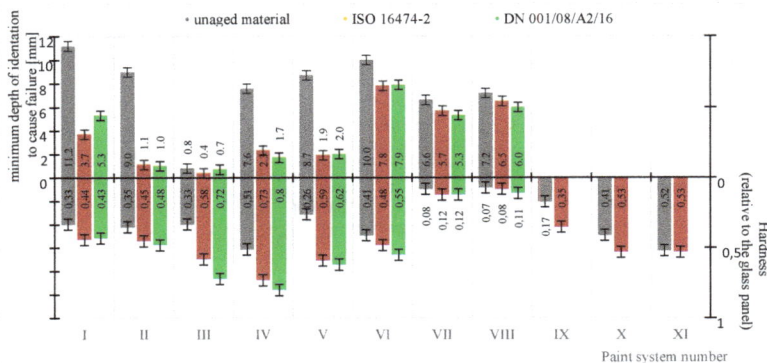

Fig. 1. *Graphical summary of the minimum indentation depth causing damage to the coating [mm] according to ISO 1520 [11] and hardness (about the glass constant) determined by the König pendulum damping method according to ISO 1522 [10] for selected tested paint systems.*

There is noticeable a certain correlation between an increase in hardness (and therefore brittleness) and a decrease in such parameters as adhesion or resistance to indentation – the more significant the increase in hardness, the greater the negative change in such parameters as adhesion or resistance to indentation. A high hardness value is a recommended parameter due to the increase of, among others, scratch resistance. However, other properties of coatings related to their elastic properties to be complete cannot be neglected – as in every case, a compromise solution has to be found.

Summary

As a result of selected mechanical tests carried out for complete paint systems used in rolling stock, it is concluded that, as a result of environmental tests conducted in the laboratory with a synergistic combination of light radiation, temperature, and moisture, there is a significant deterioration of mechanical properties such as resistance to indentation and, to a lesser extent, a decrease in the adhesion of the paint system (interlayer), while the hardness (brittleness) of the paint systems increases.

Although two different aging methods were used to evaluate the selected mechanical properties, differing primarily in the type of radiation (full light spectrum for the xenon UV-VIS-IR lamps consistent with natural light vs. UV including UV-C in the form of discrete radiation for the table lamp for the own method) and the different cycles of varying temperature and humidity, the similar character of changes were obtained and the results in specific cases were practically convergent, which allows the conclusion that the intensity of the contribution of individual aging factors on the mechanical properties is of secondary importance here and it is only essential that each of these factors is taken into account in the chosen aging method (interconnection of degradation mechanisms).

When evaluating the selected mechanical parameters of the painting systems, no clear advantage of the paint systems in the combination of base paint + clear coat over the surface paint finish was found. Based on the conducted research, it can be concluded that the paint systems used in the railway industry are characterized by high mechanical properties, which, however,

Terotechnology XII   Materials Research Forum LLC
Materials Research Proceedings **24** (2022) 59-65   https://doi.org/10.21741/9781644902059-10

deteriorate considerably as a result of environmental stress. The average decrease in adhesion for all 11 tested paint systems was approximately 0.3 adhesion units. In contrast, the reduction in resistance to indentation reached the value of a 50% decrease (the results were the same for both aging methods and referred to the average values determined before aging). The average increase in hardness was 42% for the aging method according to ISO 16474-2 [1], while for the aging method described in DN 001/08/A2/16 [2], the hardness increase was 54%. Considering the above data, it is reasonable to consider conducting selected tests of coatings after aging tests to fully assess their quality, allowing to estimate the time of their satisfactory use until the required repainting. In the railway industry, this is quite important because the downtime of trains in painting shops and the renovation of the system itself incur high financial expenses.

References

[1]   ISO 16474-2:2013 Paints and varnishes – Methods of exposure to laboratory light sources – Part 2: Xenon-arc lamps.

[2]   DN 001/08/A2/16, „Wyroby lakierowe stosowane w pasażerskim taborze szynowym w lokomotywach, wagonach i zespołach trakcyjnych", Instytut Kolejnictwa 2016.

[3]   Directive 2004/42/CE of the European Parliament and of the Council of 21 April 2004 on the limitation of emissions of volatile organic compounds due to the use of organic solvents in certain paints and varnishes and vehicle refinishing products and amending Directive 1999/13/EC.

[4]   M. Garbacz. Wpływ starzenia laboratoryjnego na wybrane właściwości fizyko-chemiczne systemów malarskich stosowanych w transporcie szynowym. Wyd. Naukowe Inst. Kolejnictwa, Warszawa, 2021, ISBN 978-83-943246-9-8.

[5]   D. Kotnarowska, M. Sirak. Destrukcja nawierzchniowych powłok akrylowych starzonych promieniowaniem UV. Autobusy 6 (2017) 816-821.

[6]   C.H. Hare. The Degradation of Coatings by Ultraviolet Light and Electromagnetic Radiation, Anatomy of Paint. Journal of Protective Coatings & Linings, May 1992.

[7]   M. Garbacz. Badania starzeniowe materiałów stosowanych w taborze szynowym z symulacją światła słonecznego i warunków pogodowych. Prace Instytutu Kolejnictwa 157 (2018) 5-15.

[8]   D. Kotnarowska. Analysis of polyurethane top-coat destruction influence on erosion kinetics of polyurethane-epoxy coating system. Eksploatacja i Niezawodność – Maintenance and Reliability 21 (2019) 103–114, https://doi.org/10.17531/ein.2019.1.12

[9]   Solar Spectral Irradiance, International Commission on Illumination (CIE). Publication No. 85,1, January 1989.

[10]   ISO 1522:2006 Paints and varnishes – Pendulum damping test.

[11]   ISO 1520:2006 Paints and varnishes – Cupping test.

[12]   ISO 2409:2020 Paints and varnishes – Cross-cut test.

[13]   ISO 16276-2:2007 Corrosion protection of steel structures by protective paint systems – Assessment of, and acceptance criteria for, the adhesion/cohesion (fracture strength) of a coating – Part 2: Cross-cut testing and X-cut testing.

Terotechnology XII
Materials Research Proceedings 24 (2022) 66-74

Materials Research Forum LLC
https://doi.org/10.21741/9781644902059-11

Virtual Methods of Testing Automatically Generated Camouflage Patterns Created Using Cellular Automata

PRZYBYŁ Wojciech[1, a*], MAZURCZUK Robert[1,b], SZCZEPANIAK Marcin[1,c], RADEK Norbert[2,d] and MICHALSKI Marek[3,e]

[1]Military Institute of Engineer Technology, ul. Obornicka 136, 50-961 Wrocław, Poland

[2]Kielce University of Technology, Faculty of Mechatronics and Mechanical Engineering, Al. 1000-lecia P. P. 7, 25-314 Kielce, Poland

[3]F. H. Barwa, ul. Warkocz 3-5, 25-253 Kielce, Poland

[a]przybyl@witi.wroc.pl, [b]mazurczuk.r@witi.wroc.pl, [c]szczepaniak@witi.wroc.pl, [d]norrad@tu.kielce.pl, [e]mmichalski@barwa.kielce.pl

Keywords: Military, Camouflage, Image Processing, Virtual Reality, Circular Automata

Abstract. The article presents a method of designing and assessing the camouflage effectiveness of a novel camouflage pattern used in military uniforms. "Artificial intelligence", namely cellular automata, which, based on terrain imaging, allowed the authors to obtain dedicated, tailor-made, or universal camouflage patterns, was used as the design method. On the other hand, the study on camouflage effectiveness was based on tests with the participation of observers in a virtual environment using a personal computer.

Introduction

Camouflage is one of the most effective and, simultaneously, more economical methods of concealing the forces, and thus gaining an offensive and defensive advantage over the enemy, both in defense and in an attack [i] Given the technological progress, especially in the field of sensors, on today's battlefield the reconnaissance can conducted via hyperspectral imaging, i.e. from a wide range of the electromagnetic spectrum, covering not only the visible spectrum but also near- and far-infrared (thermal vision) and microwave (radar) radiation wavebands.

On the other hand, multispectral methods are also employed for this purpose. These are based on the acquisition of data from selected characteristic frequencies (channels), e.g. where spectral characteristics of live flora include the minimum locally 670 nm (VIS red) and maximum 800 nm (NIR). So a good battle dress uniform (BDU), an effective camouflage coating, or proper vehicle camouflage painting should fit into the background in a wide range of electromagnetic spectra.

The Eye as a Tool

However, the human eye remains the essential reconnaissance instrument. It is equipped with a retina, on the surface of which two types of sensors are located: rod cells responsible for monochromatic and night vision (scotopic vision) and cone cells[ii] responsible for color vision and functioning in daylight (photopic vision). The former (amounting to about 120 million), due to their location outside of the central part of the retina, do not provide acute vision but are about 100 times more sensitive to light than the cone cells and react well to movement. On the other hand, the cone cells (about 6 million), thanks to their huge concentration – mainly in the central part of the retina[iii], near the macula (macula lutea) – ensure sharp vision, especially up to 20 degrees in the visual axis of the human eye.

Materials Research Forum LLC

https://doi.org/10.21741/9781644902059-11

Moreover, due to their sensory characteristics, the cone cells have a particularly good but selective color sensitivity[iv]. Hence, there are cone receptors sensitized to visible shortwave radiation (short-preferring (S), with peak spectral sensitivity at approx. 420 nm) responsible for the perception of blue color, those sensitive to colors with medium wavelengths, such as green (mid-preferring (M), with the peak at approx. 530 nm)) and the ones sensitive to colors with long wavelengths (long-preferring (L), with the peak at around 565 nm)), such as red.

Stimulating these three types of cones in the right proportions allows the eye to perceive (trichromatism). In the case of sufficiently intense and balance stimulation of all three types of cone cells, humans are capable of seeing the color white (white light).

Considering the limitations and inadequacies of the human sense of sight – trichromatism, and not detecting a continuous signal in the visible range (400-700 nm), the CIE 1931 XYZ color space was developed to correspond to human perception. To obtain a perceptually uniform space, one in which the colors could be compared easily, the CIE XYZ model underwent mathematical transformations, resulting in the CIELAB model, where "L" is light, "a" is a color from green to crimson, and "b" corresponds to the colors blue to yellow. This model is used as the primary device-independent model for describing and comparing colors.

Camouflage Patterns

We can therefore distinguish central and peripheral vision. In terms of the camouflage itself, its goal is to neutralize distinctive features of the object to be camouflaged. These physical properties make the object stand out from the background and focus the observer's attention. They include, among others, the color, shape, size, gloss (shine), contrast, and interrupted continuity of the background. Three main types of patterns can be distinguished (Fig. 1):

– the mimetic pattern, whose function is to cause the object to blend into its background of the terrain (crypsis). Detection is minimized while the observer makes an eye sweep;

– the disruptive (dazzle) pattern is to break the shape of the given object so that the observer has greater difficulty focusing their attention on the given object and, what follows, recognizing or even identifying this object.

– the mixed pattern, i.e., mimetic-disruptive pattern, where the mimetic camouflage is obtained by applying intricate details referring to the terrain background, and the disruptive element is its proper aggregation.

Fig 1. *Types of camouflage: a) mimetic – bird [v]; b) dazzle – USS Hobson [vi]; c) mixed – Danish M84 [vii]*

Camouflage patterns are usually dedicated to a selected environment: forest, desert, an agricultural area, urban surroundings, etc. One extreme case is the photorealistic camouflage used by nature observers or hunters. Attempts were also made to design and implement a universal camouflage pattern to provide adequate protection in diverse environments, e.g., forests and rocky areas. However, their universality was revealed to be questionable.

Terotechnology XII
Materials Research Proceedings **24** (2022) 66-74

Materials Research Forum LLC
https://doi.org/10.21741/9781644902059-11

Currently, in several research centers worldwide, including WITI (Fig. 2), work is underway on technologies that would enable the creation of adaptive camouflage, which would automatically adjust itself to its background.

Fig. 2. *Electrochromic cells change their color with the voltage applied*

Generation of Camouflage Pattern

The main goal of the camouflage pattern designed by the authors was its effectiveness and applicability, which meant the possibility of using it in specific military applications (uniforms, camouflage nets, mobile camouflage, camouflage painting of vehicles). However, several limitations have been imposed on the design, including the guidelines on the nature of camouflage and the technological requirements resulting from the assumed perceptual model of the human eye. These included:

- reduction of the number of colors to only a few;
- creating a micro-pattern with mimetic functions;
- obtaining a macro-pattern disrupting the silhouette outline;
- avoidance of a noticeably repetitive pattern;
- the pattern has to blend well with the given environment.

To meet the above requirements, the authors chose a method based on processing representative images of the selected environment. The work used only images that meet the above requirements. At least five photographs were selected for a given type of environment (such as a Central European forest) and season (e.g., summer). Only the areas that could serve as the proper background were cut out of these images for further processing, and so, for example, forest areas, sky, or ground were discarded. In contrast, fragments showing forest vegetation were left (Fig. 3).

Fig. 3. *Fragment of the photograph selected for further processing*

The next step was to reduce the color palette too, but a few – in the works, it was decided to use 4, as this is the most common number of colors for uniforms and camouflage covers. After

Terotechnology XII
Materials Research Proceedings **24** (2022) 66-74

Materials Research Forum LLC
https://doi.org/10.21741/9781644902059-11

analyzing various methods, the global algorithm with the minimal variance (the so-called Otsu method of automatic image thresholding) was selected and tested, which enabled the selection of colors representative for the entire set of representative frames. The method is based on dividing the color space into areas with a minimum variance inside each area and with maximum variance between the areas. Afterward, a color representative for each area is indicated. This can be done, among others, using the median or weighted average method to obtain the colors that are most different from one another and with the greatest possible optical contrast between them.

This method yielded not only four representative color coordinates in the CIELAB color space model but also the percentage share of individual colors, and the processed images with a reduced number of colors assumed the form of textures containing a large number of fine spots (micropattern) and no macro pattern (Fig. 4).

Fig. 4. *Image with the colors reduced to four*

To obtain the macro pattern, the authors reviewed the techniques of grouping pixels and decided that cellular automata (CA)would be employed for further investigations.

Similarly, as artificial neural networks, cellular automata facilitate replacing complex descriptions of complicated circuits with a system of numerous simple circuits with simple rules [viii] Cellular automata are constructed of neighboring cells, forming a network of d-dimensions, where each cell at a given moment (cycle) can assume one of the defined states at a given moment (cycle). The way states alter between successive cycles is determined by simple rules relating to the state of the cell and its neighbors (Fig. 5).

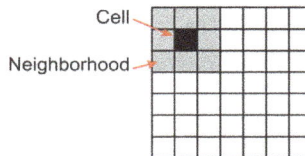

Fig. 5. *Cellular automata system model*

In the adopted camouflage design methodology, what follows the stage of reducing the number of colors is graphics described as pairs of two-dimensional arrays:

- a color matrix with dimensions of n×3, where *n* specifies the number of colors used; in the authors' case, it was 4, and the number 3 corresponds to the color coordinates of the CIELAB color space model, where L can assume the values from 0-100, and *a* and *b* – from 128 to 127;
- pixel matrix with dimensions *x*×*y*, where *x* and *y* are the width and height of the graphic expressed in pixels, respectively, the colors of the graphic are described by the color index from the color matrix; in the authors' case, there were four colors, from 0 to 3.

The image model adopted in this manner made it possible to directly translate them into the cellular automata model, in which a 2-dimensional network of cellular automata was incorporated. The network corresponded to the pixel matrix, with states defined by the index from the color matrix.

Among the distinct types of neighborhoods (Fig. 6a and Fig. 6b), the von Neumann neighborhood was selected (Fig. 6c), and in the case of cells at the margins, they were supplemented with cells from the opposite side of the network that was to surround them.

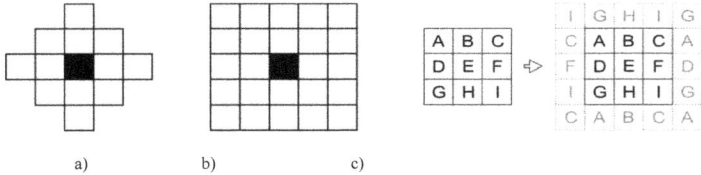

Fig. 6. a) the Moor and b) von Neumann neighborhood; c) surrounding the cells at the margins

As the result of the review and testing of diverse types of CA, the majority-vote CE type, with r = 1, and 4 states, was used in work. The transition rule (function) determined the new state as the one which obtained the highest number (sum) of states in the previous cycle for the cell and its neighborhood (Fig. 7).

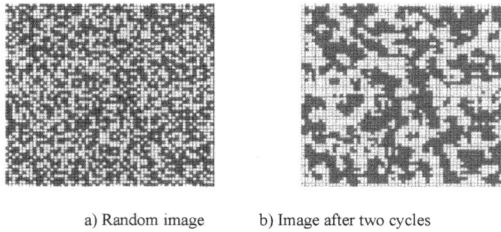

a) Random image   b) Image after two cycles

Fig. 7. Operation of the majority vote-type cellular automata

Further image processing using CA, involved an averaged image created out of n (min. 5) images with a reduced number of colors k ($k = 4$). The images were unified in terms of the scale of the elements in the photo so that in each image, 1 px corresponds to the same dimension in each image (px/cm = constant). All images have also been cropped to one size $x \times y$.

Then, a 3-dimensional array with dimensions $[y \times x \times n]$ was created from those layers, and it corresponded to the size of the frame and the number of images and contained images (pixels) in the form of successive layers (Fig. 8a). To obtain an averaged image, the 3-dimension array $[y \times x \times n]$ was reduced to a 2-dimension matrix $[y \times x]$, taking the value for the $[y, x]$ cell as the most frequent value for all the layers n of the array for this index in the matrix (Fig. 8b).

Terotechnology XII   Materials Research Forum LLC
Materials Research Proceedings **24** (2022) 66-74   https://doi.org/10.21741/9781644902059-11

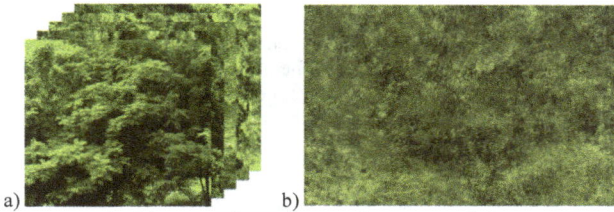

Fig. 8. *3-dimensional table created out of several images and a) its averaged version b)*

After several cycles of cellular automata operation, the authors received a macro pattern. CA operation was stopped when the size of the larger macro pattern spots reached the intended maximum size for a given application. Per the defensive standard [ix], this threshold was set at 0.3 of the dimension of the camouflaged object. For example, for uniforms – 0.3×200 cm = 60 cm, and for medium-sized armed personnel carrier – 0.3×800 cm = 240 cm. The generated pattern is shown in Figure 9.

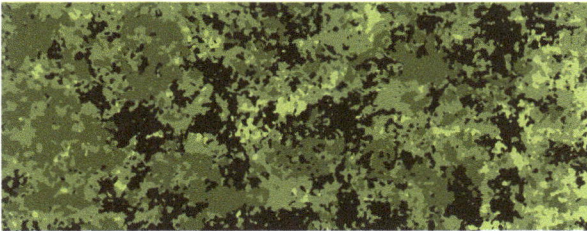

Fig. 9. *Camouflage pattern generated using CA*

The next step was to verify the camouflage effectiveness of the newly created pattern. For this purpose, the camouflage had to be applied to the military facility/object for which it was intended (e.g., a uniform, camouflage net, etc.) and then assessed. Initial tests could be conducted in a virtual environment using camouflage patterns and 3D models of military objects (Fig. 10), and the final tests, once verifying the camouflage on real objects during field tests using observers.

Fig. 10. *Application of the camouflage pattern onto a 3D object*

Terotechnology XII   Materials Research Forum LLC
Materials Research Proceedings **24** (2022) 66-74   https://doi.org/10.21741/9781644902059-11

Virtual Environment

Bearing in mind the constraints of human vision, the concepts of a virtual environment [x] were presented to assess camouflage effectiveness of the newly designed camouflage patterns, which, based on modern computer systems and displays, could largely replace the onerous and quite expensive field research (Fig. 11), in particular at the initial stage of designing camouflage patterns

***Fig. 11.** Graphic representation of field tests*

Virtual environment studies are based on presenting the observers with images containing the background (most often real images) from a given distance with a model of a camouflage pattern on the background (Fig. 12) and performing a typical detection-recognition-identification (DRI) assessment during which the observer states, whether the object has been detected (D), recognized (R) or identified (I). Detection is understood to mean the presence of an object, recognition – the ability to determine the type of object spotted (e.g., a tank), and identification, where the observer can provide details of the kind of object spotted (e.g., an M1 Abrams main battle tank).

Fig. 12. Typical virtual test display

Additionally, the time measurement for each observer from the start of the observation to the occurrence of either D, R, or I, for each image under observation.

One of the measures of central tendencies is taken as the test results. Usually, it is the average, or in the case of a large spread – the median of the distance, and the average or median of the detection, recognition and identification times for a given pattern against a given background. This is recorded for at least 6 observers.

To ensure the conditions for virtual tests resemble those in the natural environment the most, quite a few conditions have to be met and the researchers have to have at their disposal the following:

- images that have to meet most of the requirements [xi] in terms of quality – appropriate spatial and tonal resolution, correct model and color space, good sharpness, lighting, white balance, and be with the attached metadata on the environmental and atmospheric conditions, location (geographical), direction (NSEW) and time of when the image was taken, distance from the background (observation border), as well as its photographic parameters (focal length of the lens, size of the matrix, shutter, diaphragm);
- models of objects with the tested camouflage applied to them, and the camouflage colors are in a tightly controlled color scheme;
- a test station ensuring the correct colors representation (calibration and color management system) and appropriate simulation of the observation distance, which results from the image geometry and which in virtual testing means the distance from the display to the observer, and one which will also enable testing in accordance with selected methodologies and registration of test parameters.

In addition, it is also necessary to visually synchronize the model and the background to ensure that their sizes, lighting, and color space are consistent. The benefits of research studies performed in virtual environments include, first of all:

- becoming independent from the availability of training areas (proving grounds) with the required background, for which the camouflage pattern is dedicated. The desired terrain may not be accessible due to e.g., military operations and weather conditions.
- independence from weather conditions and the season;
- allowing research to be conducted in safe and comfortable conditions where observers can focus on reconnaissance;
- access to a convenient and immediate preview of the test results, thanks to the analytical module, which is a part of the virtual research management system;
- availability of military equipment and soldiers necessary to carry out field research is no longer an issue.

Summary

The proposed method of generating and testing camouflage patterns enables the design of novel patterns containing both the micro-pattern responsible for blending the camouflaged object into the background, thus reducing the likelihood of detection and the macro-pattern, which disrupts the silhouette outline and makes the object recognition challenging. The discussed methods can be used to design camouflages for various military objects: uniforms, camouflage covers, mobile camouflages, or vehicles with camouflage painting. The investigated pattern-generating method allows obtaining patterns both strictly dedicated to specific environments and seasons, as well as more universal patterns, depending on the provided initial images.

Together with the virtual testing method, which enables economic and quick testing, several individual solutions can be evaluated, not only limited to different types of cellular automata but also using morphological transformations, genetic algorithms, or artificial neural networks to find the most optimal solutions. In the future, it would be worthwhile to perform further cognitive science research in the field of image perception and recognition to replace human observers with their artificial counterparts, which would reduce the testing costs even further and which would lead to obtaining even more effective camouflage solutions using various camouflage patterns. The presented methods do not have to be constrained to the visible spectrum, and if the researchers had at their disposal an appropriate library of images from other electromagnetic spectra, that would allow them to design and examine solutions for the different ranges as well.

References

[1] M. Laprus. Leksykon wiedzy wojskowej. Wydawnictwo MON, Warszawa, 1979.

[2] H. Hofer, J. Carroll, D. R. Williams, Photoreceptor mosaic, in: L.R. Squire (ed.), Encyclopaedia of Neuroscience, Academic Press, 2009, 661-668. https://doi.org/10.1016/B978-008045046-9.01814-3

[3] D.M. Dacey. Parallel pathways for spectral coding in primate retina. Annual Review of Neuroscience 23 (2000) 743-775. https://doi.org/10.1146/annurev.neuro.23.1.743

[4] B.B. Lee. The evolution of concepts of colour vision, Neurociencias 4 (2008) 209–224.

[5] BirdsVarious0013 [online]. [accessed: 2022-01-31]. Available from: https://www.textures.com/download/BirdsVarious0013/40251

[6] NH 53549 USS Hobson (DD-464) [online]. [accessed: 2022-01-31]. Available from: https://www.history.navy.mil/our-collections/photography/numerical-list-of-images/nhhc-series/nh-series/NH-53000/NH-53549.html.

[7] Danish M84 Kampuniform Pattern [online]. [accessed: 2022-01-31]. 2020. Available from: https://www.joint-forces.com/kit-camo/32901-danish-m84-kampuniform-patern?fbclid=IwAR2c7C6Yj4pphLowqrJwyHFIhCEUGknfnJ2vuzjaMVknqw8S6SFLZQ3aF8U

[8] S. Wolfram. Physica, North-Holand, Amsterdam, 1984.

[9] Polska Norma Obronna NO-10-A800:2007, Malowanie maskujące uzbrojenia i sprzętu wojskowego, Wymagania ogólne – NO-10-A800.

[10] W. Przybył, I. Plebankiewicz, A. Januszko. A Concept of Virtual Reality in Military Camouflage Application. Materials Research Proceedings 17 (2020) 86-93. https://doi.org/10.21741/9781644901038-13

[11] W. Przybył, I. Plebankiewicz, A. Januszko, C. Śliwiński, W. Malej. Images of the environment and models of military objects in virtual method of assessing camouflaging efficiency. Selected issues of mechanical engineering technologies (2020) 234-242.

Terotechnology XII
Materials Research Proceedings **24** (2022) 75-82

Materials Research Forum LLC
https://doi.org/10.21741/9781644902059-12

Nonlinear Analysis of Beams made of High Strength Concrete Prestressed with Unbonded Tendons

KOWALSKI Damian[1,a*]

[1] Częstochowa University of Technology, Faculty of Civil Engineering, Department of Construction Processes Engineering, Dąbrowskiego 69, 42-201 Częstochowa, Poland

[a]damian.kowalski@pcz.pl

Keywords: Reinforced Concrete Modelling, Numerical Calculations, Prestressed Structures, Finite Element Analysis

Abstract. The paper presents numerical analysis methods of High-Strength Concrete (HSC) beams prestressed with unbonded tendons. Furthermore, it compares obtained results with experimental data from the literature. Prestressing tendons have been modelled in a discrete form, using one-dimensional finite elements. A temperature drop inflicted prestress force. Contact issues have been considered, i.e. friction and pressure at the interface between the cable and the duct wall. In the work, it was found that it is possible to obtain satisfactory accuracy of results with the model in use. Accurate P-Δ (load-deflection) curves were achieved matching experimental data.

Introduction

Main objective of this research was to test the possibility and the feasibility of nonlinear concrete modelling for beams made of High Strength Concrete (HSC), prestressed with unbonded tendons. A damage-plasticity model of concrete and a bilinear elastoplastic model for prestressing tendons were utilized. Both these models proved their accuracy for reinforced concrete structural elements modelling, yet there are no tests of these models, known to the author, performed for prestressed structures.

The experimental data have been taken from [1]. A scheme of used test setup is shown in Fig. 1. The tested beams were subjected to 4-point bending. This allows obtaining a section free from transversal forces due to external load.

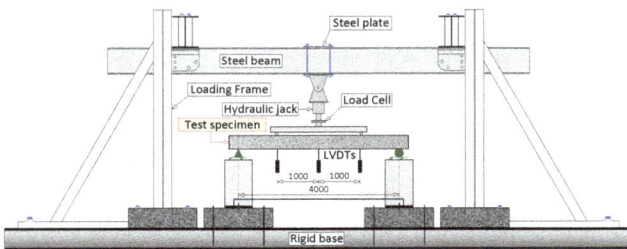

Fig. 1. Scheme of test set-up of beams.

From a variety of beams presented in [1], two of them have been chosen (namely B8 and B9 in original paper). The results were then compared with the results for the respective beams with bonded tendons (B2 and B5 in source paper). The results for bonded tendons were invoked from [2]. All the beams were partially prestressed and contained, except for the tendon, also passive reinforcement. The difference between pair of beams B2&B8 and the pair B5&B9 was a concrete class. Compressive strength of the former pair is ~73 MPa while the latter pair was made of ~96 MPa concrete. Basic geometrical data of investigated beams are presented in Fig. 2.

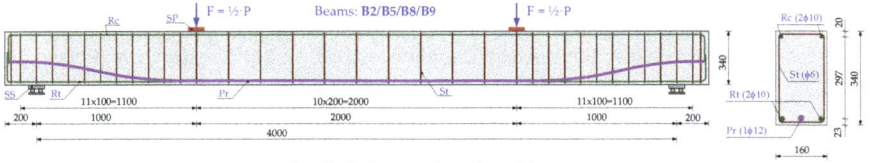

Fig. 2. Scheme of analyzed beams.

The values of concrete parameters not mentioned in the respective article were calculated based on [3]. The biaxial compressive strength of concrete was determined concerning uniaxial strength, according to [4]. The essential factors were gathered in Table 1.

Table 1. Basic concrete parameters

Sample	Young's modulus, E_{cm}, [GPa]	Uniaxial compressive strength, f_{uc}, [MPa]	Biaxial compressive strength, f_{bc}, [MPa]	Poisson's ratio, v	Intersection point abscissa between compression cap and Drucker-Prager yield function, σ_V^C, [MPa]
B2	40.27	75	86.25	0.2	-57.5
B5	43.50	97	111.55	0.2	-74.4
B8	39.78	72	82.8	0.2	-55.2
B9	43.23	95	109.25	0.2	-72.8

Methods

Finite element model. Both concrete matrix and steel plates were modelled using 8-node hexahedron (brick) finite elements with 3 DOFs (degrees of freedom) in each node. Due to symmetry, only half of the beam has been modelled (see Fig.3). Based on Timoshenko's theory, the prestressing tendons were modeled like sets of 1-dimensional 2-node beam elements. In the case of unbonded tendons, a simplified rectangle-shaped channel was assumed (see Fig.4). This is due to microplane formulation requirement for regular-shaped finite elements of the concrete matrix. Passive reinforcement has been modelled using simple 1-dimensional 2-node link elements having 1 DOF at each node, associated with longitudinal stiffness.

Prestress force in the cables was applied to the tendon in the form of a temperature drop, according to Eq. 1.

$$\Delta t = -\frac{F}{\alpha E A}$$

(1)

where:

Terotechnology XII   Materials Research Forum LLC
Materials Research Proceedings **24** (2022) 75-82   https://doi.org/10.21741/9781644902059-12

Δt – temperature drop in the cables during prestressing,
F – prestressing force,
α – thermal expansion coefficient,
E – Young's modulus,
A – cable cross-sectional area.

Prestress force has been taken as 0.9 of the values given in the paper [1]. It was established based on many trial-and-error simulations on models created for all the prestressed beams described in this research.

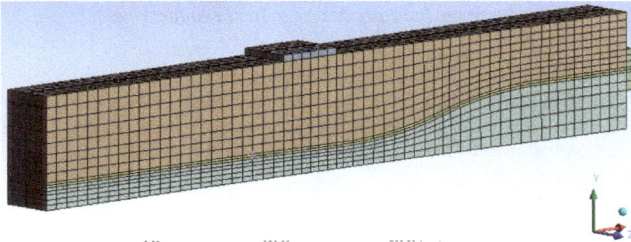

Fig. 3. *Finite element model of a sample beam.*

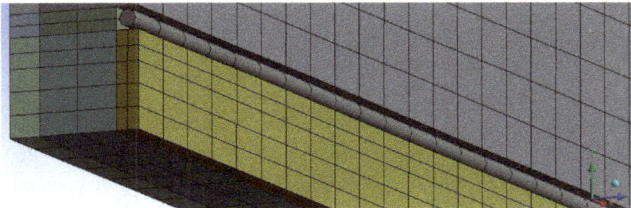

Fig. 4. *Unbonded tendon model in the square channel.*

Concrete model. A state-of-the-art concrete model was utilized and implemented in ANSYS Mechanical software; namely, the microplane coupled damage-plasticity (widely described in many papers: before the year 2000 [5-9], after the year 2000 [10-17] and the last five years [18-24]) with enhanced gradient regularization. The regularization uses 2 additional degrees of freedom per node, though it reduces solution sensitivity to finite element mesh size. Parameters for the model were taken by trial and error simulations and sensitivity analyses.

Reinforcing steel model. A bilinear model of material, elastoplastic, with kinematic hardening, was adopted. The kinematic hardening takes account of the Bauschinger effect. Fig. 5 presents a finite element model of reinforced bars and prestressing tendons.

Contact parameters. As mentioned above, the tendon channel has been modeled as simplified rectangular in transversal cross-section. Contact has been defined as fully nonlinear, with stiffness updates every load step. The friction coefficient between tendon and concrete surface was taken to equal 0.2.

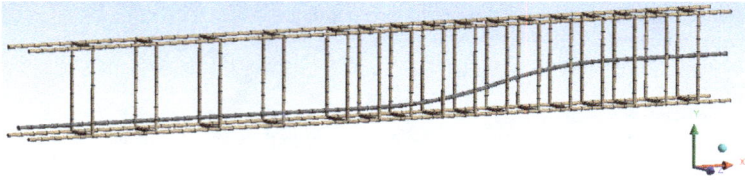

Fig. 5. *Finite elements model of active (prestressed)
and passive (non-prestressed) steel reinforcement.*

Results

Fig. 4 presents load – mid-span displacement curves of all four tested beams. An excellent agreement of the simulation results with experimental data was achieved. This proves the usefulness of the applied model for nonlinear simulations of prestressed concrete structures, including HSC. A typical map of plastic strains is depicted in Fig. 6. It reflects cracks distribution over the concrete matrix.

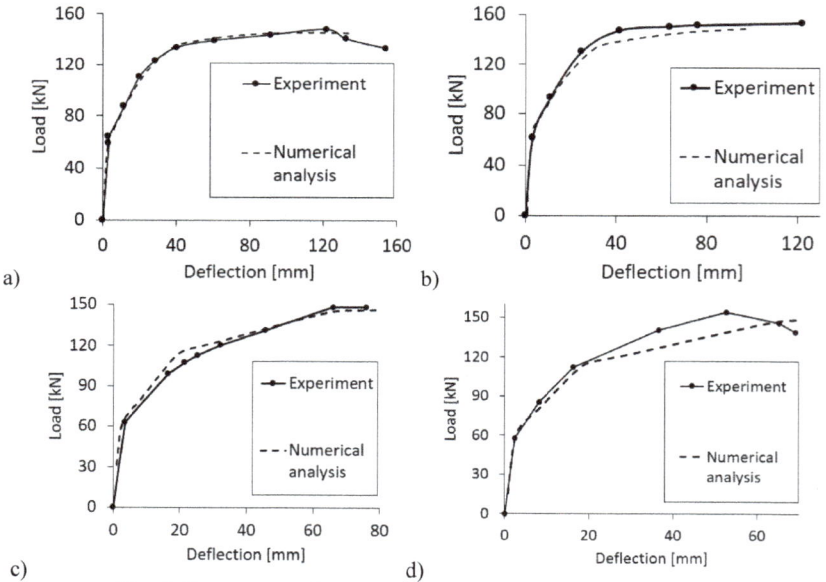

Fig. 6. *Comparison of numerical and experimental load-deflection
curves of beams B2 (a), B5 (b), B8 (c) and B9 (d).*

Materials Research Forum LLC

https://doi.org/10.21741/9781644902059-12

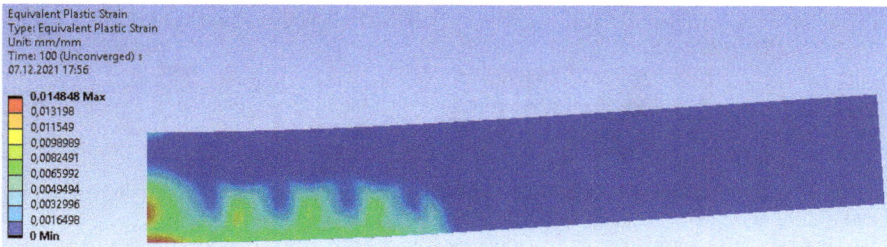

Fig. 7. *Typical cracks distribution in tested beam (expressed by means of plastic strains).*

Summary

The presented analysis is, of course, critical primarily in the area of concrete applications [25, 26] but also for all predictive analyzes [27-29], also in other scientific areas like materials science [30-33]. It may also creatively influence production engineering [34] when analyzing heavy-duty materials exposed to many factors, including exposure to corrosion [35-37], bio-corrosion in biotechnology [38-40] and agriculture [41-43], or high loads in superheaters [44, 45] and turbines [46, 47].

The paper presents the results of a nonlinear analysis of beams made of High Strength Concrete (HSC), prestressed with unbonded tendons and compares them to the available experimental and numerical data. The comparison shows that the curves match very well for bonded tendons, where there are no contact issues involved. On the other hand, the differences become slightly more significant when unbonded tendons are engaged. A possible reason for this might be modeling the tendons with 1-dimensional finite elements and associated with its contact formulation. Nevertheless, the accuracy, in this case, is still satisfying. However, further research is needed directed towards more accurate tendon modeling.

References

[1] O.F. Hussien, T.H.K. Elafandy, A.A. Abdelrahman, S.A. Abdel Baky, E.A. Nasr. Behavior of bonded and unbonded prestressed normal and high strength concrete beams. HBRC Journal 8 (2012) 239–251. https://doi.org/10.1016/j.hbrcj.2012.10.008

[2] D. Kowalski, R. Gąćkowski, J. Selejdak. Validation of microplane coupled damage-plasticity model with gradient regularization (MCDPMwGR) on prestressed and non-prestressed concrete beams. (submitted)

[3] EN 1992-1-1:2004 Eurocode 2: Design of concrete structures - Part 1-1: General rules and rules for buildings.

[4] Fédération Internationale du Béton., Comité Euro-International du Béton., and Fédération Internationale de la Précontrainte. Model Code 2010 : First complete draft, Fédération internationale du béton, 2010.

[5] Z.P. Bazant, P.G. Gambarova. Crack shear in concrete: crack band microplane model. J. Struct. Eng. 110 (1984) 2015-2035.
https://doi.org/10.1061/(ASCE)0733-9445(1984)110:9(2015)

[6] Z.P. Bažant, P.C. Prat. Microplane model for brittle-plastic material: I. Theory. J. Eng. Mech. 114 (1988) 1672-1688.
https://doi.org/10.1061/(ASCE)0733-9399(1988)114:10(1672)

[7] J. Lubliner, J. Oliver, S. Oller, E. Oñate. A plastic-damage model for concrete. Int. J. Solids Struct. 25 (1989) 299-326.

https://doi.org/10.1016/0020-7683(89)90050-4

[8] I. Carol, P.C. Pratt. New explicit microplane model for concrete : theoretical aspects and numerical implementation. Int. J. Solids Struct. 29 (1992) 1173-1191.

https://doi.org/10.1016/0020-7683(92)90141-F

[9] Z.P. Bazant, Y. Xiang, P.C. Prat. Microplane model for concrete. I: Stress-strain boundaries and finite strain. Journal of Engineering Mechanics 122 (1996) 245-254.

https://doi.org/10.1061/(ASCE)0733-9399(1996)122:3(245)

[10]Z.P. Bažant, F.C. Caner, I. Carol, M.D. Adley, S.A. Akers. Microplane model M4 for concrete. I: Formulation with work-conjugate deviatoric stress. J. Eng. Mech. 126 (2000) 944 953.

https://doi.org/10.1061/(ASCE)0733-9399(2000)126:9(944)

[11]P. Grassl, M. Jirásek. Damage-plastic model for concrete failure. Int. J. Solids Struct. 43 (2006) 7166-7196.

https://doi.org/10.1016/j.ijsolstr.2006.06.032

[12]P. Grassl, D. Xenos, U. Nyström, R. Rempling, K. Gylltoft. CDPM2: A damage-plasticity approach to modelling the failure of concrete. Int. J. Solids Struct. 50 (2013) 3805 3816.

https://doi.org/10.1016/j.ijsolstr.2013.07.008

[13]I. Zreid, M. Kaliske. Regularization of microplane damage models using an implicit gradient enhancement. Int. J. Solids Struct. 51 (2014) 3480-3489.

https://doi.org/10.1016/j.ijsolstr.2014.06.020

[14]H. Jiang, J. Zhao. Calibration of the continuous surface cap model for concrete. Finite Elements in Analysis and Design 97 (2015) 1-19.

https://doi.org/10.1016/j.finel.2014.12.002

[15]D. Xenos, P. Grassl. Modelling the failure of reinforced concrete with nonlocal and crack band approaches using the damage-plasticity model CDPM2. Finite Elements in Analysis and Design 117-118 (2016) 11-20.

https://doi.org/10.1016/j.finel.2016.04.002

[16]J. Zhang, J. Li, J.W. Ju. 3D elastoplastic damage model for concrete based on novel decomposition of stress. Int. J. Solids Struct. 94-95 (2016) 125-137.

https://doi.org/10.1016/j.ijsolstr.2016.04.038

[17]I. Zreid, M. Kaliske. Microplane modeling of cyclic behavior of concrete: a gradient plasticity-damage formulation. Proc. Appl. Math. Mech. 16 (2016) 415-416.

https://doi.org/10.1002/pamm.201610196.

[18]W. Demin, H. Fukang. Investigation for plastic damage constitutive models of the concrete material. Procedia Eng. 210 (2017) 71-78.

https://doi.org/10.1016/j.proeng.2017.11.050

[19]B. R. Indriyantho, I. Zreid, M. Kaliske. Modeling of a Concrete Dam under Earthquake Loading by A Nonlocal Microplane Approach. Procedia Eng. 171 (2017) 1010-1018.

https://doi.org/10.1016/j.proeng.2017.01.435

[20] B. Paliwal, Y. Hammi, R.D. Moser, M.F. Horstemeyer. A three-invariant cap-plasticity damage model for cementitious materials. Int. J. Solids Struct. 108 (2017) 186-202. https://doi.org/10.1016/j.ijsolstr.2016.12.015

[21] M. Szczecina, A. Winnicki. Relaxation Time in CDP Model Used for Analyses of RC Structures. Procedia Eng. 193 (2017) 369-376.

https://doi.org/10.1016/j.proeng.2017.06.226

[22] I. Zreid, M. Kaliske. A gradient enhanced plasticity–damage microplane model for concrete. Computational Mechanics 62 (2018) 1239-1257.

https://doi.org/10.1007/s00466-018-1561-1

[23] A. Wosatko, A. Winnicki, M. A. Polak, J. Pamin. Role of dilatancy angle in plasticity-based models of concrete. Arch. Civil Mech. Eng. 19 (2019) 1268-1283.

https://doi.org/10.1016/j.acme.2019.07.003

[24] B.R. Indriyantho, I. Zreid, R. Fleischhauer, M. Kaliske. Modelling of high velocity impact on concrete structures using a rate-dependent plastic-damage microplane approach at finite strains. Materials 13 (2020) art. 5165.

https://doi.org/10.3390/ma13225165 [25] M. Ulewicz, A. Pietrzak. Properties and structure of concretes doped with production waste of thermoplastic elastomers from the production of car floor mats. Materials 14 (2021) art. 872. https://doi.org/10.3390/ma14040872

[26] J. Jura, M. Ulewicz. Assessment of the Possibility of Using Fly Ash from Biomass Combustion for Concrete. Materials 14 (2021) art. 6708. https://doi.org/10.3390/ma14216708

[27] J. Pietraszek, A. Gadek-Moszczak, N. Radek. The estimation of accuracy for the neural network approximation in the case of sintered metal properties. Studies in Computational Intelligence 513 (2014) 125-134. https://doi.org/10.1007/978-3-319-01787-7_12

[28] J. Pietraszek, E. Skrzypczak-Pietraszek. The uncertainty and robustness of the principal component analysis as a tool for the dimensionality reduction. Solid State Phenom. 235 (2015) 1-8. https://doi.org/10.4028/www.scientific.net/SSP.235.1

[29] J. Pietraszek, R. Dwornicka, A. Szczotok. The bootstrap approach to the statistical significance of parameters in the fixed effects model. ECCOMAS Congress 2016 - Proceedings of the 7th European Congress on Computational Methods in Applied Sciences and Engineering 3, 6061-6068. https://doi.org/10.7712/100016.2240.9206

[30] Ł.J. Orman Ł.J., N. Radek, J. Pietraszek, M. Szczepaniak. Analysis of enhanced pool boiling heat transfer on laser-textured surfaces. Energies 13 (2020) art. 2700. https://doi.org/10.3390/en13112700

[31] N. Radek, J. Pietraszek, A. Gadek-Moszczak, Ł.J. Orman, A. Szczotok. The morphology and mechanical properties of ESD coatings before and after laser beam machining, Materials 13 (2020) art. 2331. https://doi.org/10.3390/ma13102331

[32] N. Radek, J. Konstanty, J. Pietraszek, Ł.J. Orman, M. Szczepaniak, D. Przestacki. The effect of laser beam processing on the properties of WC-Co coatings deposited on steel. Materials 14 (2021) art. 538. https://doi.org/10.3390/ma14030538

[33] A. Szczotok, J. Pietraszek, N. Radek. Metallographic Study and Repeatability Analysis of γ' Phase Precipitates in Cored, Thin-Walled Castings Made from IN713C Superalloy. Archives of Metallurgy and Materials 62 (2017) 595-601. https://doi.org/10.1515/amm-2017-0088

[34] J. Pietraszek, N. Radek, A.V. Goroshko. Challenges for the DOE methodology related to the introduction of Industry 4.0. Production Engineering Archives 26 (2020) 190-194. https://doi.org/10.30657/pea.2020.26.33

[35] K. Jagielska-Wiaderek, H. Bala, P. Wieczorek, J. Rudnicki, D. Klimecka-Tatar. Corrosion resistance depth profiles of nitrided layers on austenitic stainless steel produced at elevated temperatures, Archives of Metallurgy and Materials 54 (2009) 115-120.

[36] D. Klimecka-Tatar, H. Bala, B. Slusarek, K. Jagielska-Wiaderek. The effect of consolidation method on elctrochemical corrosion of polymer bonded Nd-Fe-B type magnetic material, Archives of Metallurgy and Materials   54 (2009) 247-256.

[37] R. Włodarczyk, A. Dudek, Z. Nitkiewicz. Corrosion analysis of sintered material used for low-temperature fuel cell plates, Archives of Metallurgy and Materials 56 (2011) 181-186. https://doi.org/10.2478/v10172-011-0021-0

[38] E. Skrzypczak-Pietraszek, A. Szewczyk, A. Piekoszewska, H. Ekiert. Biotransformation of hydroquinone to arbutin in plant in vitro cultures - Preliminary results. Acta Physiologiae Plantarum 27 (2005) 79-87. https://doi.org/10.1007/s11738-005-0039-x

[39] E. Skrzypczak-Pietraszek. Phytochemistry and biotechnology approaches of the genus exacum. In: The Gentianaceae - Volume 2: Biotechnology and Applications, 2015, 383-401. https://doi.org/10.1007/978-3-642-54102-5_16

[40] E. Skrzypczak-Pietraszek, K. Reiss, P. Żmudzki, J. Pietraszek. Enhanced accumulation of harpagide and 8-O-acetyl-harpagide in Melittis melissophyllum L. agitated shoot cultures analyzed by UPLC-MS/MS. PLoS ONE 13 (2018) art. e0202556. https://doi.org/10.1371/journal.pone.0202556

[41] T. Lipiński, A. Wach. Influence of outside furnace treatment on purity medium carbon steel, METAL 2014 – 23rd Int. Conf. on Metallurgy and Materials (2014), Ostrava, Tanger 738-743.

[42] T. Lipiński. Corrosion resistance of 1.4362 steel in boiling 65% nitric acid, Manufacturing Technology 16 (2016) 1004-1009.

[43] T. Lipiński. Roughness of 1.0721 steel after corrosion tests in 20% NaCl, Production Engineering Archives 15 (2017) 27-30. https://doi.org/10.30657/pea.2017.15.07

[44] A. Szczotok, R. Przeliorz. Phase transformations in CMSX-4 nickel-base superalloy, IOP Conference Series: Materials Science and Engineering 35 (2012) art. 012005. https://doi.org/10.1088/1757-899X/35/1/012005

[45] K. Trzewiczek, A. Szczotok, A. Gadek-Moszczak. Evaluation of the state for the material of the live steam superheater pipe coils of V degree. Advanced Materials Research 874 (2014) 35-42. https://doi.org/10.4028/www.scientific.net/AMR.874.35

[46] J. Pietraszek, A. Szczotok, N. Radek. The fixed-effects analysis of the relation between SDAS and carbides for the airfoil blade traces. Archives of Metallurgy and Materials 62 (2017) 235-239. https://doi.org/10.1515/amm-2017-0035

[47] A. Szczotok, N. Radek, R. Dwornicka. Effect of the induction hardening on microstructures of the selected steels. METAL 2018 – 27th Int. Conf. Metall. Mater. (2018), Ostrava, Tanger 1264-1269.

Terotechnology XII
Materials Research Proceedings **24** (2022) 83-89

Materials Research Forum LLC
https://doi.org/10.21741/9781644902059-13

The use of Polymer Recyclates in the Technology of Concrete Composites Production

PIETRZAK Alina[a*]

Czestochowa University of Technology / Faculty of Civil Engineering, Poland

[a] alina.pietrzak@pcz.pl

Keywords: Concrete, Concrete Composite, Polymer Waste

Abstract. The paper presents a short literature review related to the possibility of using various types of post-production or post-use waste in concrete technology. The main focus was on polymer waste, namely polyethylene terephthalate, polyethylene and polypropylene, and rubber waste.

Introduction

In modern construction, concrete, a composite material with a cement matrix, is used on a large scale. Its use allows for the construction of durable and aesthetic objects. The concrete mix components, i.e. cement, aggregate and water, are environmentally friendly. Unfortunately, the growing demand for this type of material causes enormous exploitation of the above-mentioned natural resources, which in turn causes a heavy burden on the surrounding environment. Over the past few decades, there has been a growing interest in reusing by-products from various industries. The range of used mineral additives for concrete has been extended with new raw materials and waste types. It takes into account the components that make up the concrete mix. According to the literature review, domestic and industrial waste [1], fly ash, silica dust, ground blast furnace slag, biofuel ash [2-8], sanitary waste and ceramics [9, 10], CRT glass is used for the production of concrete mix. [11, 12], recycled glass sand [13] and slag aggregate [14]. The article focuses on using various types of polymer recyclates in the production of concrete.

Polyethylene terephthalate materials

The literature reports show that polyethylene terephthalate (PET) is one of the most studied waste management [15-22]. The use of polyethylene terephthalate waste in concrete is presented in Table 1. The recyclate obtained from PET waste was used for concrete as a replacement for aggregate or in the form of fibers as reinforcement. The method of PET recyclate preparation and shape significantly impacts the produced concrete's parameters. Recycling with a smooth, spherical surface has a lesser effect on concrete workability than recycling with a non-uniform shape. The addition of PET waste material to concrete, as demonstrated by most authors, reduces the compressive strength, tensile and bending strength of concrete and the modulus of elasticity, regardless of the tested consistency and water-cement ratio. Research on polymer recyclates utilization as substrates involved in the production technology of mortars and concrete mixtures is significant in the ecological aspect.

Materials Made of Polyethylene and Polypropylene

For over fifty years, modifiers in the form of polymers have been used in concrete to improve its quality. The plastic fibers added to the fresh concrete mix act as micro-reinforcement, reducing plastic shrinkage and limiting the formation of shrinkage cracks in the hardened concrete. However, after the concrete has achieved the designed strength and the appropriate modulus of elasticity, these fibers cease to function, then the stress is transferred by the concrete itself or the main reinforcement. The research was also conducted on using waste recyclate made of

Materials Research Forum LLC

https://doi.org/10.21741/9781644902059-13

polyethylene and polypropylene as a replacement for aggregate or concrete reinforcement (Table 2).

Table 1. *Utilization of polyethylene terephthalate waste in concrete*

recycling methods	size, shape recyclate	properties of recyclate	dosing	ref.
shredded PET bottles combined at 250°C with powdered GBFS blast furnace slag	round and smooth particles ≤ 0.15 cm	PD = 1390 kg/m3	round and smooth particles ≤ 0.15 cm	[15]
PET bottles shredded in the melting process	fibers of length 30 and 40mm	SG = 1.34 g/cm^3 TS = 450	fiber 0.3 - 1.5% by volume	[16]
mechanically shredded PET bottles combined at 250°C with powdered river sand	round and smooth particles 0.15-4.75 mm	PD = 1390 kg/m^3 BD = 840 kg/m^3 FM = 4,1	fine aggregate 25-75% by volume	[17]
3 types of artificial aggregates Pc, Pf - shredded, Pp - thermal treatment	Pc $\leq$ 11.2 mm Pf $\leq$ 4 mm Pf, Pc - scaly shape Pp $\leq$ 4 mm - irregular granules	P$_c$: BD = 261 kg/m^3 SG = 1.3 g/cm^3 P$_f$: BD = 438 g/cm^3 SG = 1.28g/cm^3 P$_p$: BD = 739 kg/m^3 SG = 1.31g/cm^3	Pc - coarse grained and fine aggregate 7.5% by volume Pf, Pp - fine aggregate 7.5 and 15% by volume	[18]
3 types of artificial aggregates Pc, Pf - shredded, Pp - thermal treatment	Pc $\leq$ 11.2 mm Pf $\leq$ 4 mm Pf, Pc - scaly shape Pp $\leq$ 4 mm - irregular granules	P$_c$: BD = 351 kg/m^3 SG = 1.34 g/cm^3 P$_f$: BD = 555 kg/m^3 SG = 1.34 g/cm^3 P$_p$: BD = 827 kg/m^3 SG = 1.34 g/cm^3	Pc - coarse grained and fine aggregate 5, 10 and 15% by volume Pf, Pp - fine aggregate 5, 10 and 15% by volume	[19]
shredded PET bottles	fibers length 25 mm	SG = 1.34 g/cm^3	fiber 0.5-3% by weight of cement	[20]
shredded PET particles	< 7 mm	PD=464 kg/m^3 SG=1.11 g/cm^3	fine aggregate (sand) 5, 10 and 15%	[21]
shredded PET bottles (melted and extruded fibers)	fibers A, length 40 mm (straight, smooth profile) B fibers 52 mm long (crimped, notched profile)	A: SG = 1.34 g/cm^3 TS = 550 B: SG = 1.34 g/cm^3 TS = 274	fiber 1% by volume	[22]

Explanations: PD - density (kg / m^3); BD - bulk density (kg / m^3); SG - specific weight (g / cm^3); FM - accuracy module; TS - tensile strength (MPa); MP - melting point (°C); MoE - modulus of elasticity (GPa)

Terotechnology XII
Materials Research Proceedings 24 (2022) 83-89

Materials Research Forum LLC
https://doi.org/10.21741/9781644902059-13

Recycling mixtures of this material are characterized by different quality and mechanical properties [23, 24]. Because the properties of fibers made of pure synthetic material differ significantly from the properties of fibers obtained from recycled material, fibers made of pure polypropylene or polyethylene were used much more frequently in the research [25-27]. Concretes containing polypropylene fibers in an amount up to 1% show higher compressive, splitting, or bending strength compared to concrete without the addition of plastic (normal concrete). Increasing the content of synthetic fibers above this level deteriorates the mechanical properties of modified concretes.

Table 2. The use of PE and PP waste in concrete

recycling methods	size, shape recyclate	properties of recyclate	dosing	ref.
industrial waste from mechanically shredded carpets	PP and nylon fibers fiber I, length 12-25mm fiber II, length 3–25mm	-	fiber in quantity 1 and 2% by volume	[28]
shredded HDPE waste	small particles	-	0.5 - 5% for fine aggregate of the total weight	[29]
industrial waste of carpets, shredded mechanically	PP and nylon fibers 12 – 25 mm	-	$0.07 – 1.4\%$ volume	[30]
mix of HDPE, PP waste and PVC recycling by grinding	grain size ≤10 mm irregular shape	HDPE: BD = 534 kg/m^3 SG = 1.0 g/cm^3 PVC: BD = 684 kg/m^3 SG = 1.0 g/cm^3	too thick aggregate in a ratio of 1: 0.274 by volume	[31]
ground HDPE waste	grain size ≤ 2.36 mm heated coarse-grained aggregate combined with HDPE powder	BD = 945 - 962 g/cm^3 SG = 1.04 g/cm^3	modifier was added in an amount of 2, 4 and 6% to replace the same amount of cement and sand	[32]
fibers from HDPE waste from pots, buckets, cans, kitchen utensils	fiber length 20 - 100 mm, the ratio of the fiber length to its diameter 20 - 100	SG = 0.9 g/cm3	0.6% by volume	[33]
LDPE waste - bags	grain size ≤2.36 mm	SG = 0.93 g/cm^3 FM = 5.92	for fine aggregate 0.4 - 1% of the total weight	[34]

Explanations: PD - density (kg / m^3); BD - bulk density (kg / m^3); SG - specific weight (g / cm^3); FM - accuracy module; TS - tensile strength (MPa); MP - melting point (°C); MoE - modulus of elasticity (GPa)

Rubber Waste

There have also been reports on the utilization of rubber waste in concrete production [35-44]. With the increase in the content of rubber waste used as a sand substitute, concrete mixes were characterized by lower [40, 41] or higher [36] workability. Concrete containing the addition of rubber recyclate showed lower values of mechanical parameters than concretes produced without the addition of waste [37, 44]. Such an approach may be inspiring for other industrial branches struggling with the problem of recyclates, e.g. packaging [45], steel industry [46],

Summary
Limiting the consumption of mineral resources is particularly important in the construction sector, which consumes large amounts of mineral resources, especially felt and aggregate. The consumption of these raw materials can be reduced by replacing them with, for example, recycled materials. Such activities are consistent with the idea of sustainable development, which is defined as development that meets the needs of the present generation without reducing the ability of future generations to meet their needs. However, replacing mineral materials with recyclates from various types of waste in producing new building materials has several technical and technological limitations, so a series of laboratory tests must precede it. Numerous researchers have attempted to determine the possibilities of using various types of waste, including dust from biomass combustion, ceramics, glass cullet, and some synthetic materials to produce concrete. In the case of plastics, as shown in the literature review, due to their different physicochemical properties (diversified composition of modifying additives and a diverse chemical structure of the primary polymer), their use is tough and it requires each time separate tests for a selected group of plastics. The cited review of the literature presents both research carried out with the use of polymer recyclates produced for the needs of the study and with the use of materials derived from the recycling process. Most studies using recycled plastics concern polyethylene terephthalate and polypropylene. At the same time, there are few reports on the use of other plastics (e.g. PC, PUR) or mixtures of various synthetic materials in the production of concrete.

References

[1] K. Kishore, N. Gupta. Application of domestic & industrial waste materials in concrete: A review, Materials Todays Proc. 26 (2020) 2926-2931.
https://doi.org/10.1016/j.matpr.2020.02.604

[2] Y. Aggarwal, R. Siddique. Microstructure and properties of concrete using bottom ash and waste foundry sand as partial replacement of fine aggregates, Constr. Build. Mater. 54 (2014) 210-223.

[3] J. Halbiniak., Projektowanie składu betonowego z dodatkiem popiołów lotnych oraz ich wpływa na tempo przyrostu wytrzymałości, Budownictwo o Zoptymalizowanym Potencjale Energetycznym 2 (2012) 29-36.

[4] J. Jura, M. Ulewicz. Application of fly ash and CRT glass waste in cement mortars. Scientific Review – Engineering and Environmental Sciences 27 (2018) 348-354.

[5] J. Jura. Influence of type of biomass burned on the properties of cement mortar containing the fly ash, Construction of Optimized Energy Potential (CoOEP) 9 (2020) 77-82.

[6] J. Jura, M. Ulewicz. Assessment of the Possibility of Using Fly Ash from Biomass Combustion for Concrete, Materials 14 (2021) art. 6708. https://doi.org/10.3390/ma14216708

[7] J. Popławski. Influence of biomass fly-ash blended with bituminous coal fly-ash on properties of concrete, Construction of Optimized Energy Potential (CoOEP) 9 (2020) 89-96.

Terotechnology XII   Materials Research Forum LLC
Materials Research Proceedings **24** (2022) 83-89   https://doi.org/10.21741/9781644902059-13

[8] A. Pietrzak, The effect of ashes generated from the combustion of sewage sludge on the basic mechanical properties of concrete, Construction of Optimized Energy Potential (CoOEP) 1 (2019) 29–35.

[9] A. Halicka, P. Ogrodnik, B. Zegardli. Using ceramic sanitary ware waste as concrete aggregate, Constr. Build. Mater. 48 (2013) 295–305.

[10] M. Ulewicz, J. Halbiniak. Application of waste from utilitarian ceramics for production of cement mortar and concrete, Physicochem. Probl. Miner. Process. 52 (2016) 1002–1010.

[11] P. Walczak, J. Małolepszy, M. Reben, K. Rzepa. Mechanical properties of concrete mortar based on mixture of CRT glass cullet and fluidized fly ash, Procedia Eng. 108 (2015) 453 – 458.

[12] A. Pietrzak, M. Ulewicz. The Influence of Addition of CRT Glass Cullet on Selected Parameters of Concrete Composites, Earth and Environmental Science, 2nd Int. Conf. Sustainable Energy and Environmental Development (SEED'17), Kraków, Polska 2019.

[13]. S.C. Bostanci,, M. Limbachiya, H. Kew. Portland-composite and composite cement concretes made with coarse recycled and recycled glass sand aggregates: engineering and durability properties, Constr. Build. Mater. 128 (2016) 324–340.

[14] S-J. Choi Kim, Y.-U., T-G. Oh, B-G. Cho. Compressive Strength, Chloride Ion Penetrability, and Carbonation Characteristic of Concrete with Mixed Slag Aggregate, Materials 13 (2020) art.940. https://doi.org/10.3390/ma13040940

[15] Y. W. Choi, D. J Moon, J. S. Chung, S. K. Cho. Effects of waste PET bottlers aggregate on the properties of concrete, Cem. Concr. Res. 35 (2005) 776–781.

[16] T. Ochi, S. Okubo, K. Fukui. Development of recycled PET fiber and its application as concrete-reinforcing fiber, Cement Concr. Compos. 29 (2007) 448–455.

[17] Y. W. Choi, D. J. Moon, Y. J. Kim, M. Lachemi. Characteristics of mortar and concrete containing fine aggregate manufactured from recycled waste polyethylene terephthalate bottles, Constr. Build. Mater. 23 (2009) 2829–2835.

[18] R. Silva, J. De Brito, N. Saikia. Influence of curing conditions on the durability-related performance of concrete made with selected plastic waste aggregates, Cement Concr. Compos. 35 (2013) 23–31.

[19] N. Saikia, J. de Brito. Waste polyethylene terephthalate as an aggregate in concrete. Mater. Res. 16 (2013) 341–350.

[20] R. Nibudey, P. Nagarnaik, D. Parbat, A. Pande, Strength and fracture properties of post consumed waste plastic fiber reinforced concrete, Int. J. Civ. Struct. Environ. Infrastruct. Eng. Res. Dev. (IJCSEIERD) 3 (2013) 9–16.

[21] E. Rahmani, M. Dehestani, M. H. A. Beygi, H. Allahyari, I. M. Nikbin. On the mechanical properties of concrete containing waste PET particles, Constr. Build. Mater. 47 (2013) 1302-1308.

[22] F. Fraternali, S. Spadea, V. P. Berardi. Effects of recycled PET fibres on the mechanical properties and seawater curing of Portland cement-based concretes, Constr. Build. Mater. 61 (2014) 293–302.

[23] I. Borovanska, T. Dobreva, R. Benavente, S. Djoumaliisky, G. Kotzev, Quality assessment of recycled and modified LDPE/PP 404 blends, J. Elastom, Plast. 44 (2012) 479-497

[24] T. Kojnoková, L. Markovičová, F. Nový. The changes of LD-PE films after exposure in different media, Prod. Eng. Arch. 26 (2020) 185-189. https://doi.org/10.30657/pea.2020.26.32

[25] S. Kakooei, H. M. Akil, M. Jamshidi, J. Rouhi. The effects of polypropylene fibers on the properties of reinforced concrete 408 structures, Constr. Build. Mater. 27 (2012) 73–77.

[26] A. Sivakumar, M. Santhanam. A quantitative study on the plastic shrinkage cracking in high strength hybrid fibre 410 reinforced concrete, Cement Concr. Compos. 29 (2007) 575–581.

[27] A. Pietrzak, M. Ulewicz. The impact of the length of polypropylene fibers on selected properties of concrete, Acta Sci. Pol. 412 Architectura 18 (2019) 21-25.

[28] Y. Wang, A.-H. Zureick, B.S. Cho, D. Scott. Properties of fibre reinforced concrete using recycled fibres from carpet industrial waste. J. Mater. Sci. 29 (1994) 4191-4199.

[29] T. R. Naik, S. S. Singh, C. O. Huber, B. S. Brodersen. Use of post-consumer waste plastics in cement-based composites. Cem. Concr. Res. 26 (1996) 1489–1492.

[30] Y. Wang, H. Wu, V. C. Li. Concrete reinforcement with recycled fibers, J. Mater. Civ. Eng. 12 (2000) 314–319.

[31] M. Elzafraney, P. Soroushian, M. Deru. . Development of energy-efficient concrete buildings using recycled plastic aggregates, J. Archit. Eng. 11 (2005) 122–130.

[32] K. Kumar, P. Prakash, Use of waste plastic in cement concrete pavement, Adv. Mater. Res. J. 15 (2006) 15, 1–21.

[33] B. Khadakbhavi, D.V.V. Reddy, D. Ullagaddi. Effect of aspect ratios of waste HDPE fibres on the properties of fibres on fiber reinforced concrete, Res. J. Eng. Technol. 3 (2010) 13–21.

[34] M. Chaudhary, V.Srivastava, V. Agarwal. Effect of waste low density polyethylene on mechanical properties of concrete, J. Acad. Ind. Res. 3 (2014) 123.

[35] A. Sofi. Effect of waste tyre rubber on mechanical and durability properties of concrete – A review, Ain Shams Eng. J. 9 (2018) 2691-2700.

[36] M.M. Balaha, A.A.M. Badawy, M. Hashish. Effect of using ground tire rubber as fine aggregate on the 424 behaviour of concrete mixes, Indian J. Eng. Mater. Sci. 14 (2007) 427-435.

[37] Gesoglu, Mehamet, Guneyisi, Erhan, Strength development and chloride penetration in rubberized concretes with and 426 without silica fume, Mater. Struct. 40 (2007) 953–964.

[38] C. Albano, N. Camacho, J. Reyes, J.L. Feliu, M. Herna´ndez. Influence of scrap rubber to Portland I concrete composites: 428 destructive and non-destructive testing, Compos. Struct. 71 (2005) 439–446.

[39] N. Holmes, K. Dunne, J. O'Donnell. Longitudinal shear resistance of composite slabs containing crumb rubber 430 in concrete toppings, Constr. Build. Mater. 55 (2014) 365–378.

[40] M. Bravo, J. de Brito. Concrete made with used tire aggregate: durability-related performance, J. Clean. Prod. 25 (2012) 42–50.

[41] O. Onuaguluchi, D. K. Panesar. Hardened properties of concrete mixtures containing pre-coated crumb rubber and silica 434 fume, J. Clean. Prod. 82 (2014) 125–131.

[42] M.M. Taha, A.S. El-Dieb, M.A. AbdEl-Wahab, M.E. Abdel-Hameed. Mechanical, fracture, and microstructural investigations of rubber concrete, J. Mater. Civ. Eng. ASCE 20 (2008) 640-649. https://doi.org/10.1061/(ASCE)0899-1561(2008)20:10(640)

[43] Batayneh, K. Malek, Marie, Iqbal, Asi, Ibrahim, Promoting the use of crumb rubber concrete in developing countries, Waste 438 Manage. 28 (2008) 2171–2176.

[44] A. Pietrzak, M. Ulewicz. Properties and Structure of Concretes doped with Production 3 Waste of Thermoplastic Elastomers from the Production of Car 4 Floor Mats, Materials 14 (2021) art.872. https://doi.org/10.3390/ma14040872

[45] S.T. Dziuba, M. Ingaldi. Segragation and recycling of packaging waste by individual consumers in Poland, Int. Multidisciplinary Scientific GeoConference Surveying Geology and Mining Ecology Management, SGEM 3 (2015) 545-552.

[46] R. Ulewicz, J. Selejdak, S. Borkowski, M. Jagusiak-Kocik. Process management in the cast iron foundry, METAL 2013 - 22nd Int. Conf. Metallurgy and Materials (2013), Ostrava, Tanger 1926-1931.

[47] A. Pacana, A. Gazda, D. Malindzak, R. Stefko. Study on improving the quality of stretch film by Shainin method, Przemysl Chemiczny 93 (2014) 243-245. https://doi.org/10.12916/przemchem.2014.243

Terotechnology XII
Materials Research Proceedings 24 (2022) 90-95

Materials Research Forum LLC
https://doi.org/10.21741/9781644902059-14

Sustainable Buildings – Thermal Sensations Case Study

DĘBSKA Luiza[1,a *] and KAPJOR Andrej[2,b]

[1]Faculty of Environmental, Geomatic and Energy Engineering, Kielce University of Technology, 25-314 Kielce, Poland

[2]Faculty of Mechanical Engineering, University of Zilina, Univerzitna 1, 01026 Zilina, Slovakia

[a]ldebska@tu.kielce.pl, [b]Andrej.Kapjor@fstroj.uniza.sk

Keywords: Thermal Comfort, Relative Humidity, Thermal Sensations, Sustainable Building, BMI

Abstract. Currently, thermal comfort is becoming one of the most important aspects of human life. It is related to the time people spend in a closed environment. That is why it is so important to study this issue in terms of their actual thermal sensations, in order to better understand and clarify the scope of the relevant parameters, in particular by looking for such data for buildings built with a view to sustainable development. They are to meet all standards, ensuring thermal comfort for people staying in them. The article discusses thermal sensations, thermal preferences, and humidity for the group aged 21-25 in the winter in the Energis building belonging to the Kielce University of Technology. The methods that were used for this purpose include an environmental measure that takes microclimate parameters and questionnaires assessing the conditions in the room under study.

Introduction

Modern, sustainable construction encourages designers to provide the highest-quality internal conditions in a closed space so that everyone can feel comfortable in it. Thermal comfort depends on many factors, such as air temperature, clothing, airflow speed, carbon dioxide level, relative humidity, etc. can cause fatigue and weariness. Therefore, it is so important that intelligent buildings can provide the greatest possible comfort for working and learning people. Taking into account the above factors, an analysis of the available literature was carried out, in which not much research related to modern buildings of an educational nature was found, but many studies related to buildings being trained in various climate zones. An example of such research is the research [1, 2] carried out in China, which in the first case showed that students prefer cooler temperatures, with the selected temperature range from about 21°C to about 27°C, while in the second case, 90% of people believed that they felt thermal comfort. The intelligent building, which was examined by the authors [3, 4], provided information in both cases that it meets the thermal expectations of people because the conditions that prevailed at that time were closest to their expectations of comfort. Other interesting studies concern thermal comfort in tropical climates like Australia [5] where the temperature range, according to the respondents, was from 19°C to 26°C, Malaysia and Japan [6] where the best temperature was about 25°C-26°C degrees, and Bangladesh [7] in which students identified the temperature of 27°C as the most neutral in their climate zone. As mentioned earlier thermal comfort depends on many factors including also individual preferences, medical treatment as well as heat transfer issues [8-11]. A thorough theoretical analysis of the heat distribution would be very complex [12-14] and would require either time-consuming FEM modeling or the use of semi-empirical methods [15-17].

The main purpose of the study was to learn about the thermal sensations of the group in order to check whether the intelligent building meets the thermal expectations of people in the winter.

Terotechnology XII
Materials Research Proceedings 24 (2022) 90-95

Materials Research Forum LLC
https://doi.org/10.21741/9781644902059-14

Methodology

The study was conducted in the winter in the Energis building at the Kielce University of Technology. The Energis building was created with sustainable development in mind, so as to be self-sufficient in energy, using renewable energy sources, such as heat pumps, photovoltaic and solar panels, and windmills. In addition, the building has a BMS (Building Manager System) system programmed, the main assumption of which is to control microclimate parameters in lecture halls or employees' rooms, additionally it provides control of lighting inside the building and electricity yield and its further use to power teaching devices. Energis is shown in Fig. 1.

Fig. 1. Energis building, Kielce University of Technology.

The study was performed using a microclimate meter that collected data from the internal parameters of the room and questionnaires, which were completed by the study group. First, the meter was placed in the central part of the room. After approximately 15 minutes, the results were written down. Secondly, the questionnaires were distributed and completed by the respondents. The task of the students was to assess thermal sensations, and moisture, and determine their well-being and productivity on the basis of the questions contained in the questionnaire. 20 people aged 21 to 25 participated in the study. After the analysis, 4 questionnaires were rejected because these people felt sick and could not be taken into account in the further evaluation of the results. Men accounted for 62.5% and women accounted for 37.5% of the studied group. The lecture hall is presented in Fig. 2.

Fig. 2. Lecture hall in Energis building.

Results

The conditions that prevailed during the performed test were:
- air and black ball temperatures: 23.8 [°C] and 23.5 [°C];

Materials Research Forum LLC
https://doi.org/10.21741/9781644902059-14

- relative humidity and air speed 25.51 [%] and 0.08 [m/s];
- light intensity and CO_2 concentration: 475.7 [lx] and 719 [ppm].

The obtained parameters are within the applicable ranges for closed rooms. The responses of the respondents are shown in the next figures. Fig. 3 shows the real thermal sensations of people.

No person has identified room conditions as cool or cold. Definitely 43.75% of people considered the conditions in the room as comfortable and as negative – 31.25%. On the other hand, 25% of people declared that the room was too warm. The chart below presents the acceptability of temperature by the participants of the study.

Half of the group consider the prevailing temperature as comfortable, and the other half that it is acceptable. It means that all the people who were participating in the current study expressed positive ideas about the indoor microclimate. No one complained about the conditions in the room, which is quite strange, because a quarter of the population there said that it was too hot (Fig. 3), but none expressed any complaints about it regarding the acceptability of the environment they were in.

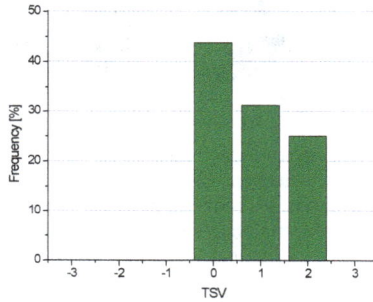

Fig. 3. *Thermal sensation of respondents (TSV): -3 – too cold, -2 – too cool, -1 – pleasantly cool, 0 – comfortable, 1 – pleasantly warm, 2 – too warm, 3 – too hot.*

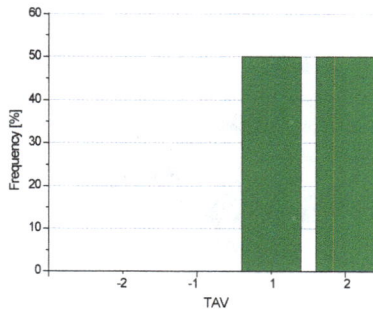

Fig. 4. *Temperature accepted by the respondents (TAV): -2 – definitely unpleasant,, -1 – unpleasant, 1 – acceptable 2 – comfortable.*

Materials Research Forum LLC
https://doi.org/10.21741/9781644902059-14

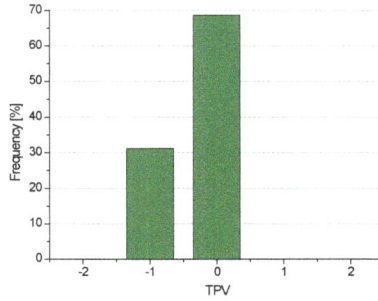

Fig. 5. *Thermal preferences vote (TPV):*
-2 – definitely cooler, -1 – cooler, 0 – no change, 1 – warmer, 2-definitely warmer.

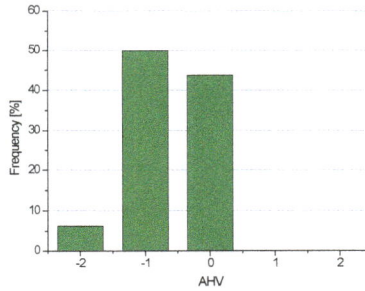

Fig. 6. *Assessment of humidity vote (AHV):*
-2 – too dry, -1 – quite dry, 0 – pleasantly, 1 – quite humid, 2 – too humid.

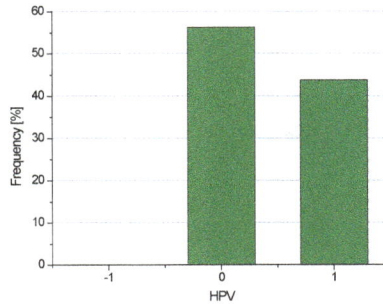

Fig. 7. *Humidity preferences vote (HPV):*
-1 – more dry, 0 – no change, 1 – more humid.

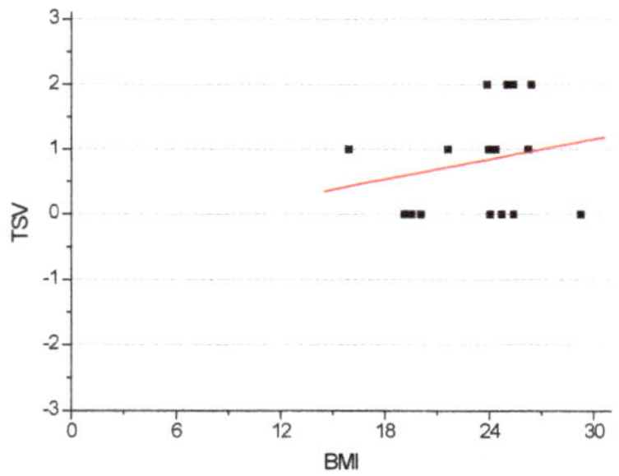

Materials Research Forum LLC
https://doi.org/10.21741/9781644902059-14

Fig. 8. *The relationship between the BMI index and the TSV.*

The next question in the questionnaire dealt with another important issue – Fig. 5 shows people's heat preferences, i.e. what temperature would be better, warmer, or cooler. It turned out that 68.75% of people would leave the temperature unchanged, while 31.25% would like it to be cooler. Fig. 6 below shows the assessment of humidity in the lecture hall.

An overall estimate shows that around 56.25% of people in the room are too dry or too dry. Contrary to 43.75% who rated the humidity as pleasant. The purpose of Fig. 7 is to show the willingness to change the humidity in the lecture hall, where the test took place, to a more humid or drier one.

According to the respondents, the humidity could remain unchanged at 56.25%. Changes in humidity to the more humid one were expressed by 43.75% of people. Fig. 8 below shows the relationship between the BMI and Thermal Sensation Vote (TSV).

Summary

The surveyed group was 75% satisfied with temperature – only ca. 31% would prefer it to be cooler. The humidity was judged as quite dry or too dry, so it could be wetter in the room. The sustainable building met people's expectations during the test, except for relative humidity which could be higher.

References

[1] Z. Fang, S. Zhang, Y. Cheng, A. M. L. Fong, M. O. Oladokun, Z. Lin, H. Wu. Field study on adaptive thermal comfort in typical air conditioned classrooms, Build. Environ. 133 (2018) 73–82. https://doi.org/10.1016/j.buildenv.2018.02.005

[2] J. Jiang, D. Wang, Y. Liu, Y. Di, J. Liu. A field study of adaptive thermal comfort in primary and secondary school classrooms during winter season in Northwest China. Build. Environ. 175 (2020) art. 106802. https://doi.org/10.1016/j.buildenv.2020.106802

[3] L. Dębska, J. Krakowiak. Thermal environment assessment in selected Polish educational buildings. Cold Climate HVAC & Energy (2021). E3S Web of Conferences 246 (2021) art.15004. https://doi.org/10.1051/e3sconf/ 202124615004.

[4] L. Dębska. Assessment of the Indoor Environment in the Intelligent Building. Civ. Environ. Eng. 17 (2021) 572-582. https://doi.org/10.2478/cee-2021-0058

[5] R. de Dear, J. Kim, C. Candido, M. Deuble. Adaptive thermal comfort in Australian school classrooms. Build. Res. Inf. 43 (2015) 383-398. https://doi.org/10.1080/09613218.2015.991627

[6] S.A. Zaki, S.A. Damiati, H.B. Rijal, A. Hagishima, A.A. Razak. Adaptive thermal comfort in university classrooms in Malaysia and Japan. Build. Environ. 122 (2017) 294-306. https://doi.org/10.1016/j.buildenv.2017.06.016

[7] S.J. Talkudar, T.H. Talukdar, M.K. Singh, A. Baten, S. Hossen. Status of thermal comfort in naturally ventilated University classrooms of Bangladesh in hot and humid summer season. J. Build. Eng. 32 (2020) art. 101700. https://doi.org/10.1016/j.jobe.2020.101700

[8] A. Szczotok, J. Pietraszek, N. Radek. Metallographic Study and Repeatability Analysis of γ' Phase Precipitates in Cored, Thin-Walled Castings Made from IN713C Superalloy. Archives of Metallurgy and Materials 62 (2017) 595-601. https://doi.org/10.1515/amm-2017-0088

[9] Ł.J. Orman, N. Radek, J. Pietraszek, M. Szczepaniak. Analysis of enhanced pool boiling heat transfer on laser – textured surfaces, Energies 13 (2020), art. 2700. https://doi.org/10.3390/en13112700

[10] N. Radek, J. Pietraszek, A. Gądek-Moszczak, Ł.J. Orman, A. Szczotok. The morphology and mechanical properties of ESD coatings before and after laser beam machining, Materials 13 (2020) art. 2331. https://doi.org/10.3390/ma13102331

[11] G. Majewski, Ł.J. Orman, M. Telejko, N. Radek, J. Pietraszek, A. Dudek. Assessment of thermal comfort in the intelligent buildings in view of providing high quality indoor environment, Energies 13 (2020) art. 1973. https://doi.org/10.3390/en13081973

[12] T. Styrylska, J. Pietraszek. Numerical modeling of non-steady-state temperature-fields with supplementary data. Zeitschrift fur Angewandte Mathematik und Mechanik 72 (1992) T537-T539.

[13] M. Zmindak, L. Radziszewski, Z. Pelagic, M. Falat. FEM/BEM techniques for modelling of local fields in contact mechanics, Communications - Scientific Letters of the University of Zilina 17 (2015) 37-46.

[14] I. Nová, K. Fraňa, T. Lipiński. Monitoring of the Interaction of Aluminum Alloy and Sodium Chloride as the Basis for Ecological Production of Expanded Aluminum. Physics of Metals and Metallography 122 (2021) 1288-1300. https://doi.org/10.1134/S0031918X20140124

[15] J. Pietraszek. Response surface methodology at irregular grids based on Voronoi scheme with neural network approximator. 6th Int. Conf. on Neural Networks and Soft Computing JUN 11-15, 2002, Springer, 250-255. https://doi.org/10.1007/978-3-7908-1902-1_35

[16] J. Pietraszek, A. Gadek-Moszczak, N. Radek. The estimation of accuracy for the neural network approximation in the case of sintered metal properties. Studies in Computational Intelligence 513 (2014) 125-134. https://doi.org/10.1007/978-3-319-01787-7_12

[17] J. Pietraszek, R. Dwornicka, A. Szczotok. The bootstrap approach to the statistical significance of parameters in the fixed effects model. ECCOMAS 2016 – Proc. 7th European Congress on Computational Methods in Applied Sciences and Engineering 3, 6061-6068. https://doi.org/10.7712/100016.2240.9206

Terotechnology XII
Materials Research Proceedings 24 (2022) 96-101

Materials Research Forum LLC
https://doi.org/10.21741/9781644902059-15

Traditional and Intelligent Buildings – Perceptions of Thermal Comfort

KRAWCZYK Natalia[1,a] *

[1]Faculty of Environmental, Geomatic and Energy Engineering, Kielce University of Technology, Poland

[a]nkrawczyk@tu.kielce.pl

Keywords: Thermal Comfort, Predicted Mean Vote, Thermal Sensation Vote, Microclimate

Abstract. The article presents the perception of thermal comfort in two buildings, intelligent and traditional. 32 people aged 18 to 22 and one women aged 52 participated in the study. Two indicators were analyzed, PMV (Predicted Mean Vote) and PPD (Predicted Percentage of Dissatisfied). The analysis consisted in comparing the actual feelings of the respondents with the results based on Fanger's model. The assessment of air humidity and thermal preferences are also shown.

Introduction

Thermal comfort is a state in which we are neither too cold nor too warm. It is difficult to choose the right parameters in a room because one person may be too warm and for another, it may be too cold in the same air parameters and in the same room. Everyone's perception of our surroundings is different. This is only a subjective judgment. The perception of thermal comfort is influenced by air temperature, relative humidity, carbon dioxide concentration, air speed, light intensity, seasons of the year, noise, etc. Each of these factors may affect our well-being or discomfort. The consequence of not ensuring thermal comfort may also be, for example, a headache, which makes us feel less productive. The assessment of thermal comfort is a very important issue. It helps us determine the parameters in which a person feels indifferent. For modern construction, it is assumed that the requirements for controlling the parameters of the internal environment with the use of BMS (Building Management System) systems will meet the highest requirements compared to traditional construction.

In the 1970s, O. Fanger, on the basis of the applicable standards: ISO 7730 [1] together with PN-EN 16798-1: 2019 [2], developed two indicators, PMV (Predicted Mean Vote) and PPD denoting the predicted number of dissatisfied people (Percentage of Dissatisfied People). The PMV index is expressed on a seven-point scale, where "-3" means cold and "+3" hot.

Currently, many researchers are involved in research on thermal comfort in school buildings, homes, universities, and offices, and less in intelligent buildings. The research on thermal comfort in intelligent buildings was carried out by Krawczyk and Krakowiak [3], who compared the results of an intelligent building with a traditional building. The research shows a difference between the PMV index and the actual results from the questionnaires. Their results showed that PMV in summer was below zero, and vice versa in winter. Homoda et al. [4] found that not only temperature is an important factor influencing thermal comfort. Relative humidity should also be taken into account. The research on thermal comfort in schools was carried out by Jindal [5] on 130 students. He examined the thermal environment and students' perceptions, finding that students feel best at temperatures ranging from 15.5 to 33.7°C. The authors [6, 7] dealt with research on intelligent buildings. It has to be mentioned that thermal sensations depend not only on the air temperature but also on a number of factors for example individual preferences, and heat transfer issues [8-11]. It may be evaluated by a time-consuming FEM analysis or, alternatively, by a surrogate model derived from DOE methods [12, 13] and a non-parametric approach [14, 15].

Air-conditioning and heating devices significantly affect the conditions in the room, therefore the designed air-conditioning and heating systems should provide appropriate parameters, depending on the purpose of the room.

Material and Method

In this study, a study was carried out for two buildings, a school building and an intelligent building of Energis of the Kielce University of Technology. The results for the first building were taken from the article [3]. However, for the second building, research was carried out in January 2022. Air temperature, relative air humidity and air velocity for the traditional building are 27.7°C, 47.60% and 0.05 m/s [3], while for the intelligent one: 25.1°C, 29.72% and 0.09 m/s.

The school building has no ventilation in the classrooms. On the other hand, the building of the Kielce University of Technology has mechanical ventilation. The intelligent building houses 22 lecture halls, laboratories, and rooms for lecturers and other employees. It is equipped with solar collectors, heat pumps, and photovoltaic cells, which are placed on the roof of the building. Thanks to photovoltaic panels and solar collectors, the building uses solar energy to heat water and generate electricity to illuminate the building. The school building is shown in Fig. 1 below, and the building of the Kielce University of Technology is shown in Fig. 2.

Fig. 1. Photo of the school building.

Fig. 2. Photo of the Energis building.

The school building was built in 1903, and the supply of fresh air is possible only through doors and windows. The Energis building, on the other hand, was built in 2012 and was designed with sustainable development in mind.

The study was conducted using two methods. The first method was to measure the microclimate parameters with a Testo 400 gauge. It was placed at the level of the thermal center of gravity. Air temperature, black sphere temperature, air velocity, mean radiation temperature, relative humidity, and carbon dioxide concentration were measured. The values were read after the measurements stabilized for 15 minutes. The second method consisted in completing the questionnaires on thermal impressions of the microclimate. This allowed for the assessment of thermal comfort in the room in which they stayed. In addition, the survey contained a question about clothing, thanks to which the average level of insulation of the garment was determined. The thermal resistance of the office chair was added to the value of thermal resistance, which is 0.1 clo. The total value of the thermal resistance for the first room was 0.62 clo, and for the second - 0.73 clo. At the end of the questionnaire, there was a record with data on the age, sex, weight, and height of the

Terotechnology XII   Materials Research Forum LLC
Materials Research Proceedings **24** (2022) 96-101   https://doi.org/10.21741/9781644902059-15

respondents. 32 people aged 18 to 22 and one women aged 52 participated in the buildings in question.

Results and Discussion

The same number of people took part in both rooms. In the first room, the subjects were 17 to 18 years old and one woman was 52 years old, and in the second room, they were from 20 to 22 years old. The respondents were asked about their thermal feelings, which are included on a seven-point scale: "-3" - too cold, "-2" - too cool, "-1" - pleasantly cool, "0" - comfortable, "1" - pleasantly warm, "2" - too warm, "3" - too hot. Fig. 3 below shows the frequency of the answers given on the subject of thermal impressions.

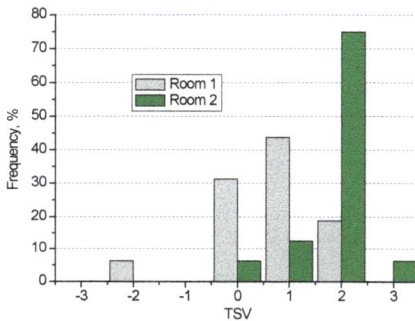

Fig. 3. Frequency of responses regarding thermal sensations vote (TSV).

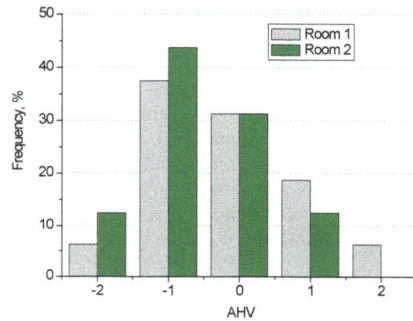

Fig. 4. Frequency of responses regarding air humidity vote (AHV).

In the figure above, the results for a traditional building are marked in gray, and the results for an intelligent building are green. It can be noticed that for a traditional building, students most often chose the answer "pleasantly warm", which amounted to 43.75%. They then chose the answer "comfortable" - 31.25%. Consecutively "too warm" - 18.75%, while the least chosen answer by students was "too cool" (6.25%). For the Energis building it looks like this: the most common answer was "too warm" (75%), then "pleasantly warm" (12.50%), and the options "comfortable" and "too hot" had the same number of answers - 6,25%. Looking at all the answers, it was found that the respondents did not feel well in the rooms they studied. For both buildings, the percentage of choosing the answer -3, -2, + 2, + 3 was greater than 10.00% and amounted to 25% for a traditional building, and 81% for an intelligent building. The respondents were also asked about the assessment of indoor humidity. In the questionnaire, when asked about the assessment of air humidity, there were the following answers: "-2" - too dry, "-1" - fairly dry, "0" - pleasantly, "1" - quite damp, "2" - too humid. The graph for the assessment of air humidity is presented in Fig. 4. The most frequently chosen answer for both rooms was the answer "fairly dry". It was 37.50% for the first room and 43.75% for the second room. The second most frequently chosen answer was "pleasantly". For both buildings, it accounts for 31.25% of all the answers provided. The respondents chose the next answer, it is "quite damp". It is 18.75% for the first room and 12.50% for the second. The options "too dry" and "too humid" are 6.25% each for the first room. For the second room, the answer "too dry" is 12.50%. It can be concluded that the respondents assessed the humidity in the examined rooms in a similar way. Fig. 5 shows respondents' preferences for humidity.

Terotechnology XII
Materials Research Proceedings **24** (2022) 96-101

Materials Research Forum LLC
https://doi.org/10.21741/9781644902059-15

Respondents had a choice of individual answers: "-1" - drier, "0" - no change, "1" - more humid. It can be seen from the chart above that the respondents would like their indoor air to be more humid. It looks like this. For room 1 "no change" and "more humid" are at the same level, they constitute 43.75%. When asked about their preferences regarding humidity, the respondents also selected the "more dry" option, which is 12.50%. When assessing the air humidity for the second room, more than half of the respondents (56.25%) considered the prevailing conditions "pleasantly". 43.75% of the respondents chose the answer "no change".

From this chart, it can be seen that the respondents want changes in indoor air humidity. The comparison between the TSV obtained from the questionnaires and the PMV calculated from the Fanger model is presented in Fig. 6.

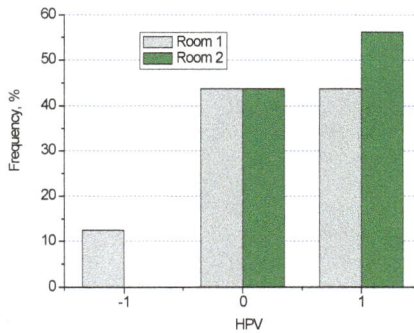

Fig. 5. Frequency of responses regarding the humidity preferences vote (HPV).

Figure 6 shows the gray color of the PMV results calculated on the basis of the Fanger model, while the green color represents the TSV obtained from the questionnaires. It can be seen that for the traditional building (Room 1) and the intelligent building (Room 2), the PMV and TSV values do not coincide. The adopted range of thermal comfort ranges from -0.5 to +0.5. Taking the answers obtained from the questionnaires, both buildings mix in this respect, but looking at the calculated results, they do not fit anymore. PMV for the first room is 0.16 and for the second room -0.1. On the other hand, the TSV is outside the scope of the ISO 7730 standard. The TSV for the traditional one is 0.69, and for the intelligent one is 1.81. Such a discrepancy between these indicators may depend on many factors that were not included in Fanger's model. Figure 7 below presents the thermal sensations of dissatisfied people in the examined rooms.

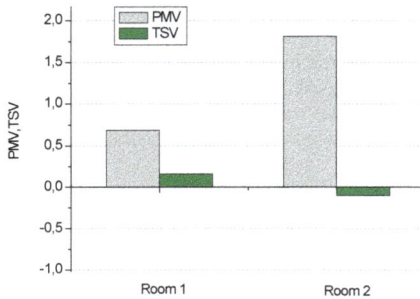

Fig. 6. A comparison of PMV calculated on the basis of the Fanger model and TSV obtained using questionnaires for both the buildings.

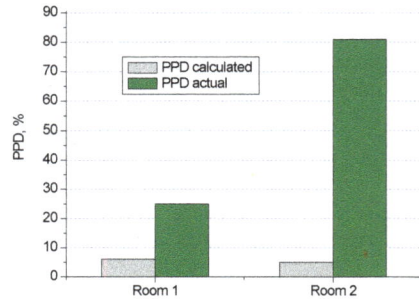

Fig. 7. Comparison of actual survey results with calculated results for PPD.

The above figure presents the results concerning the percentage of people dissatisfied with the prevailing indoor conditions. The results calculated on the basis of the Fanger model are marked in grey, and the results from the questionnaires are marked in green. The PPD is in no way overlapping with each other. For building 1, PPD was calculated at 6%, and 25% from the questionnaires. For building 2, the PPD z calculated on the basis of the Fanger model is equal to 5%, and on the basis of the surveys, it is equal to 81%. Hence the conclusion that the actual PPD results are much higher than those calculated on the basis of the Fanger model.

Conclusions

The measurement results of the microclimate parameter from the Testo 400 measuring device were compared with the actual results based on the questionnaires. The research for the traditional building was carried out in summer, and for the intelligent building in winter. Overall, the study confirmed that the TSV results (based on the questionnaires) were not consistent with the PMV results (calculated using the Fanger model). Therefore, the best solution is to modify the Fanger model. This discrepancy may result from other parameters that are not included in the model but have an impact on people's feelings and their responses to the surveys. Differences can also be seen in the PPD indicator. According to the ISO 7730 standard, the PPD ratio should not exceed 10%. Unfortunately, for a traditional and intelligent room, the percentage of dissatisfied people exceeded 10%. This proves that the Fanger model needs to be modified. Additionally, the survey showed that respondents prefer higher relative humidity. Therefore, it is necessary to take care of the thermal conditions of people staying in educational rooms.

References

[1] ISO International Organisation for Standardization, Ergonomics of the thermal environment – Analytical determination and interpretation of thermal comfort using calculation of the PMV and PPD indices and local thermal comfort criteria, International Standard ISO 7730, 2005.

[2] PN-EN 16798-1:2019, Energy Performance of Buildings-Ventilation for Buildings-Part 1: Indoor Environmental Input Parameters for Design and Assessment of Energy Performance of Buildings Addressing Indoor Air Quality, Thermal Environment, Lighting and Acustics, 2019.

Terotechnology XII   Materials Research Forum LLC
Materials Research Proceedings **24** (2022) 96-101   https://doi.org/10.21741/9781644902059-15

[3] N. Krawczyk, J. Krakowiak. The comparison of thermal comfort test results in selected traditional and modern buildings, E3S Web of Conferences 286 (2021) art. 02008. https://doi.org/10.1051/e3sconf/202128602008

[4] R.Z. Homoda, K.S.H. Saharia, H.A.F. Almurib, F.H. Nagi. RLF and TS fuzzy model identification of indoor thermal comfort based on PMV/PPD, Build. Environ. 49 (2012) 141-153. https://doi.org/10.1016/j.buildenv.2011.09.012

[5] A. Jindal. Thermal comfort study in naturally ventilated school classrooms in composite climate of India, Build. Environ. 142 (2018) 34-46. https://doi.org/10.1016/j.buildenv.2018.05.051

[6] N. Krawczyk, Comparison of thermal comfort in a traditional and intelligent building, E3S Web of Conferences 336 (2022) art. 00019. https://doi.org/10.1051/e3sconf/202233600019

[7] A. Białek, L. Dębska, N. Krawczyk. Assessment of light intensity and productivity in the intelligent building – case study, E3S Web of Conferences 336 (2022) art. 00011. https://doi.org/10.1051/e3sconf/202233600011

[8] G. Majewski, M. Telejko, Ł.J. Orman. Preliminary results of thermal comfort analysis in selected buildings, E3S Web of Conf. 17 (2017) art. 56. https://doi.org/10.1051/e3sconf/20171700056

[9] Ł.J. Orman, N. Radek, J. Pietraszek, M. Szczepaniak. Analysis of enhanced pool boiling heat transfer on laser – textured surfaces, Energies 13 (2020) art. 2700. https://doi.org/10.3390/en13112700

[10] N. Radek, J. Pietraszek, A. Gądek-Moszczak, Ł.J. Orman, A. Szczotok. The morphology and mechanical properties of ESD coatings before and after laser beam machining, Materials 13 (2020) art. 2331. https://doi.org/10.3390/ma13102331

[11] G. Majewski, Ł.J. Orman, M. Telejko, N. Radek, J. Pietraszek, A. Dudek. Assessment of thermal comfort in the intelligent buildings in view of providing high quality indoor environment, Energies 13 (2020) art. 1973. https://doi.org/10.3390/en13081973

[12] J. Pietraszek, E. Skrzypczak-Pietraszek. The uncertainty and robustness of the principal component analysis as a tool for the dimensionality reduction. Solid State Phenom. 235 (2015) 1-8. https://doi.org/10.4028/www.scientific.net/SSP.235.1

[13] J. Pietraszek, N. Radek, A.V. Goroshko. Challenges for the DOE methodology related to the introduction of Industry 4.0. Production Engineering Archives 26 (2020) 190-194. https://doi.org/10.30657/pea.2020.26.33

[14] J. Pietraszek, A. Gadek-Moszczak, N. Radek. The estimation of accuracy for the neural network approximation in the case of sintered metal properties. Studies in Computational Intelligence 513 (2014) 125-134. https://doi.org/10.1007/978-3-319-01787-7_12

[15] J. Pietraszek, R. Dwornicka, A. Szczotok. The bootstrap approach to the statistical significance of parameters in the fixed effects model. ECCOMAS Congress 2016 - Proceedings of the 7th European Congress on Computational Methods in Applied Sciences and Engineering 3, 6061-6068. https://doi.org/10.7712/100016.2240.9206

Terotechnology XII
Materials Research Proceedings 24 (2022) 102-108

Materials Research Forum LLC
https://doi.org/10.21741/9781644902059-16

Corrosion of the S235JR Carbon Steel after Normalizing and Overheating Annealing in 2.5% Sulphuric Acid at Room Temperature

LIPIŃSKI Tomasz[1,a] * and PIETRASZEK Jacek[2,b]

[1]University of Warmia and Mazury in Olsztyn, The Faculty of Technical Sciences Department of Materials and Machines Technology, 10-957 Olsztyn, ul. Oczapowskiego 11, Poland

[2] Cracow University of Technology, Faculty of Mechanical Engineering, Department of Applied Computer Science, 31-864 Kraków, al. Jana Pawła II 37, Poland

[a]tomekl@uwm.edu.pl, [b]jacek.pietraszek@pk.edu.pl

Keywords: Steel, Structural Steel, Carbon Steel, Corrosion, Corrosion Rate, Profile Roughness

Abstract. The low-carbon steels offer economical properties of mean hardness, strength, and low corrosion resistance, but the steel can be welded without restrictions, usually. The structural low carbon steels have a ferritic-perlitic microstructure. The microstructure depends on the manufacturing technology and heat treatments of steel. This steel is not intended for heat treatment. However, due to the technological process, which is welding, the material often overheats. This overheating causes microstructure growth. The effect of larger grains of the steel microstructure is the reduction of its functional properties. Corrosion resistance is an essential factor in structural steel's quality and application. The purpose of this article is to investigate corrosion resistance using weight loss and profile roughness parameters of typical structural steel in grade S235JR in 2.5% sulphuric acid solution in distilled water. Samples were tested after normalizing and superheating annealing. Corrosion tests show that continued corrosion characterizes tested steel in both corrosive environments. Roughness parameters for every one of the research times determine the size of steel corrosion.

Introduction

Low-carbon steel is a very popular construction material. The mechanical, physical, and chemical properties of low carbon steel are influenced by different factors, including chemical composition, manufacturing technology, the morphology of microstructure, inclusions, etc. Steels from the low carbon group are a wide range of industrial applications, mainly as a good welded construction material. The microstructure and properties of these steels are still tested to improve the quality. The structural low carbon steels have a ferritic-perlitic microstructure in normal conditions [1-3].

Steel structures with low-carbon structural steel can be built quickly by welding at a low price, but the main problem is their corrosion protection. Corrosion resistance is an essential factor in structural steel's quality and application. Corrosion processes can extract metal atoms from the metal lattice, which atoms during the process pass to corrosion products. Particularly dangerous is the corrosion causing the local diffusion of metal atoms. The problem is enormous because low-carbon structural steel is sensitive to corrosion. The corrosion rate depends on different environments [1, 4-8]. The risks of losing steel properties due to corrosion occur especially clearly in aggressive atmospheres. In the atmosphere, gaseous pollutants classified as compounds of nitrogen, ammonia, and sulfur, including hydrogen sulfide, occur more and more often [9]. They occur in large amounts, usually around industrial plants. The aggressive factors include sulfur

Terotechnology XII

Materials Research Proceedings **24** (2022) 102-108

Materials Research Forum LLC

https://doi.org/10.21741/9781644902059-16

compounds. The activation of corrosive processes by sulfur compounds causes high chemical activity, which is dangerous for metal alloys. Sulfur dioxide dissolved in water creates aggressive SO_3^{2-} ions. In the hydrochloric acid environment, the corrosion rate of carbon structural steel increases intensively with increasing acid concentration and with increasing carbon content in the steel. In addition, the corrosion products of some metal alloys, including iron alloys, promote the rapid oxidation of sulfurous anhydride, contributing to the intensification of corrosion [4, 10, 11].

Another problem affecting the corrosion rate of steel is its grain size. As a result of the production process, e.g. welding, hot-forming, etc., the material may overheat, resulting in grain growth.

One of the corrosive environmental factors is sulfur compounds. They occur mainly in the vicinity of industries polluted by sulfur. Because of this, structural steel has often been tested with sulfur acid on corrosion resistance. Regarding the importance of the problem, this study was carried out to investigate the corrosion resistance of the overheated low carbon structural steel in 2.5% sulphuric acid reaching water at ambient temperature.

The subject of the article may be of interest to a broad audience related to corrosive, aggressive environments at room temperatures, e.g. biotechnologists [12], water supply system conservators [13, 14], and housing infrastructure conservators [15, 16]. The problems of such corrosion are also crucial for the machine industry [17], in particular for the quality management systems implemented in this industry [18, 19] and the latest automation systems for management supported by dense IT networks [20].

Materials and Methods

The experiment was performed on low carbon S235JR (1.0038) steel designation according to EN 10025-2:2004, plate - thickness t = 5.00 mm. The actual chemical composition tested steel is presented in Table 1.

Table 1. Chemical composition of the S235JR steel

Mean chemical compositions [wt. %]								
C	Si	Mn	P	S	Cr	Cu	Ni	N
0.19	0.22	0.90	0.03	0.04	0.03	0.02	0.02	0.01

Before experiments, the specimens, after being mechanically cut off with an area of 13 cm2 (40 x 10 x 5 mm), were successively polished with water paper to Ra = 0.32 μm and next cleaned with 95% alcohol.

The samples with ferritic-perlitic microstructure were tested in accordance with the standard dedicated for stainless steel PN EN ISO 3651-1 [21] corrosive media were represented by 2,5% sulphuric acid.

The corrosion rare of the S235JR steel measured in mm/year was calculated with the use of the below formula (1), measured in g/m2 were calculated with the use the below formula (2):

$$r_{corm} = \frac{8760 \cdot m}{S \cdot t \cdot \rho} \tag{1}$$

$$r_{corg} = \frac{10000 \cdot m}{S \cdot t} \tag{2}$$

where:

t – time of soaking in a corrosive solution of 2.5% sulphuric acid water solution [hours],

S – the surface area of the sample (the starting value was assumed) [cm²],

Terotechnology XII   Materials Research Forum LLC
Materials Research Proceedings 24 (2022) 102-108   https://doi.org/10.21741/9781644902059-16

M – average mass loss in solution (measured as the difference between initial mass and mass
after corrosion time) [g],

P – sample density [g/cm^3].

Two variants of heat treatment were used for the tests. In the first variant, the alloy was normalized
at 860°C for 8 minutes and cooled in air. In the second variant, the alloy was superheated, annealed
at 1100°C for 30 minutes, and cooled in air.

The corrosion resistance of the S235JR steel in 2.5% sulphuric acid was tested using weight
loss. The mass of samples was measured by Kern ALT 3104AM digital laboratory precision scales
with an accuracy of 0.0001 g.

Profile roughness parameters were analyzed according to the PN-EN 10049:2014-03 standard
(Measurement of roughness average Ra and peak count RPc on flat metallic products) by the
Diavite DH5 profilometer.

Results and Discussion

The microstructure of raw overheated S235JR steel is presented in Fig. 1. The effect of overheating
on the microstructure presents the enlarged, equaxial grains ferrite phase (white area) at the
background of the perlite (a gray area).

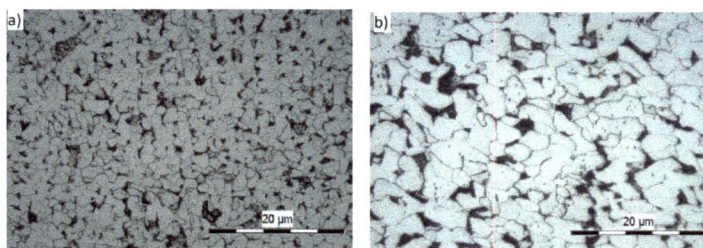

*Fig. 1. Microstructure of S235JR: light etched ferrite and dark etched perlite: a) after
normalizing annealing, b) after overheating*

Influence time of soaking the S235JR structural steel in 2.5% sulphuric acid at ambient
temperature on the relative mass loss (RML) with regression equation and correlation coefficient
r is presented in Fig. 2.

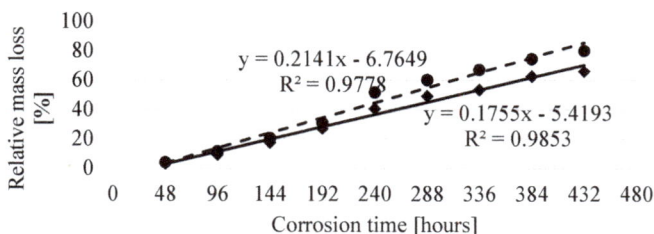

*Fig. 2. Influence time of soaking the S235JR structural steel in 2.5% sulphuric acid on
the relative mass loss (RML) at ambient temperature; solid line alloy after normalizing
annealing, dashed line alloy after overheating*

Terotechnology XII Materials Research Forum LLC
Materials Research Proceedings **24** (2022) 102-108 https://doi.org/10.21741/9781644902059-16

Influence time of soaking the S235JR structural steel in 2.5% sulphuric acid at ambient temperature on the corrosion rate measured in mm per year with the regression equation and correlation coefficient r is presented in Fig. 2 and the corrosion rate measured in gram per m^2 with the regression equation and correlation coefficient r is presented in Fig. 3.

Profile roughness parameters of S235JR steel after corrosion tests in 2,5% sulphuric acid the regression equations and correlation coefficients r are presented in Fig. 4 with Ra – arithmetical mean roughness value [μm], Rp – maximum roughness depth [μm] for time range: 48, 96, 144, 192, 240, 288, 336, 384 and 432 hours of soaking is presented in Fig. 4 and for Rq - mean peak width [μm], Rt - total height of the roughness profile [μm]in Fig. 5.

Relative mass loss (RML – Fig. 1) in the initial period of samples soaking is similar for both states of heat treatment. As the soaking time increases, the difference becomes more and more apparent. Steel after superheating annealing loses its corrosion resistance (Fig. 1) faster than steel after normalizing. This is due to the larger grains obtained after overheating the steel. Analyzing the corrosion rates (Fig. 2 and Fig. 3), it was found that in the soaking process up to 192 hours, the corrosion rate increased for the steel after overheating compared to the normalized steel. From 240 hours of soaking, the corrosion rate of the steel after overheating stabilizes compared to the corrosion rate of normalized steel. It follows that coarse grain steel in the initial period corrodes at a higher rate than fine grain steel. This difference stabilizes over time. Nevertheless, a higher corrosion rate of coarse grain steel than that of fine grain steel was observed.

The analyzed roughness parameters indicate that steel increases the surface roughness with increasing soaking time (Fig. 4 and Fig. 5).

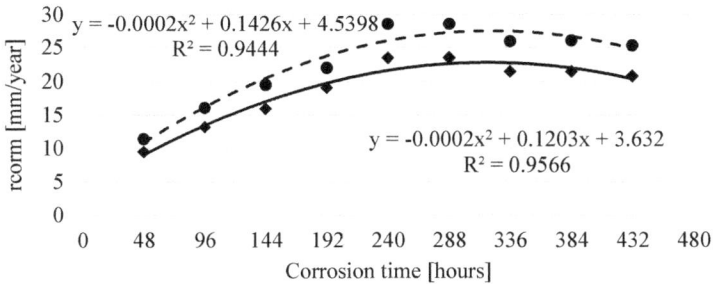

Fig. 3. *Influence time of soaking the S235JR structural steel in 2.5% sulphuric acid at ambient temperature on the corrosion rate measured in mm per year; solid line alloy after normalizing annealing, dashed line alloy after overheating*

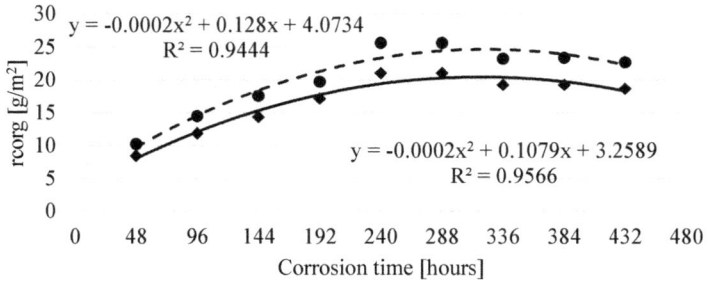

Fig. 4. *Influence time of soaking the S235JR structural steel in 2.5% sulphuric acid at ambient temperature on the corrosion rate measured in gram per m²; solid line alloy after normalizing annealing, dashed line alloy after overheating*

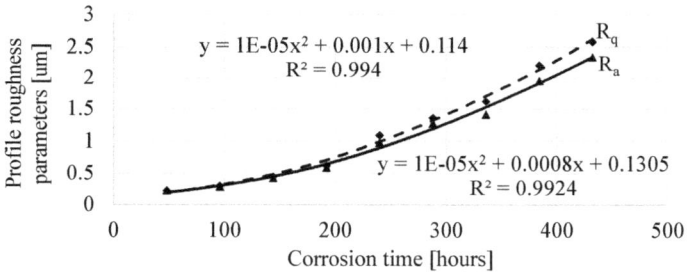

Fig. 5. *Profile roughness of S235JR steel after corrosion tests in 2.5% sulphuric acid at ambient temperature for different corrosion times: Ra - arithmetical mean roughness; alloy after overheating*

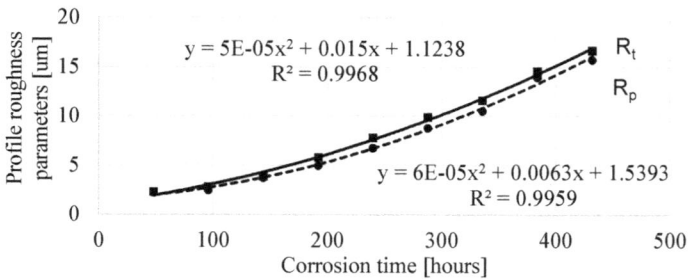

Fig. 5. *Profile roughness of S235JR steel after corrosion tests in 2.5% sulphuric acid at ambient temperature for different corrosion times: Rp - maximum roughness depth [μm], Rt - total height of the roughness profile [μm]; alloy after overheating*

Terotechnology XII   Materials Research Forum LLC
Materials Research Proceedings 24 (2022) 102-108   https://doi.org/10.21741/9781644902059-16

Conclusions

1. The research results show that weight loss of S235JR steel soaking in 2.5% sulphuric acid at ambient temperature depends on the type of heat treatment (microstructure morphology).
2. The roughness of the sample increases, but the corrosion rate measured as a corrosion velocity decreases with time.
3. When steel is overheated, it partially loses its corrosion resistance.

References

[1] L.L. Machuca, R. Jeffrey, R.E. Melchers. Microorganisms associated with corrosion of structural steel in diverse atmospheres, Int. Biodeterior. Biodegradation 114 (2016) 234-243. http://dx.doi.org/10.1016/j.ibiod.2016.06.015

[2] T. Lipiński. Investigation of corrosion rate of X55CrMo14 stainless steel at 65% nitrate acid at 348 K. Production Engineering Archives 27 (2021) 108-111. https://doi.org/10.30657/pea.2021.27.13

[3] K. Knop. The use of quality tools to reduce surface defects of painted steel structures. Manufacturing Technology 21 (2021) 805-817. https://doi.org/10.21062/mft.2021.088

[4] Uhlig H.H., Revie R.W. Corrosion and corrosion control 3rd Ed. John Wiley and Sons, 1985.

[5] Z. Blikharskyy, J. Selejdak, Y. Blikharskyy, R. Khmil. Corrosion of reinforce bars in RC constructions, System Safety: Human-Technical Facility-Environment 1 (2019) 277-283. https://doi.org/10.2478/czoto-2019-0036

[6] M.A. Azam, N.E. Safie, H.H. Hamdan. Effect of sulfur content in the crude oil to the corrosion behavior of internal surface of API 5L X65 petroleum pipeline steel, Manufacturing Technology 21 (2021) 561-574. https://doi.org/10.21062/mft.2021.066

[7] T. Lipiński. Corrosion effect of 20% NaCl solution on basic carbon structural S235JR steel, Engineering for Rural Development 16 (2017) 1069-1074. https://doi.org/10.22616/ERDev2017.16.N225

[8] E. Naveen, B.V. Ramnath, C. Elanchezhian, M.S.S. Nazirudeen. Influence of organic corrosion inhibitors on pickling corrosion behaviour of sinter-forged C45 steel and 2% Cu alloyed C45 steel. Journal of Alloys and Compounds 695 (2017) 3299-3309. https://doi.org/10.1016/j.jallcom.2016.11.133

[9] N.G. Thompson, M. Yunovich, D. Dunmire. Cost of corrosion and corrosion maintenance strategies, Corros. Rev. 25 (2007) 247-261. https://doi.org/10.1515/CORRREV.2007.25.3-4.247

[10] D. Bricín, A. Kříž. Influence of the Boriding Process on the Properties and the Structure of the Steel S265 and the Steel X6CrNiTi18-10, Manufacturing Technology 21(1) (2021) 37-44. https://doi.org/10.21062/mft.2021.003

[11] P. Szabracki. Effect of aging on the microstructure and the intergranular corrosion resistance of X2CrNiMoN25-7-4 duplex stainless steel, Solid State Phenom. 203-204 (2013) 59-62. https://doi.org/10.4028/www.scientific.net/SSP.203-204.59

[12] E. Skrzypczak-Pietraszek, A. Urbańska, P. Żmudzki, J. Pietraszek. Elicitation with methyl jasmonate combined with cultivation in the Plantform™ temporary immersion bioreactor highly increases the accumulation of selected centellosides and phenolics in Centella asiatica (L.) Urban

shoot culture, Engineering in Life Sciences 19 (2019) 931-943.
https://doi.org/10.1002/elsc.201900051

[13] M. Dobrzański. The influence of water price and the number of residents on the economic efficiency of water recovery from grey water, Technical Transactions 118 (2021) art. e2021001. https://doi.org/10.37705/TechTrans/e2021001

[14] K. Wójcicka. The efficiency of municipal sewage treatment plants inspiration for water recovery. Technical Transactions 118 (2021) art. e2021023.
https://doi.org/10.37705/TechTrans/e2021023

[15] G. Majewski, Ł.J. Orman, M. Telejko, N. Radek, J. Pietraszek, A. Dudek. Assessment of thermal comfort in the intelligent buildings in view of providing high quality indoor environment. Energies 13 (2020) art. 1973. https://doi.org/10.3390/en13081973

[16] A. Sikora. Rzeszów as an example of a 'new town' tailored for the modern era. Technical Transactions 118 (2021) art. e2021006. https://doi.org/10.37705/TechTrans/e2021006

[17] S. Marković, D. Arsić, R.R. Nikolić, V. Lazić, B. Hadzima, V.P. Milovanović, R. Dwornicka, R. Ulewicz, R. Exploitation characteristics of teeth flanks of gears regenerated by three hard-facing procedures, Materials 14 (2021) art. 4203. https://doi.org/10.3390/ma14154203

[18] A. Pacana, R. Ulewicz. Analysis of causes and effects of implementation of the quality management system compliant with iso 9001, Polish Journal of Management Studies 21 (2020) 283-296. https://doi.org/10.17512/pjms.2020.21.1.21

[19] R. Ulewicz, F. Novy, R. Dwornicka. Quality and work safety in metal foundry, METAL 2020 – 29th Int. Conf. Metall. Mater. (2020), Ostrava, Tanger 1287-1293.
https://doi.org/10.37904/metal.2020.3649

[20] J. Pietraszek, N. Radek, A.V. Goroshko. Challenges for the DOE methodology related to the introduction of Industry 4.0, Production Engineering Archives 26 (2020) 190-194.
https://doi.org/10.30657/pea.2020.26.33

[21] PN EN ISO 3651-1, Determination of resistance to intergranular corrosion of stainless steels. Part 1: Austenitic and ferritic-austenitic (duplex) stainless steels. Corrosion test in nitric acid medium by measurement of loss in mass (Huey test).

Terotechnology XII
Materials Research Proceedings 24 (2022) 109-117

Materials Research Forum LLC
https://doi.org/10.21741/9781644902059-17

The Influence of the Solvent on the Corrosion Resistance of Vinyltrimethoxysilane Based Coatings Deposited on X20Cr13 Steel

KUCHARCZYK Aleksandra [1,a], ADAMCZYK Lidia [1,b*]
and MIECZNIKOWSKI Krzysztof [2,c]

[1]Czestochowa University of Technology, Faculty of Production Engineering and Materials Technology, Department of Material Engineering, Czestochowa, Poland, EU

[2]University of Warsaw, Faculty of Chemistry, Department of Inorganic and Analytical Chemistry, Warsaw, Poland, EU

[a]aleksandra.kucharczyk@pcz.pl, [b]lidia.adamczyk@pcz.pl, [c]kmiecz@chem.uw.pl

Keywords: VTMS, Sol-Gel Method, Stainless Steel, Alcohol, Anticorrosive Coatings

Abstract. The paper presents the results of research on the anti-corrosion properties of coatings based on vinyltrimethoxysilane, alcohol, and acetic acid in sulfate solutions with or without chloride ions (pH = 2). The experiment investigated the influence of a solvent (methanol, ethanol, butanol, propanol) on anti-corrosion properties, surface morphology, and adhesion to the steel substrate. The coatings were deposited on X20Cr13 stainless steel using the sol-gel dip method.

Introduction

Stainless steel is used in many industries: from the production of heavy machinery and energy to precision mechanics and electronics [1, 2]. The protective mechanism of stainless steel relies on a naturally generated (passive) oxide layer that prevents corrosion in mildly corrosive environments. In the presence of aggressive halogen ions (Cl-, Br-, I-), the oxide layer undergoes pitting corrosion [3-5]. In order to increase the protection against corrosion, coatings, e.g. chromate or phosphate, are used. The toxicity and carcinogenicity of chromium and phosphorus limited the use of this type of anti-corrosion coating [6-8]. The search for newer, better, more environmentally friendly coating systems has begun. One of the best alternatives in technology is the sol-gel method [9, 10].

The sol-gel process involves hydrolysis and condensation of M (OR) n metal alkoxides. Tralkoxysilanes such as VTMS vinyltrimethoxysilane is well known as coupling agents and crosslinkers. These are compounds that have such functional groups as vinyl, amino, carboxyl, glycidyl, hydroxyl, or acrylic. The speed of the hydrolysis and condensation reactions depends on the parameters: pH, temperature, and concentration of the reagents. Changing these parameters may change the structure and properties of the silane layer to a large extent [11-15].

The precursors of the synthesis reaction in the sol-gel method are various metal alkoxides, salts, or nitrates [16]. The preparation of sol solutions is based on dissolving the appropriate alkoxide in an organic solvent, most often anhydrous alcohol [17]. The simplest alcohols such as methanol, ethanol, propanol, or butanol are widely used raw materials in chemical synthesis [18, 19]. The hydrolysis and condensation (polycondensation) reactions take place simultaneously in the entire volume of the solution. The speed of the sol-gel process can be controlled, for example, by using a suitable catalyst (acid or basic) [20]. In this study, the influence of the solvent used as a component of the VTMS / Alkohol / AcOH coating on its adhesion and corrosion protection of X20Cr13 steel was assessed. Over the last dozen or so years, many articles have been written on the protection of metal surfaces with silanes, and these publications do not deal with the issue of the influence of the solvent on the process of protecting metals covered with silane coatings against corrosion. The paper presents research on alcohol: methyl, ethyl, butyl, and propyl alcohol.

The results presented in this article may be of interest to those industries where there is a risk of corrosion [21-24] and biocorrosion [25-28]. In the case of using special coatings [29-31], especially those subjected to laser treatment [32-34], the issue of texture may be important so that the anti-corrosion layer has proper adhesion. Improvement of anti-corrosion properties will significantly affect the risk of occurrence of particular failure scenarios [35-38] and thus modify the quality management schemes [39, 40]. Undoubtedly, it will also have an impact on the design of new devices and machines [41], especially those operating in difficult conditions [42-44] or military treatment [45-47], taking into account the increased corrosion resistance in accordance with the Robust Design concept [48, 49], and thus will be an inspiration for data analysis methods [50-52] in the field of production engineering and quality management [53].

Materials and Methods

Materials. The chemical composition of X20Cr13 stainless steel (in wt.%) was as follows: C-0.17; Cr-12.6; Si-0.34 ;, Ni-0.25; Mn-0.30; V-0.04; P-0.024; S <0.005; the rest is Fe. The following reagents were used: vinyltrimethoxysilane VTMS (by Sigma Aldrich), methanol (by Chempur), ethanol (by POCH Basic), propanol (by Sigma Aldrich), butanol (by Chempur), acetic acid AcOH (by Chempur) with the analytical purity grade and deionized water. The volume ratio of the resulting VTMS: Alcohol: AcOH coatings was 4.84: 2.16: 3.0.

Measurements were carried out on electrodes made of X20Cr13 stainless steel samples. Metal with a diameter of 5 mm was mounted in polymethyl methacrylate frames. The initial treatment of metal samples consisted of wet polishing with polishing papers (600, 1000, 2000), rinsing with deionized water and ethyl alcohol, and then drying at room temperature. In order to decrease the surface of the samples, each time before applying the coating, they were washed with acetone.

Preparations of VTMS/Alkohol/AcOH Coatings. Preparation with the use of a solvent was included in the publication [54]. Four sols were prepared according to the following procedure: 3.16 mol dm^{-3} VTMS was dispersed in alcohol:

1) methyl MtOH,

2) ethyl EtOH,

3) propyl IPOH,

4) butyl BtOH.

The amount of 0.1 mol dm^{-3} AcOH was gradually added to the solutions. The solutions were continued to mix for 2 days. After the solution had changed its consistency, the samples were immersed for approximately 20 minutes. The methodology of applying silane coatings was developed in the publication [55]. The coated samples were dried for 1 day at room temperature in a silica gel desiccator.

Experimental Conditions. The electrochemical tests were carried out using the CHI 706 measurement station (CH Instruments, Austin, Texas, USA) in a three-electrode system, in which the auxiliary electrode was a platinum electrode and the reference electrode was a saturated calomel electrode NEK. In order to determine the protective properties of VTMS/Alcohol/AcOH coatings, potentiodynamic polarization curves of uncoated and coated steel electrodes were recorded in the following solutions: 0.5 mol dm^{-3} Na$_2$SO$_4$ (pH = 2) and 0.5 mol dm^{-3} Na$_2$SO$_4$ + 0.5 mol dm^{-3} NaCl (pH = 2), the potential range was from -0.8 V to 1.6 V, the polarization rate was 10 mVs^{-1}. The surface appearance of the coatings deposited on the tested steel was assessed using

Terotechnology XII
Materials Research Proceedings **24** (2022) 109-117

Materials Research Forum LLC
https://doi.org/10.21741/9781644902059-17

the Olympus GX41 optical microscope. The adhesion tests were carried out using a simple Scotch[TM] tape sticking and peeling test after each coating application.

Experimental Part

Fig. 1 shows the morphology of VTMS/Alkohol/AcOH coatings deposited on the surface of X20Cr13 steel. The morphology of all shells is smooth and transparent. The coatings evenly cover the entire surface of the electrodes, with no visible pits when the coating is removed. The adhesion to the substrate of X20Cr13 stainless steel was checked using Scotch[TM] tape, immediately after the deposition of the VTMS/Alcohol/AcOH coatings. The produced coatings are characterized by good adhesion to the steel substrate.

In order to characterize the anti-corrosion properties of the produced VTMS/Alcohol/AcOH coatings against general corrosion, potentiodynamic curves were recorded in an acidified solution of 0.5 mol dm^{-3} Na_2SO_4 at pH = 2, in the potential range of -0.8 ÷ 1.6 V for X20Cr13 steel. uncoated and coated. As can be seen from Fig. 2, the produced coatings inhibit anodic processes and shift the corrosion potential of the steel by approx. 0.5 V towards positive values. The addition of the solvent leads to a reduction (1 - 4 times) of the cathode and anode currents density (Fig. 2c).

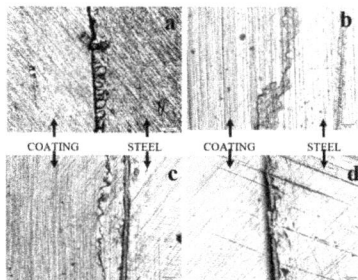

Fig. 1. *VTMS/Alcohol/AcOH coating with a concentration of 3,16 mol dm^{-3} VTMS in solution and containing alcohol: methyl MtOH (a), ethyl EtOH (b), propyl PyOH (c), butyl BtOH (d). Olympus GX41 (x100)*

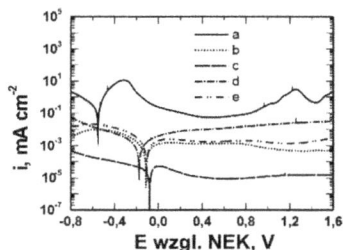

Fig. 2. *Potentiodynamic polarization curves recorded in a solution of 0.5 mol dm^{-3} Na_2SO_4 pH = 2 for X20Cr13 steel not coated (a) and coated with VTMS/Alcohol/AcOH coatings with a VTMS concentration of 3.16 mol dm^{-3} and containing alcohol: methyl MtOH (b), ethyl EtOH (c), propyl PyOH (d), butyl BtOH (e). Polarization speed 10 mVs^{-1}, solutions in contact with air, 24 ° C.*

In order to assess the inhibition capacity of the local corrosion coatings produced, analogous potentiodynamic curves were performed for the sulfate solution containing the addition of 0.5 mol dm^{-3} chloride ions (Fig. 3). The corrosion potential of X20Cr13 steel covered with VTMS/Alcohol/AcOH coatings is shifted towards the positive values by approx. 0.5 V (Fig.3c) in relation to the value of the corrosion potential recorded for the uncoated steel (E_{cor} = -0.597 V) (Fig. 3a). Lower values of cathode and anode current densities were also observed for steels covered with these coatings. For all VTMS/Alcohol/AcOH coatings deposited on X20Cr13 steel, no breakthrough potential of the passive layer (pitting nucleation potential) was observed. Thus, VTMS/Alcohol/AcOH coatings effectively protect the steel against the penetration of a solution containing Cl⁻ ions and inhibit pitting nucleation processes. Fig. 4 shows the surface photos of the samples after corrosion tests

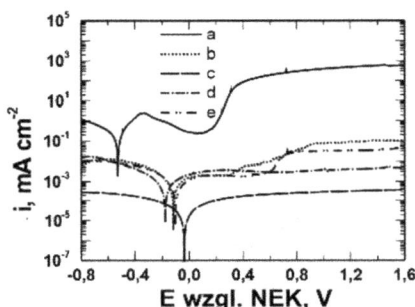

Fig. 3. Potentiodynamic polarization curves recorded in a solution of 0.5 mol dm^{-3} Na₂SO₄ + 0.5 mol dm^{-3} NaCl pH = 2 for X20Cr13 steel not coated (a) and coated with VTMS/Alcohol/AcOH coatings with a VTMS concentration of 3.16 mol dm^{-3} and containing alcohol: methyl MtOH (b), ethyl EtOH (c), propyl PyOH (d), butyl BtOH (e). Polarization speed 10 mVs⁻¹, solutions in contact with air, 24°C.

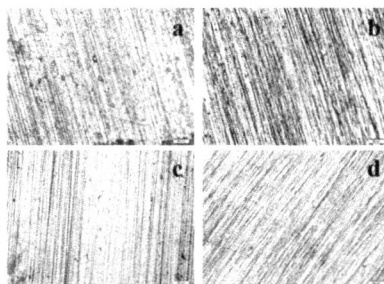

Fig. 4. Morphology of the sample surface after corrosion tests, coating: VTMS/MtOH/AcOH (a), VTMS/EtOH/AcOH (b), VTMS/PyOH/AcOH (c), VTMS/BtOH/AcOH (d). Olympus GX41 (x200)

Fig. 4 shows the morphology of samples with VTMS/Alcohol/AcOH coating after corrosion tests in a solution of 0.5 mol dm^{-3} Na₂SO₄ + 0.5 mol dm^{-3} NaCl pH = 2. The photos show steel surfaces without pitting. Thus, the obtained coatings effectively inhibit the corrosive processes.

Conclusions
The tests of VTMS/Alcohol/AcOH coatings have shown that the selection of an appropriate solvent has a significant impact on the anti-corrosive properties of VTMS coatings. The produced coatings show a homogeneous surface without visible defects in the structure, good adhesion to the substrate, and extend the time of steel resistance to the action of chloride and sulfate ions in an acidic environment. The VTMS/EtOH/AcOH coating shows the best ability to block the transport of chloride ions responsible for pitting corrosion of steel.

References
[1] L. Adamczyk, A. Pietrusiak. Zastosowanie powłok kompozytowych na bazie kwasu 4-aminobezoesowego (4-ABA) i poli(3,4-etylenodioksytiofenu) (PEDOT) w ochronie stali X20Cr13 przed korozją. Ochrona Przed Korozją 11 (2009) 465-468.

[2] L. Adamczyk, A. Dudek. The influence of components concentration in the electrodeposition process on the protective properties of 3,4-ethylenedioxythiophene (EDOT) with 4-(pyrrole-1-yl) benzoil acid (PYBA), polyoxyethylene-10-laurylether (BRIJ) and lithium perchlorate. METAL 2019 Int. Conf, Tanger, Brno 2019, 973-978. https://doi.org/10.37904/metal.2019.881

[3] M. Vayer, I. Reynaud, R. Erre. XPS characterisations of passive films fordem on martensitic stainless steel: qualitative and quantitative investigations. J. Mater. Sci. 35 (2000) 2581-2587. https://doi.org/10.1023/A:1004719213960

[4] L. Adamczyk, A. Pietrusiak, H. Bala. Protective properties of PEDOT/ABA coatings deposited from micel lar solution on stainless steel. Arch. Metall. Mater. 56 (2011) 883-889.

[5] M. Slemnik. Effect of testing temperature on corrosion behaviour of different heat treated stainless steel in the active-passive region. Anti-Corrosion Methods and Mater. 55 (1) (2008) 20-26. https://doi.org/10.1108/00035590810842799

[6] A.A. Burkov, M.A. Kulik. Wear-resistant and anticorrosive coatings based on chrome carbide Cr7C3 obtained by electric Spark deposition. Prot. Met. Phys. Chem. Surf. 56 (2020) 1217-1221. https://doi.org/10.1134/S2070205120060064

[7] B. Del Amo, R. Romagnoli, C. Deya, J.A. Gonzalez. High performance water-based paints with non-toxic anticorrosive pigments. Progress in Organic Coatings 45 (4) (2002) 389-397. https://doi.org/10.1016/S0300-9440(02)00125-X

[8] S. Bera, G. Udayabhanu, R. Narayan, T.K. Rout. High performance chrome free coating for white rust protection of zinc. Materials Science and Technology 32 (4) (2016) 338-347. https://doi.org/10.1179/1743284715Y.0000000081

[9] Y.M. Liza, R.C. Yasin, S.S. Maidani, R. Zainul. Sol Gel: Principle and technique (a review). INA-Rxiv 2018. https://doi.org/10.31227/osf.io/2cuh8

[10] I. De Graeve, A. Franquet, C. Le Pen, H. Terryn, I. Vereecken. Silanefilms for the pre-treatment of aluminum: film formation mechanism and caring. In: L. Fedrizzi, H. Terryn, A. Simoes (Eds.) Innovative Pre-Treatment Techniques to Prevent Corrosion of Metallic Surfaces. Elsevier, 2007, 1-18.

[11] V.A. Ogarev, S.L. Selector. Organosilicon promoters of adhesion andtheir influence on the corrosion of metals. Progress in Organic Coatings 21(1992) 135-187. https://doi.org/10.1016/0033-0655(92)87003-S

Terotechnology XII
Materials Research Proceedings **24** (2022) 109-117

Materials Research Forum LLC
https://doi.org/10.21741/9781644902059-17

[12] S. Zheng, J. Li. Inorganic-organic sol gel hybrid coatings for corrosion protection of metals. J. of Sol-Gel Sci. Technol. 54 (2010) 174-187. https://doi.org/10.1007/s10971-010-2173-1

[13] H. Wang, R. Akid. A room temperature cred sol-gel anticorrosion pre-treatment for Al. 2024-T3 alloys. Corros. Sci. 49 (12) (2007) 4491-4503. https://doi.org/10.1016/j.corsci.2007.04.015

[14] S.R. Davis, A.R. Brough, A. Atkinson. Formation of silica/epoxy hybrid Network polymers. J. Non-Cryst. Solids 315 (2003) 197-205. https://doi.org/10.1016/S0022-3093(02)01431-X

[15] Y.H. Han, A. Taylor, K. M. Knowles. Characterization of organic-inorganic hybrid coatings deposited on aluminum substrates. Surf. Coat. Technol. 202 (9) (2008)1859-1868. https://doi.org/10.1016/j.surfcoat.2007.08.006

[16] M. Guglielmi, G. Carturan, Precursors for sol-gel preparations. J. Non-Crystalline Solids 100 (1988) 16. https://doi.org/10.1016/0022-3093(88)90004-X

[17] H. Schmidt, G. Jonschker, S. Goedlicke, M. Menning. The sol-gel process as a Basic technology for nanoparticle- dispersed inorganic-organic composites. J. Sol-Gel Sci. Technol. 19 (2000) 39-51. https://doi.org/10.1023/A:1008706003996

[18] M. Starowicz, J. Banaś. Anodowe właściwości żelaza w etalonowych i n-propanolowych roztworach elektrolitów. Ochrona Przed Korozją 55 (2012) 110-113.

[19] M. Starowicz, K. Banaś, J. Banaś. Anodowe właściwości żelaza w metanolowych roztworach elektrolitów. Ochrona Przed Korozją 52 (2009) 543-545.

[20] D.L. Meixner, P.N. Dyer. Influence of sol-gel synthesis parameters on the microstructure of particulate silica xerogels, J. Sol- Gel Sci. Technol. 14 (1999) 223-232. https://doi.org/10.1023/A:1008774827602

[21] D. Klimecka-Tatar, G. Pawlowska, R. Orlicki, G.E. Zaikov. Corrosion characteristics in acid, alkaline and the ringer solution of Fe68-xCoxZr10Mo5W2B 15 metallic glasses, J. Balk. Tribol. Assoc. 20 (2014) 124-130.

[22] D. Klimecka-Tatar, G. Pawłowska, K. Radomska. The effect of Nd12Fe77Co5B6 powder electroless biencapsulation method on atmospheric corrosion of polymer bonded magnetic material METAL 2014 – 23rd Int. Conf. Metallurgy and Materials (2014), Ostrava, Tanger 985-990.

[23] T. Lipiński. Corrosion resistance of 1.4362 steel in boiling 65% nitric acid, Manufacturing Technology 16 (2016) 1004-1009.

[24] T. Lipiński. Roughness of 1.0721 steel after corrosion tests in 20% NaCl, Production Engineering Archives 15 (2017) 27-30. https://doi.org/10.30657/pea.2017.15.07

[25] A. Dudek. Surface properties in titanium with hydroxyapatite coating, Optica Applicata 39 (2009) 825-831.

[26] A. Dudek, R. Wlodarczyk. Structure and properties of bioceramics layers used for implant coatings, Solid State Phenom. 165 (2010) 31-36. https://doi.org/10.4028/www.scientific.net/SSP.165.31

[27] E. Skrzypczak-Pietraszek. Phytochemistry and biotechnology approaches of the genus exacum. In: The Gentianaceae – Volume 2: Biotechnology and Applications, 2015, 383-401. https://doi.org/10.1007/978-3-642-54102-5_16

[28] A. Dudek, M. Klimas. Composites based on titanium alloy Ti-6Al-4V with an addition of inert ceramics and bioactive ceramics for medical applications fabricated by spark plasma sintering (SPS method), Materialwissenschaft und Werkstofftechnik 46 (2015) 237-247. https://doi.org/10.1002/mawe.201500334

[29] Ł.J. Orman. Boiling heat transfer on meshed surfaces of different aperture, AIP Conference Proc. 1608 (2014) 169-172. https://doi.org/10.1063/1.4892728

[30] L. Dąbek, A. Kapjor, Ł.J. Orman. Distilled water and ethyl alcohol boiling heat transfer on selected meshed surfaces, Mechanics and Industry 20 (2019) art. 701. https://doi.org/10.1051/meca/2019068

[31] D. Nowakowski, A. Gądek-Moszczak, P. Lempa. Application of machine learning in the analysis of surface quality - the detection the surface layer damage of the vehicle body, METAL 2021 – 30th Int. Conf. Metallurgy and Materials (2021), Ostrava, Tanger 864-869. https://doi.org/10.37904/metal.2021.4210

[32] Ł.J. Orman Ł.J., N. Radek, J. Pietraszek, M. Szczepaniak. Analysis of enhanced pool boiling heat transfer on laser-textured surfaces. Energies 13 (2020) art. 2700. https://doi.org/10.3390/en13112700

[33] N. Radek, J. Pietraszek, A. Gadek-Moszczak, Ł.J. Orman, A. Szczotok. The morphology and mechanical properties of ESD coatings before and after laser beam machining, Materials 13 (2020) art. 2331. https://doi.org/10.3390/ma13102331

[34] N. Radek, J. Konstanty, J. Pietraszek, Ł.J. Orman, M. Szczepaniak, D. Przestacki. The effect of laser beam processing on the properties of WC-Co coatings deposited on steel. Materials 14 (2021) art. 538. https://doi.org/10.3390/ma14030538

[35] J. Korzekwa, M. Bara, J. Pietraszek, P. Pawlus. Tribological behaviour of Al2O3/inorganic fullerene-like WS2 composite layer sliding against plastic, Int. J. Surf. Sci. Eng. 10 (20160 570-584. https://doi.org/10.1504/IJSURFSE.2016.081035

[36] G. Filo, J. Fabiś-Domagała, M. Domagała, E. Lisowski, H. Momeni. The idea of fuzzy logic usage in a sheet-based FMEA analysis of mechanical systems, MATEC Web of Conf. 183 (2018) art.3009. https://doi.org/10.1051/matecconf/201818303009

[37] J. Fabis-Domagala, M. Domagala, H. Momeni. A concept of risk prioritization in FMEA analysis for fluid power systems, Energies 14 (2021) art. 6482. https://doi.org/10.3390/en14206482

[38] J. Fabis-Domagala, M. Domagala, H. Momeni. A matrix FMEA analysis of variable delivery vane pumps, Energies 14 (2021) art. 1741. https://doi.org/10.3390/en14061741

[39] R. Ulewicz, J. Selejdak, S. Borkowski, M. Jagusiak-Kocik. Process management in the cast iron foundry, METAL 2013 – 22nd Int. Conf. Metallurgy and Materials (2013), Ostrava, Tanger 1926-1931.

[40] J. Fabiś-Domagała, G. Filo, H. Momeni, M. Domagała. Instruments of identification of hydraulic components potential failures, MATEC Web of Conf. 183 (2018) art.03008. https://doi.org/10.1051/matecconf/201818303008

[41] M. Ingaldi, S.T. Dziuba. Modernity evaluation of the machines used during production process of metal products, METAL 2015 – 24th Int. Conf. Metallurgy and Materials (2015), Ostrava, Tanger 1908-1914.

[42] M. Zenkiewicz, T. Zuk, J. Pietraszek, P. Rytlewski, K. Moraczewski, M. Stepczyńska. Electrostatic separation of binary mixtures of some biodegradable polymers and poly(vinyl chloride) or poly(ethylene terephthalate), Polimery/Polymers 61 (2016) 835-843. https://doi.org/10.14314/polimery.2016.835

[43] E. Radzyminska-Lenarcik, R. Ulewicz, M. Ulewicz. Zinc recovery from model and waste solutions using polymer inclusion membranes (PIMs) with 1-octyl-4-methylimidazole, Desalination and Water Treatment 102 (2018) 211-219. https://doi.org/10.5004/dwt.2018.21826

[44] M. Dobrzański. The influence of water price and the number of residents on the economic efficiency of water recovery from grey water, Technical Transactions 118 (2021) art. e2021001. https://doi.org/10.37705/TechTrans/e2021001

[45] B. Szczodrowska, R. Mazurczuk. A review of modern materials used in military camouflage within the radar frequency range, Technical Transactions 118 (2021) art.e2021003. https://doi.org/10.37705/TechTrans/e2021003

[46] M. Morawski, T. Talarczyk, M. Malec. Depth control for biomimetic and hybrid unmanned underwater vehicles, Technical Transactions 118 (2021) art. e2021024. https://doi.org/10.37705/TechTrans/e2021024

[47] A. Kubecki, C. Śliwiński, J. Śliwiński, I. Lubach, L. Bogdan, W. Maliszewski. Assessment of the technical condition of mines with mechanical fuses, Technical Transactions 118 (2021) art. e2021025. https://doi.org/10:37705/TechTrans/e2021025

[48] J. Pietraszek, E. Skrzypczak-Pietraszek. The uncertainty and robustness of the principal component analysis as a tool for the dimensionality reduction. Solid State Phenom. 235 (2015) 1-8. https://doi.org/10.4028/www.scientific.net/SSP.235.1

[49] J. Pietraszek, N. Radek, A.V. Goroshko. Challenges for the DOE methodology related to the introduction of Industry 4.0. Production Engineering Archives 26 (2020) 190-194. https://doi.org/10.30657/pea.2020.26.33

[50] J. Pietraszek, A. Gadek-Moszczak, N. Radek. The estimation of accuracy for the neural network approximation in the case of sintered metal properties. Studies in Computational Intelligence 513 (2014) 125-134. https://doi.org/10.1007/978-3-319-01787-7_12

[51] J. Pietraszek, R. Dwornicka, A. Szczotok. The bootstrap approach to the statistical significance of parameters in the fixed effects model. ECCOMAS 2016 – Proc. 7th European Congress on Computational Methods in Applied Sciences and Engineering 3, 6061-6068. https://doi.org/10.7712/100016.2240.9206

[52] J. Pietraszek, A. Szczotok, N. Radek. The fixed-effects analysis of the relation between SDAS and carbides for the airfoil blade traces. Archives of Metallurgy and Materials 62 (2017) 235-239. https://doi.org/10.1515/amm-2017-0035

Materials Research Forum LLC
https://doi.org/10.21741/9781644902059-17

[53] A. Pacana, D. Siwiec, L. Bednárová. Method of choice: A fluorescent penetrant taking into account sustainability criteria, Sustainability 12 (2020) art. 5854. https://doi.org/10.3390/su12145854

[54] A. Kucharczyk, L. Adamczyk. Wpływ stężenia składników w procesie osadzania zanurzeniowego na własności ochronne powłok sianowych wytworzonych na stali nierdzewnej. Ochrona Przed Korozją 10 (2020) 327-331. https://doi.org/10.15199/40.2020.10.2

[55] A. Kucharczyk, L. Adamczyk, K Miecznikowski. The Influence of the Type of Electrolyte in the Modyfying Solution on the Protective Properties of Vinyltrimethoxysilane/Ethanol-Based Coatings Formed on Stainless Steel X20Cr13. Materials 14 (2021) art. 6209. https://doi.org/10.3390/ma14206209

Terotechnology XII

Materials Research Proceedings **24** (2022) 118-125

Materials Research Forum LLC

https://doi.org/10.21741/9781644902059-18

Evaluation of the Technological Modernity of the Machines used in the Metallurgical Industry

INGALDI Manuela [1,a *] and MAZUR Magdalena[1,b]

[1] Department of Production Engineering and Safety, Faculty of Management, Czestochowa University of Technology, Al. Armii Krajowej 19b, 42-218 Czestochowa, Poland

[a] manuela.ingaldi@pcz.pl, [b] magdalena.mazur@pcz.pl

Keywords: ABC Technology Method, Machines, Modernity, Technology

Abstract. The analysis of the modernity of production machines is an important issue due to two aspects. First of all, modernity will have a direct impact on the efficiency of using both the machine and the entire production process (the entire production line), as well as the productivity of this process. On the other hand, this meaning will refer to the qualitative results of the processes that are carried out by the analyzed device. The level of modernity of machines is the starting point in the analysis of the effectiveness of their use and planned organizational changes in the production line. The aim of the paper was to evaluate the modernity of the milling machine CNC, which is used in the production of various types of steel products. The ABC technology method was the research instrument. Individual parts of this machine were divided into 3 subassemblies (parts of main subassembly A, parts of supportive subassembly B, parts of collateral subassembly C) and then each part was evaluated on the basis of Parkers' five-point scale. The analysis showed that despite the fact that the machine is not modern, there is no need to replace it, because the quality level of the steel products produced with it, is high.

Introduction

Nowadays, the tasks of modern production depend to the greatest extent on the use/application of scientific and technical innovations [1,2]. The use of new solutions, both in organizational and technical terms, is the main element of the development / modernity of production systems. The assessment of the modernity level of the production devices is a multifaceted and very complex concept, which should include: place, time, material flow, technologies, and intangible resources (people). Production equipment and machines constitute one of the elements determining the used technologies and the achieved results in terms of efficiency and productivity, which is one of the elements of market competitiveness. The inability to measure the level of modernity in production with one integral measure forces analyzes to be made using a number of partial indicators. These indicators make it possible to determine the level of modernity of the manufacturing processes implemented in the enterprise. The assessment indicators include: assessment of the modernity of the used machinery park, the modernity of the used technologies and organizational solutions influencing the efficiency of the processes. In recent years, there has been a development of multi-criteria assessment methods [3]. Many authors emphasize the usefulness of these methods in the assessment of technical systems [4,5].

The basic criterion for the analysis and assessment of modernity level of the equipment in production is money. All activities performed in the production line generate costs and revenues. These two aspects are directly related to the supervision of the technical condition of devices and the assessment of the capabilities of these machines. And these are possibilities both in terms of the results (production capacity per unit of time) and the necessary inputs (work, time and energy). These two factors are the basis of actions taken in the field of maintenance, which takes into account the possibility of implementing production plans and the necessity to undertake investment activities [6,7]. The method of assessing the modernity level of the subject should be constructed in such a way as to meet the basic assumptions of the analysis: 1 - credibility of the results, 2 - achievement of the full results of the assessment process (comprehensiveness), 3 - validity, i.e. obtaining a result based on the analysis of facts.

Modernity is directly related (and can be identified) with the concept of the quality. This is due to the strong relationship between the modernity level of the machine and high quality of workmanship (product) [8,9]. Additionally, a detailed presentation distinguishes three aspects of the concept of modernity of devices: 1- technical and constructional modernity (technical level), 2 - modernity as a high level of workmanship, and 3 - functional and operational modernity (quality of use). The broadest scope includes technical and design modernity, which can be defined as the highest level of technical system parameters, in relation to other comparable products [10].

The aim of the paper was to evaluate the modernity of the chosen machine used in the metallurgical industry. The milling machine CNC of the Polish production was the research object. This machine was manufactured in 2007, so it is not so new, however is well equipped and allow to produce big range of products. As a research method, ABC technology method was chosen. It let classify individual parts of the machine according to level of the modernity of production. Such an analysis can show the modernity of the individual parts but also of entire machine and if it should be replaced by new one.

The scheme of the evaluation method shown in the article can also be used in similar situations requiring qualitative assessment, such as biotechnology [11], traffic engineering [12, 13], management [14-17] or quality assurance systems [18, 19]. Moreover, carrying out the evaluation of modernity requires the development of action scenarios, including scenarios of possible failures and their consequences [20-24]. This work can inspire the further development of many data analysis methods [25] based on both quantitative [26-29] and qualitative [30-32] data, thus providing useful tools in many areas of technology [33-36]. Effective predictions, especially explained in an accessible and understandable language, significantly improve the reception of organizational and technical solutions, increasing the general technical culture [37, 38].

Methodology

The ABC technology method is based on the assessment of the value of each of its components through the prism and significance in the total value of the entity being assessed. The analysis makes it possible to determine which elements of the device (more precisely, the evaluation for that element) require a special approach, due to the significant impact on the functionality of the device. The ABC technology method, made for individual devices, facilitates decision-making by the maintenance services and other related departments. The analysis performed in a team and based on the analysis of historical data allows for a quick response (decision) in terms of ensuring continuity of production and improvement of efficiency for the implemented activities [39-42].

All machine components are divided in three groups: Technologies of level A (subassembly A), also known as main technologies, are basic technologies, fundamental for business. They help to give special attributes to produced products. Technologies of level B (subassembly B) are the enabling technologies of a general nature, available to all companies in a given industry. The company does not show interest in development, but benefits from such progress during the purchase of the machine. Technologies of level C (subassembly C) are supporting technology which are usually part of the overall business. These technologies are associated with its own machinery or equipment and are not subject to the innovative activity of the entity using it [6, 7].

The evaluations of individual parts of the machine can be made on the basis of Parker's five-point scale [6, 7]. Level 1 concerns of easy the machine parts manufactured with use of craft technologies. Level 2 concerns of the machine parts manufactured with unchanging technologies used for years. Level 3 concerns of the machine parts manufactured with more complex technologies, requiring technical skills and knowledge. Level 4 concerns of the machine parts manufactured with modern technologies. Level 5 concerns of the machine parts manufactured with the most modern, unique technologies, not known by other producers.

The milling machine CNC was selected for the research presented in the paper, however, for legal reasons, the manufacturer and the number of the machine were not disclosed. The machine was manufactured in 2007 in Poland. It is equipped with a work table with dimensions (width x length) 500 x 1400 mm. Numerical control enables work in an automatic cycle with circular or linear interpolation in three-dimensional space. The machine's feeds in the X/Y/Z axes are respectively 800 mm, 500 mm and 500 mm. The CNC milling machine weighs 4500 kg. A fourth axis can be connected to the machine control panel. The machine has a hydro-mechanical quick-change device and hydraulic console relief. The machine is also equipped with stepless regulation of the spindle speed. The machine is used during production process of different types and shapes of steel products.

Results

The modernity research was conducted for the chosen milling machine CNC. The first stage of the analysis was to specify the different parts of the machine and their division into individual subassemblies, and then to evaluate each part. Evaluation of modernity level of individual parts of the research machine and average evaluation of each group was presented in Table 1.

Table 1. Evaluation of the modernity of the parts of the milling machine CNC [own study]

No	Parts of main subassembly A	Evaluation	No	Parts of main subassembly C	Evaluation
A1	Control system	4	C1	Fan	1
A2	Control panel	4	C2	Shields	2
A3	Programming system	3	C3	Sensors	3
A4	Stepless regulation of the spindle speed	3	C4	Wires	2
			C5	Switch key	2
A5	Hydro-mechanical quick-change device	3	C6	Control buttons	2
			C7	Lighting	2
A6	Hydraulic console relief	3	C8	Machine construction	3
A7	Stationary breaking cap	2	C9	Foundation	2
	Average	3.14		Average	2.11

No	Parts of main subassembly B	Evaluation
B1	Utilities connection system	3
B2	Main power transmission system	4
B3	Feeders	4
B4	Start-up system	2
B5	Safety barrier	2
B6	Integrated power units	2
B7	Height adjustment mechanism	3
	Average	2.86

a)   b)   c)

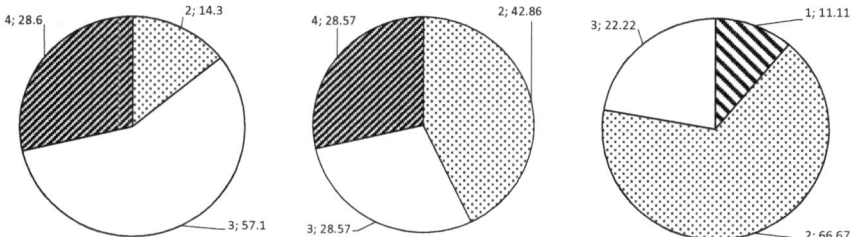

Fig. 1. *The modernity structure of subassemblies of the milling machine CNC in: a) main subassembly, b) supportive subassembly, c) collateral subassembly [own study]*

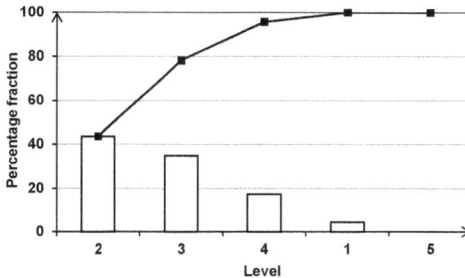

Fig. 2. *ABC analysis of the modernity level of the milling machine CNC [own study]*

The structure of modernity level of the research machine for each individual subassembly was presented in Fig. 1 in form of the pie chart, while the final result of the ABC analysis for the research machine was presented in Fig. 2, in form of the Pareto chart.

Individual parts of the milling machine CNC, according to Table 1, Fig. 1 and Fig. 2 were classified in following way:

− parts of main subassembly A: in over 57% are on level 3, in over 28% are on level 4, in over 14% are on level 2. So 3 was the most often given evaluation (4 times). The highest evaluated parts (level 4) were: control system and control panel. No part of this subassembly received evaluation 1 or 5. This group had the highest average evaluation (3.14). This means that the average part of the subassembly was manufactured with more complex technologies, requiring technical skills and knowledge.

− parts of supportive subassembly B: in almost 43% are on level 2, and in almost 29% are on level 3 and 4. So 2 was the most often given evaluation (3 times). The highest evaluated parts (level 4) were: main power transmission system, feeders. No part of this subassembly received evaluation 1 or 5. The average evaluation was 2.86, which means that parts of this subassembly were manufactured with more complex technologies, requiring technical skills and knowledge or with unchanging technologies used for years.

– parts of collateral subassembly C: in almost 57% are on level 2, in over 22% are on level 3, in over 11% are on level 1. So 2 was the most often given evaluation (6 times). The highest evaluated parts (level 3) were: sensors and machine construction. No part of this subassembly received evaluation 4 or 5. The average evaluation of all subassembly was 2.11, which means that on average, parts of this subassembly were manufactured with unchanging technologies used for years.

Average evaluation of the entire machine was 2.65, which means that most of the parts were manufactured with more complex technologies, requiring technical skills and knowledge or with unchanging technologies used for years. It can be also concluded that evaluations 2 and 3 were the most often given evaluation to individual parts of the research machine (respectively over 40% and almost 40%). Evaluations 4 and 1 had little impact on the overall evaluation of modernity of the research machine. Evaluation 5 did not appear during the analysis. It means that the research machine is not modern.

It should be emphasized, however, that the quality of the products manufactured with the use of the research machine is quite high, and the analysis of nonconformities that appeared mostly were caused by factors other than the milling machine CNC. The results of this analysis will not be shown in the paper, but they prove that despite the fact that the machine is not modern, it allows for obtaining products with appropriate parameters and quality level. This means that it does not need to be replaced in the nearest future.

Summary

One of the main and important resources of each enterprise is technical equipment in the form of machines, which is used during the production process, and which affects not only the efficiency of the production process, its productivity, but also the quality of manufactured products. One of the factor that helps in the evaluation of equipment is the modernity of machines. To evaluate it, the ABC technology method can be used, which is a simple research instrument that uses the 20/80 principle, e.g. that roughly 80% of consequences come from 20% of causes.

In the paper the indicated method was used to evaluate the modernity of the milling machine CNC, which is used in the production of various types of steel products. Average evaluation of the entire machine was 2.65, which means that most of the parts were manufactured with more complex technologies, requiring technical skills and knowledge or with unchanging technologies used for years. So it was concluded that the machine is not modern. However, the analysis showed that despite the fact that the machine is not modern, there is no need to replace it, because the quality level of the steel products produced with it, is high.

References

[1] O. Abdurazzakov , B. CS. Illés, N. Jafarov, K. Aliyev. The impact of technology transfer on innovation, Polish Journal of Management Studies 21 (2020) 9-23. https://doi.org/10.17512/pjms.2020.21.2.01

[2] A. Wiśniewska-Sałek. Managing a Sustainable Supply Chain – Statistical Analysis of Natural Resources in the Furniture Industry, Managem. Systems in Prod. Eng. 29 (2021) 227-234. https://doi.org/10.2478/mspe-2021-0028

[3] T. Trzaskalik (red.). Wielokryterialne wspomaganie decyzji. Metody i zastosowania, PWE, Warszawa, 2014.

[4] A. Kiełbus. Wielokryterialna ocena jakości wyrobów przemysłu maszynowego, PhD thesis, Politechnika Krakowska, Kraków 2007.

Terotechnology XII
Materials Research Proceedings **24** (2022) 118-125

Materials Research Forum LLC
https://doi.org/10.21741/9781644902059-18

[5] M. Krynke, K. Knop, K. Mielczarek. Analysis of the modernity and effectiveness of chosen machines in the processing of high – molecular materials, Production Engineering Archives 3/2 (2014) 18-21.

[6] M. Ingaldi. Evaluation of the machine modernity in the motor industry, Independent Journal of Management & Production 5 (2014) 993-1003. https://doi.org/10.14807/ijmp.v5i4.234

[7] J. Selejdak, R. Ulewicz, M. Ingaldi. The Evaluation of the Use of a Device for Producing Metal Elements Applied in Civil Engineering. METAL 2014: 23rd Int. Conf. on Metallurgy and Materials, Ostrava, TANGER, 2014, 1882-1887.

[8] R. Ulewicz, M. Mazur, K. Knop, R. Dwornicka. Logistic controlling processes and quality issues in a cast iron foundry, Materials Research Proceedings 17 (2020) 65-71. https://doi.org/10.21741/9781644901038-10

[9] P. Bajdor, I. Paweloszek, H. Fidlevora. Analysis and assessment of sustainable entrepreneurship practices in Polish small and medium enterprises, Sustainability 13 (2021) art. 3595. https://doi.org/10.3390/su13073595

[10] S. Francik. Metoda oceny nowoczesności techniczno-konstrukcyjnej ciągników rolniczych wykorzystująca sztuczne sieci neuronowe cz. I: założenia metody, Inżynieria Rolnicza 9 (2009) 41-47.

[11] E. Skrzypczak-Pietraszek. Phytochemistry and biotechnology approaches of the genus *Exacum*. In: The Gentianaceae – Volume 2: Biotechnology and Applications, 2015, 383-401. https://doi.org/10.1007/978-3-642-54102-5_16

[12] A. Bakowski, V. Dekŷŝ, L. Radziszewski, Z. Skrobacki, P. Świetlik. Estimation of uncertainty and variability of urban traffic volume measurements in Kielce, 11th Int. Sci. Tech. Conf. Automotive Safety (2018) 1-8. https://doi.org/10.1109/AUTOSAFE.2018.8373314

[13] A. Bakowski, V. Dekŷš, L. Radziszewski, Z. Skrobacki. Validation of traffic noise models, AIP Conf. Proc. 2077 (2019) art.020005. https://doi.org/10.1063/1.5091866

[14] S. Borkowski, R. Ulewicz, J. Selejdak, M. Konstanciak, D. Klimecka-Tatar. The use of 3x3 matrix to evaluation of ribbed wire manufacturing technology, METAL 2012 – 21st Int. Conf. Metallurgy and Materials (2012), Ostrava, Tanger 1722-1728.

[15] K. Antosz, A. Pacana. Comparative analysis of the implementation of the SMED method on selected production stands, Tehnicki Vjesnik 25 (2018) 276-282. https://doi.org/10.17559/TV-20160411095705

[16] A. Maszke, R. Dwornicka, R. Ulewicz. Problems in the implementation of the lean concept at a steel works - Case study, MATEC Web of Conf. 183 (2018) art.01014. https://doi.org/10.1051/matecconf/201818301014

[17] N. Baryshnikova, O. Kiriliuk, D. Klimecka-Tatar. Management approach on food export expansion in the conditions of limited internal demand, Polish Journal of Management Studies 21 (20200 101-114. https://doi.org/10.17512/pjms.2020.21.2.08

[18] G. Ostasz, K. Czerwińska, A. Pacana. Quality Management of Aluminum Pistons with the Use of Quality Control Points. Management Systems in Production Engineering 28 (2020) 29-33. https://doi.org/10.2478/mspe-2020-0005

[19] R. Ulewicz, D. Siwiec, A. Pacana, M. Tutak, J. Brodny. Multi-criteria method for the selection of renewable energy sources in the polish industrial sector, Energies 14 (2021) art.2386. https://doi.org/10.3390/en14092386

[20] G. Filo, J. Fabiś-Domagała, M. Domagała, E. Lisowski, H. Momeni. The idea of fuzzy logic usage in a sheet-based FMEA analysis of mechanical systems, MATEC Web of Conf. 183 (2018) art.3009. https://doi.org/10.1051/matecconf/201818303009

[21] R. Ulewicz, M. Mazur. Economic aspects of robotization of production processes by example of a car semi-trailers manufacturer, Manufacturing Technology 19 (2019) 1054-1059. https://doi.org/10.21062/ujep/408.2019/a/1213-2489/MT/19/6/1054

[22] A. Pacana, R. Ulewicz. Analysis of causes and effects of implementation of the quality management system compliant with iso 9001, Pol. J. Manag. Stud. 21 (2020) 283 296. https://doi.org/10.17512/pjms.2020.21.1.21

[23] J. Fabis-Domagala, M. Domagala, H. Momeni. A matrix FMEA analysis of variable delivery vane pumps, Energies 14 (2021) art. 1741. https://doi.org/10.3390/en14061741

[24] N. Baryshnikova, O. Kiriliuk, D. Klimecka-Tatar. Enterprises' strategies transformation in the real sector of the economy in the context of the COVID-19 pandemic, Production Engineering Archives 27 (2021) 8-15. https://doi.org/10.30657/pea.2021.27.2

[25] J. Pietraszek, N. Radek, A.V. Goroshko. Challenges for the DOE methodology related to the introduction of Industry 4.0. Production Engineering Archives 26 (2020) 190-194. https://doi.org/10.30657/pea.2020.26.33

[26] J. Pietraszek, R. Dwornicka, A. Szczotok. The bootstrap approach to the statistical significance of parameters in the fixed effects model. ECCOMAS – Proc. 7th European Congress on Computational Methods in Applied Sciences and Engineering 3, 6061-6068. https://doi.org/10.7712/100016.2240.9206

[27] T. Lipiński. Roughness of 1.0721 steel after corrosion tests in 20% NaCl, Production Engineering Archives 15 (2017) 27-30. https://doi.org/10.30657/pea.2017.15.07

[28] N. Radek, J. Pietraszek, A. Gadek-Moszczak, Ł.J. Orman, A. Szczotok. The morphology and mechanical properties of ESD coatings before and after laser beam machining, Materials 13 (2020) art. 2331. https://doi.org/10.3390/ma13102331

[29] N. Radek, J. Konstanty, J. Pietraszek, Ł.J. Orman, M. Szczepaniak, D. Przestacki. The effect of laser beam processing on the properties of WC-Co coatings deposited on steel. Materials 14 (2021) art. 538. https://doi.org/10.3390/ma14030538

[30] J. Pietraszek, E. Skrzypczak-Pietraszek. The uncertainty and robustness of the principal component analysis as a tool for the dimensionality reduction. Solid State Phenom. 235 (2015) 1-8. https://doi.org/10.4028/www.scientific.net/SSP.235.1

[31] G. Majewski, Ł.J. Orman, M. Telejko, N. Radek, J. Pietraszek, A. Dudek. Assessment of thermal comfort in the intelligent buildings in view of providing high quality indoor environment, Energies 13 (2020) art. 1973. https://doi.org/10.3390/en13081973

[32] M. Dobrzański. The influence of water price and the number of residents on the economic efficiency of water recovery from grey water, Technical Transactions 118 (2021) art. e2021001. https://doi.org/10.37705/TechTrans/e2021001

[33] A. Dudek, R. Wlodarczyk. Structure and properties of bioceramics layers used for implant coatings, Solid State Phenom. 165 (2010) 31-36. https://doi.org/10.4028/www.scientific.net/SSP.165.31

[34] A. Szczotok, J. Pietraszek, N. Radek. Metallographic Study and Repeatability Analysis of γ' Phase Precipitates in Cored, Thin-Walled Castings Made from IN713C Superalloy. Archives of Metallurgy and Materials 62 (2017) 595-601. https://doi.org/10.1515/amm-2017-0088

[35] Ł.J. Orman Ł.J., N. Radek, J. Pietraszek, M. Szczepaniak. Analysis of enhanced pool boiling heat transfer on laser-textured surfaces. Energies 13 (2020) art. 2700. https://doi.org/10.3390/en13112700

[36] S. Marković, D. Arsić, R.R. Nikolić, V. Lazić, B. Hadzima, V.P. Milovanović, R. Dwornicka, R. Ulewicz. Exploitation characteristics of teeth flanks of gears regenerated by three hard-facing procedures, Materials 14 (20210 art. 4203. https://doi.org/10.3390/ma14154203

[37] M. Grebski. Mobility of the Workforce and Its Influence on Innovativeness (Comparative Analysis of the United States and Poland), Production Engineering Archives 27 (2021) 40-46. https://doi.org/10.30657/pea.2021.27.36

[38] M. Grebski, M. Mazur. Social climate of support for innovativeness, Production Engineering Archives 28 (2022) 110-116. https://doi.org/10.30657/pea.2022.28.12

[39] M. Jagusiak-Kocik. Use of overall equipment effectiveness indicator for analysis of work time of test bench, Czasopismo Techniczne. Mechanika R113/3-M (2016) 111-117.

[40] K. Knop. Analyzing the machines working time utilization for improvement purposes, Production Engineering Archive 27 (2021) 137-147. https://doi.org/10.30657/pea.2021.27.18

[41] D. Klimecka-Tatar, G. Pawlowska, R. Orlicki, G.E. Zaikov. Corrosion characteristics in acid, alkaline and the ringer solution of Fe68-xCoxZr10Mo5W2B 15 metallic glasses, Journal of the Balkan Tribological Association 20 (2014) 124-130.

[42] D. Klimecka-Tatar, H. Bala, B. Slusarek, K. Jagielska-Wiaderek. The effect of consolidation method on electrochemical corrosion of polymer bonded Nd-Fe-B type magnetic material, Archives of Metallurgy and Materials 54/1 (2009) 247-259.

Terotechnology XII
Materials Research Proceedings 24 (2022) 126-133

Materials Research Forum LLC
https://doi.org/10.21741/9781644902059-19

The use of Computer Simulation Techniques in Production Management

KRYNKE Marek[1,a] and KLIMECKA-TATAR Dorota [1,b*]

[1]Department of Production Engineering and Safety, Faculty of Management, Czestochowa University of Technology, Al. Armii Krajowej 19b, 42-218 Czestochowa, Poland

[a] marek.krynke@pcz.pl, [b] d.klimecka-tatar@pcz.pl

Keywords: Production Management, Simulation, Optimization, FlexSim

Abstract. The purpose of this paper is to present the possibility of using computer simulation techniques to optimize production processes. The paper presents a simulation model, which is the basis for solving decision problems, in particular regarding the determination of alternative scenarios for the allocation of production resources. The simulation model was built in FlexSim. It consists of a finite set of decision variables and constraints that result from the analyzed technological process. The basic stages of creating a simulation model are discussed and the results of the simulation are presented.

Introduction

Companies around the world are constantly looking for ways to reduce costs and make the best use of available resources. Organizations are currently looking for Lean system solutions that would improve their activities by eliminating what does not bring value to the customer, while increasing the efficiency of the manufacturing process [1, 2].

Success in the global economy is often viewed in terms of competitiveness, risk and innovation [3, 4]. Success in business is based on speed in making decisions and solid information support in this process. Decision making is easy in simple systems and in a situation where there is no alternative choice [5, 6]. However, most production systems are difficult to understand. As a rule, they offer a large number of variants of action. It is difficult for an individual to analyze and make the right decision because each system has one or more of the following characteristics [7]:

– system components are subject to their own random actions,
– random actions of the environment affect the system,
– the behavior of the system is dependent on the time variable.

System components have many interactions, so there are many ways to connect paths between system components. When a decision-maker starts analyzing the system and formulates a plan to optimize its performance, then it can face extremely difficult problems [8]. In these situations, common sense thinking and the use of simple computational techniques is insufficient in view of the dynamics and random nature of the system's behavior [9]. Therefore, methods have been developed that help managers analyze processes and are commonly known as decision support systems. The decision support system acts as an analysis tool by which decision makers formulate action plans [5]. Simulation is one such tool. The simplest way to describe the role of simulation in the decision-making process is as follows: simulation is an experiment and a simplified imitation (with the help of a computer) of a specific action [10]. Process simulation requires prior construction of appropriate predictive models [11, 12], especially in the case of heavy industries [13-15], a high risk of contamination [16-20] or a risk of injuries [21-24]. The construction of predictive models requires the application of many data analysis methods to previously collected data, both parametric [25-27] and non-parametric [28-31]. The developed predictive models are components [32] that should be built into the quality management system [33, 34], taking into account various scenarios of events, including failures [35-38]. Of course, the successful implementation of such systems depends hugely on technical culture [39] and social attitudes of workers [40].

The Essence of Modelling and Simulation
Modelling means the activity of selecting an acceptable substitute called a model for the original, i.e. it is an approximate reproduction of the most important properties of the original. In other words, it is building a model that reflects the most important features of the examined or designed object from the point of view of the task it serves in a specific reality or abstraction In general, modelling tools, and simulation in particular, provide mechanisms for studying the problem presented in them, for alternative experimentation, and for predicting the results of proposed external solutions. This approach significantly increases the decision space (allows to evaluate more different ideas), does not interfere with the real system and allows to estimate the risk of actions [41-43]. The main area of application of simulation is production. The production results were easy to predict when only one operator was working on the line. The pace of work depended only on the decision of one operator [44-46]. However, once with the start of the industrial revolution, and in particular with the introduction of the assembly line, production transformed into a complex system consisting of many pieces of equipment, many operations and activities, and involving many people interacting with each other [47-50]. Optimizing the process using the simulation method means finding the best configuration of input variables that will reflect the highest efficiency and stability of the process [51, 52]. And the optimization itself usually consists in maximizing or minimizing the selected parameter [53-54].

Methodology – Case Study
This article discusses a combinatorial approach to minimize the path for an overhead crane. The study took into account the problem of selecting individual machines to fulfil the order for 5 types of products. The developed model focuses on the planning of the production process, where it is necessary to decide on which machine each type of product should be manufactured - so that the total distance of the crane (operating time and operating costs) is as low as possible. In this concept, the 3D FlexSim simulation environment with the built-in OptQuest optimization module was used to solve the problem [55]. It has been assumed that this process will take place on 5 machines, and due to the changeover of machines, each type of product will be produced on one machine. The quantities of individual products and the operating time for each type of product are summarized in Table 1. Model of simulation of the analyzed production process performed in the FlexSim program is also presented.

This problem in its basic version seems to be very simple, because with one warehouse, knowing the position of the arrangement of individual machines, it is enough to send the largest batch of semi-finished products to the nearest station. Later, the next closest position should be selected and the largest batch of semi-finished products should be shipped, etc. However, with a greater number of warehouses, or a greater number of products or means of transport, this problem becomes complicated very quickly. To find the shortest route with only one semi-finished products warehouse and 5 machines, you need to calculate the so-called the number of inversions, which is $5! = 120$ combinations. If the problem concerns more machines, e.g. 10, it will be $10! = 3,628,800$ combinations. Thus, manual calculations are unrealistic, therefore it is necessary to use the OptQuest optimizer built into FlexSim. In the base model, standard objects from the program library were used, which were programmed according to the task conditions. The flow elements simulating individual product types are generated by the Source type object. In this model, the source works in the *Arrival Sequence mode*, where five types of items (semi-finished products) are defined and their respective amounts are given (Fig. 1a). The model should define a global table with dimensions (5 x 2), where the row is the number of the port to which the machine is connected, and the columns are: product type (item) (Col 1) and processing time (Col 2) (Fig. 1b). This arrangement means that the item number l is to be routed by the RMW stock warehouse to the exit port number l to which the machine M1 is connected, etc. The transport was then assigned to the RMW warehouse via the central port. The standard FlexSim object - a crane was also used for transport.

Table 1. Data on the type of product, production volume and operating time for the discussed production process

Product type	Production volume [pcs]	Operating time [s]	Simulation model
1.	40	1500	
2.	50	1200	
3.	30	2100	
4.	25	2600	
5.	45	1400	

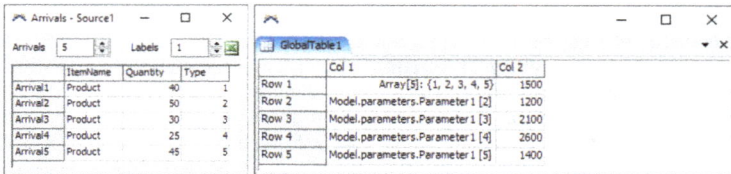

Fig. 1. a) Defining the Source object for the analyzed research problem; b) definition of the allocation of ports and production time to specific types of products [own study]

In this model, each Processor performs the function of individual M1 ÷ M5 machines. The processing time of individual products is set according to the performance of each machine. For this purpose, column 2 was defined in the global table and, similarly to the RMW warehouse, it was assigned to each Processor. Transport to the FPW finished goods warehouse is also set up in individual processors. The same crane was used for this purpose. After starting, the appropriate item type will be transported to the specific machine. However, such a choice is not optimal, because the access routes do not have to be the shortest. In order to optimize the selection of individual machines, it is necessary to define the operation of the optimizer. The optimizer will match the sequence of individual ports for the shipment of semi-finished products to individual machines until the optimal value was determined, at which the total distance of the crane will be the smallest.

Results

Based on the simulation (with input data from Table 1), the best combination of ports was obtained, with the shortest route. The result of the optimizer's work for all 120 combinations of ports, along with the length of travel, is shown in Fig. 2. As it can be seen at the presented results, the best sequence of port addressing is the combination of 3, 1, 4, 5, 2, obtained in the 31st iteration as Rank 1. This means that the first type semi-finished product in the amount of 40 items should be sent to the machine M3, semi-finished product of the type 2 in the amount of 50 pcs. for the M1 machine, semi-finished product type 3 for the machine 4, etc., machine M5 - product 4 and machine M2 - semi-finished product 5. Then the total travel route for the crane will be the shortest and will be 23.8 km. Fig. 3a shows a production flow cyclogram. A high degree of production sustainability can be observed, which translates into the end of production of the entire assortment at a similar time. On the other hand, Fig. 3b shows the degree of use of individual machines.

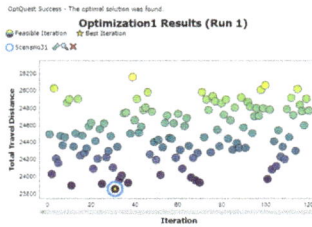

Fig. 2. The result of the optimizer work for all 120 iterations (combinations of 5 output ports for RMW warehouse) [own study]

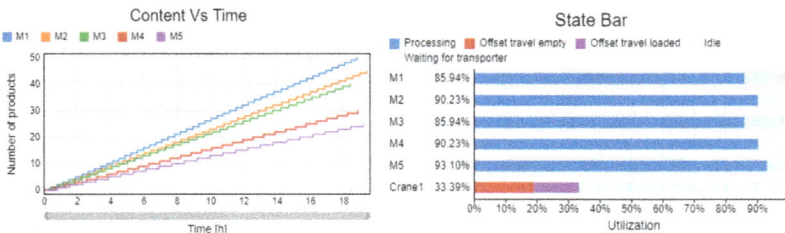

Fig. 3. a) Production flow cycle for all types of products manufactured on M1 ÷ M5 machines; b) Loads of individual machines throughout the production cycle [own study]

The diagram in Fig. 3b shows the so-called partial efficiency of machines. It is a value that represents the OEE coefficient without taking into account the quality of the manufactured products. This means that assuming that all manufactured products were 100% compliant.

Summary

The main purpose of the article was to present the possibility of using a simulation experiment to optimize the production process. The presented example shows how a discrete event simulation model can improve production process planning. The presented concept can be the basis for constructing more complex simulations. It can also be successfully used in logistics for route planning or finding the optimal location of the warehouse. The benefits of simulating real operations on average give savings from 5 to 25% as a result of lower operating costs and increasing throughput (flow) [7]. A significant part of the savings are already foreseen in the simulation process, which significantly reduces the analysis time and thus provides a quick method to try out changes and new ideas. Such analysis is especially important in situations of sharing resources and people involved in a given process or operation. The most important thing, however, is that the simulation shows the impact of plans and schedules on the actual performance, and also directs managers to choose the optimal actions. The main value of the simulation is also the fact that - based on historical and current features - it presents the dynamics of the simulated system, allowing it to be learned, studied, experimented and modified without interfering with and disturbing the real system. Alternatively, a simulation based on predicted or planned features represents the dynamics of the behavior of a system that does not yet exist and allows us to understand it before it arises in the real world.

References

[1] R. Ulewicz, R. Kucęba. Identification of problems of implementation of Lean concept in the SME sector, Eng. Manag. Prod. Serv. 8 (2016) 19–25. https://doi.org/10.1515/emj-2016-0002

[2] M. Mazur, H. Momeni. LEAN Production issues in the organization of the company - results, Production Engineering Archives 22 (2019) 50–53. https://doi.org/10.30657/pea.2019.22.10

[3] M. Krynke, K. Mielczarek, O. Kiriliuk. Cost Optimization and Risk Minimization During Teamwork Organization, Management Systems in Production Engineering 29 (2021) 145–150. https://doi.org/10.2478/mspe-2021-0019

[4] M. Krynke. Risk Management in the Process of Personnel Allocation to Jobs, 8th Int. Conf. System Safety: Human-Technical Facility-Environment (CzOTO 2019) (2020) 82–90.

[5] M. Matuszny. Building decision trees based on production knowledge as support in decision-making process, Production Engineering Archives 26 (2020) 36–40. https://doi.org/10.30657/pea.2020.26.08

[6] J. Karcz, B. Ślusarczyk. Criteria of quality requirements deciding on choice of the logistic operator from a perspective of his customer and the end recipient of goods, Production Engineering Archives 27 (2021) 58–68. https://doi.org/10.30657/pea.2021.27.8

[7] M. Beaverstock, A. Greenwood, W. Nordgren. Applied Simulation Modeling and Analysis Using FlexSim 5th Ed. FlexSim Software Products, Inc, 2017.

[8] D. Klimecka-Tatar, M. Ingaldi, M. Obrecht. Sustainable Developement in Logistic – A Strategy for Management in Terms of Green Transport, Management Systems in Production Engineering 29 (2021) 91–96. https://doi.org/10.2478/mspe-2021-0012

[9] T.D.C. Le, D.D. Nguyen, J. Oláh, M. Pakurár. Optimal vehicle route schedules in picking up and delivering cargo containers considering time windows in logistics distribution networks: A case study, Production Engineering Archives 26 (2020) 174–184. https://doi.org/10.30657/pea.2020.26.31

[10] M. Drbúl, D. Stančeková, O. Babík, J. Holubjak, I. Görögová, D. Varga. Simulation Possibilities of 3D Measuring in Progressive Control of Production, Manufacturing Technology 16 (2016) 53–58.

[11] J. Pietraszek, R. Dwornicka, A. Szczotok. The bootstrap approach to the statistical significance of parameters in the fixed effects model. ECCOMAS 2016 – Proc. 7th European

Congress on Computational Methods in Applied Sciences and Engineering 3, 6061-6068. https://doi.org/10.7712/100016.2240.9206

[12] J. Pietraszek, N. Radek, A.V. Goroshko. Challenges for the DOE methodology related to the introduction of Industry 4.0. Production Engineering Archives 26 (2020) 190-194. https://doi.org/10.30657/pea.2020.26.33

[13] M. Kekez, L. Radziszewski. Modelling of pressure in the injection pipe of a diesel engine by computational intelligence, P. I. Mech. Eng. D-J. Aut. 225 (2011) 1660-1670. https://doi.org/10.1177/0954407011411388

[14] M. Ingaldi, S.T. Dziuba. Modernity evaluation of the machines used during production process of metal products, METAL 2015 – 24th Int. Conf. Metallurgy and Materials (2015), Ostrava, Tanger 1908-1914.

[15] A. Szczotok, J. Pietraszek, N. Radek. Metallographic Study and Repeatability Analysis of γ′ Phase Precipitates in Cored, Thin-Walled Castings Made from IN713C Superalloy. Archives of Metallurgy and Materials 62 (2017) 595-601. https://doi.org/10.1515/amm-2017-0088

[16] E. Skrzypczak-Pietraszek. Phytochemistry and biotechnology approaches of the genus exacum. In: The Gentianaceae - Volume 2: Biotechnology and Applications, 2015, 383-401.https://doi.org/10.1007/978-3-642-54102-5_16

[17] M. Zenkiewicz, T. Zuk, J. Pietraszek, P. Rytlewski, K. Moraczewski, M. Stepczyńska. Electrostatic separation of binary mixtures of some biodegradable polymers and poly(vinyl chloride) or poly(ethylene terephthalate), Polimery/Polymers 61 (2016) 835-843. https://doi.org/10.14314/polimery.2016.835

[18] M. Opydo, R. Kobyłecki, A. Dudek, Z. Bis. The effect of biomass co-combustion in a CFB boiler on solids accumulation on surfaces of P91 steel tube samples, Biomass and Bioenergy 85 (2016) 61-68. https://doi.org/10.1016/j.biombioe.2015.12.011

[19] E. Radzyminska-Lenarcik, R. Ulewicz, M. Ulewicz. Zinc recovery from model and waste solutions using polymer inclusion membranes (PIMs) with 1-octyl-4-methylimidazole, Desalination and Water Treatment 102 (2018) 211-219. https://doi.org/10.5004/dwt.2018.21826

[20] M. Ulewicz, A. Pietrzak. Properties and structure of concretes doped with production waste of thermoplastic elastomers from the production of car floor mats. Materials 14 (2021) art. 872. https://doi.org/10.3390/ma14040872

[21] J. Jura, M. Ulewicz. Assessment of the Possibility of Using Fly Ash from Biomass Combustion for Concrete. Materials 14 (2021) art. 6708. https://doi.org/10.3390/ma14216708

[22] B. Szczodrowska, R. Mazurczuk. A review of modern materials used in military camouflage within the radar frequency range, Technical Transactions 118 (2021) art.e2021003. https://doi.org/10.37705/TechTrans/e2021003

[23] M. Morawski,T. Talarczyk, M. Malec. Depth control for biomimetic and hybrid unmanned underwater vehicles, Technical Transactions 118 (2021) art. e2021024. https://doi.org/10.37705/TechTrans/e2021024

[24] A. Kubecki, C. Śliwiński, J. Śliwiński, I. Lubach, L. Bogdan, W. Maliszewski. Assessment of the technical condition of mines with mechanical fuses, Technical Transactions 118 (2021) art. e2021025. https://doi.org/10.37705/TechTrans/e2021025

[25] M. Zmindak, L. Radziszewski, Z. Pelagic, M. Falat. FEM/BEM techniques for modelling of local fields in contact mechanics, Communications - Scientific Letters of the University of Zilina 17 (2015) 37-46.

[26] J. Pietraszek, E. Skrzypczak-Pietraszek. The uncertainty and robustness of the principal component analysis as a tool for the dimensionality reduction. Solid State Phenom. 235 (2015) 1-8. https://doi.org/10.4028/www.scientific.net/SSP.235.1

[27] J. Pietraszek, A. Szczotok, N. Radek. The fixed-effects analysis of the relation between SDAS and carbides for the airfoil blade traces. Archives of Metallurgy and Materials 62 (2017) 235-239. https://doi.org/10.1515/amm-2017-0035

[28] J. Pietraszek, A. Gadek-Moszczak, N. Radek. The estimation of accuracy for the neural network approximation in the case of sintered metal properties. Studies in Computational Intelligence 513 (2014) 125-134. https://doi.org/10.1007/978-3-319-01787-7_12

[29] J. Korzekwa, M. Bara, J. Pietraszek, P. Pawlus. Tribological behavior of Al_2O_3/inorganic fullerene-like WS_2 composite layer sliding against plastic, Int. J. Surf. Sci. Eng. 10 (2016) 570-584. https://doi.org/10.1504/IJSURFSE.2016.081035

[30] N. Radek, J. Pietraszek, A. Gadek-Moszczak, Ł.J. Orman, A. Szczotok. The morphology and mechanical properties of ESD coatings before and after laser beam machining, Materials 13 (2020) art. 2331. https://doi.org/10.3390/ma13102331

[31] N. Radek, J. Konstanty, J. Pietraszek, Ł.J. Orman, M. Szczepaniak, D. Przestacki. The effect of laser beam processing on the properties of WC-Co coatings deposited on steel. Materials 14 (2021) art. 538. https://doi.org/10.3390/ma14030538

[32] G. Ostasz, K. Czerwińska, A. Pacana. Quality Management of Aluminum Pistons with the Use of Quality Control Points. Management Systems in Production Engineering 28 (2020) 29-33. https://doi.org/10.2478/mspe-2020-0005

[33] A. Maszke, R. Dwornicka, R. Ulewicz. Problems in the implementation of the lean concept at a steel works - Case study, MATEC Web of Conf. 183 (2018) art.01014. https://doi.org/10.1051/matecconf/201818301014

[34] A. Pacana, R. Ulewicz. Analysis of causes and effects of implementation of the quality management system compliant with iso 9001, Pol. J. Manag. Stud. 21 (2020) 283 296. https://doi.org/10.17512/pjms.2020.21.1.21

[35] J. Fabiś-Domagała, G. Filo, H. Momeni, M. Domagała. Instruments of identification of hydraulic components potential failures, MATEC Web of Conf. 183 (2018) art.03008. https://doi.org/10.1051/matecconf/201818303008

[36] G. Filo, J. Fabiś-Domagała, M. Domagała, E. Lisowski, H. Momeni. The idea of fuzzy logic usage in a sheet-based FMEA analysis of mechanical systems, MATEC Web of Conf. 183 (2018) art.3009. https://doi.org/10.1051/matecconf/201818303009

[37] Ł.J. Orman Ł.J., N. Radek, J. Pietraszek, M. Szczepaniak. Analysis of enhanced pool boiling heat transfer on laser-textured surfaces. Energies 13 (2020) art. 2700. https://doi.org/10.3390/en13112700

[38] J. Fabis-Domagala, M. Domagala, H. Momeni. A matrix FMEA analysis of variable delivery vane pumps, Energies 14 (2021) art. 1741. https://doi.org/10.3390/en14061741

[39] M. Grebski. Mobility of the Workforce and Its Influence on Innovativeness (Comparative Analysis of the United States and Poland), Production Engineering Archives 27 (2021) 40-46. https://doi.org/10.30657/pea.2021.27.36

[40] M. Grebski, M. Mazur. Social climate of support for innovativeness, Production Engineering Archives 28 (2022) 110-116. https://doi.org/10.30657/pea.2022.28.12

[41] I. Kaczmar. Komputerowe modelowanie i symulacje procesów logistycznych w środowisku flexsim, PWN, Warszawa, 2019.

[42] D. Klimecka-Tatar. Context of production engineering in management model of Value Stream Flow according to manufacturing industry, Prod. Eng. Arch. 21 (2018) 32–35. https://doi.org/10.30657/pea.2018.21.07

[43] D. Leks, A. Gwiazda. Application of FlexSim for modelling and simulation of the production process, Selected Engineering Problems 2015 (2015) 51–56.

[44] I. Kaczmar. The use of simulation and optimization in managing the manufacturing process — case study, Gospodarka Materiałowa i Logistyka 2016 (2016) 21–28.

[45] S. Setamanit. Evaluation of outsourcing transportation contract using simulation and design of experiment, Polish Journal of Management Studies 18 (2018) 300–310. https://doi.org/10.17512/pjms.2018.18.2.24

[46] D. Krenczyk, W.M. Kempa, K. Kalinowski, C. Grabowik, I. Paprocka. Production planning and scheduling with material handling using modelling and simulation, MATEC Web Conf. 112 (2017) 9015. https://doi.org/10.1051/matecconf/201711209015

[47] K. Knop. Indicating and analysis the interrelation between terms – visual: management, control, inspection and testing, Production Engineering Archives 26 (2020) 110–120. https://doi.org/10.30657/pea.2020.26.22

[48] M. Krynke. Personnel management on the production line using the FlexSim simulation environment, Manufacturing Technology 21 (2021) 657–667. https://doi.org/10.21062/mft.2021.073

[49] J. Kyncl. Digital Factory Simulation Tools, Manufacturing Technology 16 (2016) 371–375.

[50] E. Sujová, D. Vysloužilová, H. Čierna, R. Bambura. Simulation Models of Production Plants as a Tool for Implementation of the Digital Twin Concept into Production, Manufacturing Technology 20 (2020) 527–533. https://doi.org/10.21062/mft.2020.064

[51] M. Krynke. Management optimizing the costs and duration time of the process in the production system, Production Engineering Archives 27 (2021) 163–170. https://doi.org/10.30657/pea.2021.27.21

[52] J. Kyncl, T. Kellner, R. Kubiš. Tricanter Production Process Optimization by Digital Factory Simulation Tools, Manufacturing Technology 17 (2017) 49–53.

[53] M. Krynke, K. Mielczarek, A. Vaško. Analysis of the Problem of Staff Allocation to Work Stations, Quality Production Improvement – QPI 1 (2019) 545–550. https://doi.org/10.2478/cqpi-2019-0073

[54] M. Krynke, K. Mielczarek. Applications of linear programming to optimize the cost-benefit criterion in production processes, MATEC Web Conf. 183 (2018) art. 4004. https://doi.org/10.1051/matecconf/201818304004

[55] FlexSim: User manual, 2017.

Terotechnology XII
Materials Research Proceedings 24 (2022) 134-141

Materials Research Forum LLC
https://doi.org/10.21741/9781644902059-20

Analysis of the Physicochemical Properties of The Ti6Al4V Titanium Alloy Produced by the Plastic Working Method and the SLS Method

KIERAT Oliwia[1,a], DUDEK Agata[1,b] * and MOSKAL Grzegorz[2,c]

[1]Czestochowa University of Technology, Faculty of Production Engineering and Materials Technology, Department of Material Engineering, Czestochowa, Poland, EU

[2]Silesian University of Technology, Faculty of Materials Science and Engineering, Department of Material Technologies, Katowice, Poland, EU

[a]oliwia.kierat@pcz.pl, [b]agata.dudek@pcz.pl, [c]grzegorz.moskal@polsl.pl

Keywords: Titanium Alloy, Ti6Al4V, Plastic Working, 3D Printing, Corrosion Resistance, Microstructure Analysis

Abstract. This work compared the physicochemical properties of the Ti6Al4V titanium alloy produced by plastic working and selective laser sintering. The corrosion behavior of materials was analyzed in terms of their application in medicine, particularly in implantology. For this purpose, corrosion resistance tests were carried out in Ringer's fluid. The microstructural analysis of these materials was performed – before and after corrosion tests using the KEYENCE VHX-7000 digital microscope. Phase analysis of these materials was performed using a SEIFERT T-T X-ray diffractometer.

Introduction

Ti6Al4V titanium alloy is successfully used in the automotive and aviation industries, as well as in medicine [1-3]. Taking into account biomedical applications, the use of titanium alloy in implantology deserves special attention due to a number of advantages such as corrosion resistance, high biocompatibility, low weight, and fatigue strength [4]. However, the excessively high Young's modulus of the Ti6Al4V titanium alloy compared to bone entails limitations in implantological applications, negatively affecting, among others, the connection of the implant surface with the bone tissue [1, 5]. An alternative solution to this problem is the use of 3D printing for the production of Ti6Al4V titanium alloy [6]. The following technologies are mainly used for 3D printing of powdered metals: (i) selective laser sintering (SLS), (ii) selective laser melting (SLM), and laser surfacing (LMD) [1]. Selective laser sintering uses a laser beam to sinter powdered materials, resulting in the formation of three-dimensional objects [7]. The process of selective laser melting is analogous, but in this case, the laser beam is used to melt powdered materials [1]. In the laser surfacing process (LMD), the laser creates a pool of liquid metal into which the melting powder is fed. As a result of this process, sediment is created that connects to the substrate - thus, layer by layer, a 3D object is created.

In this study, the properties of Ti6Al4V titanium alloy produced by plastic working and 3D printed by selective laser sintering were compared. In particular, the research focused on the corrosion resistance of titanium alloys produced by plastic working and selective laser sintering, their microstructural analysis, and phase composition analysis.

The presented results may be engaging in many industries due to the desired corrosion resistance, e.g. when applying special surface layers [8-10] obtained by coating deposition [11-14], their subsequent modification with laser treatment [15-18], but also in the production of devices exposed to corrosion in hydraulic heavy-duty machines [19, 20], to biocorrosion in biotechnology industry [21-23], agriculture [24-26] and surgery implants [27-29]. They can also inspire the development of classic data analysis methods [30-32], their non-parametric variants [33], and failure analysis methods [34-37]. A significant improvement in corrosion resistance will also have a large impact on the current quality management due to the extension of the service time of equipment and infrastructure [38-42].

Materials and Methods
Titanium alloy Ti6Al4V was obtained by two methods: the plastic working method and the selective laser sintering (SLS) method was used for the tests. To visualize the microstructure, the specimens were etched with a reagent containing: 2.5 ml of nitric acid, 2.5 ml of hydrofluoric acid, and 95 ml of water.

The corrosion resistance of the tested materials was tested. For this purpose, samples of titanium alloy were mounted in polymethyl methacrylate frames with epoxy resin and mechanically polished using sandpaper with a grain size of up to 2000. The corrosion resistance tests were carried out using the CH Instruments 440A (USA) measuring station in a three-electrode system, where a platinum electrode was used as the auxiliary electrode, and the reference electrode was a saturated calomel electrode, and the working electrode was the tested titanium alloy sample. In order to test the corrosion resistance, potentiodynamic polarization curves were recorded in the potential range from -1.5 to +3.0 V (measured against a saturated calomel electrode - NEK). Ringer's fluid was used as the corrosive medium. The composition of Ringer's fluid is shown in Table 1.

Table 1. Ringer's fluid composition

Ingredient	NaCl	KCl	CaCl₂
Amount [g/l]	8,6	0,3	0,333

The KEYENCE VHX-7000 digital microscope was used for microstructural analysis of the tested materials, both before and after the corrosion tests. Phase analysis was performed using a SEIFERT T-T X-ray diffractometer.

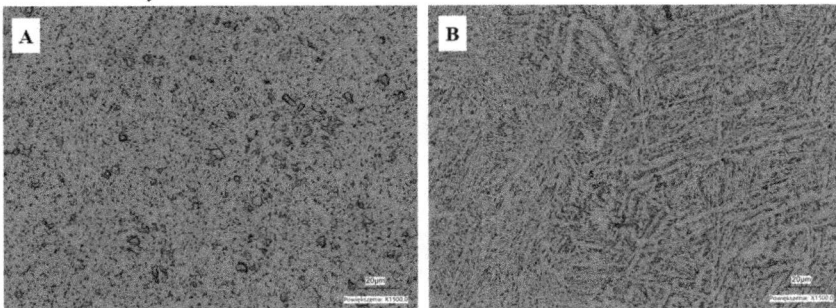

Fig. 1. Microstructure of (A) Ti6Al4V titanium alloy produced by plastic working method and (B) Ti6Al4V titanium alloy 3D printed by SLS method.

Results
Fig. 1 shows the microstructure of the Ti6Al4V titanium alloy (A) produced by plastic working and (B) 3D printed using the SLS method. Ti6Al4V titanium alloy is an $\alpha + \beta$ two-phase alloy.

Based on the obtained microstructures, it was found that the method of producing the Ti6Al4V titanium alloy significantly affects its structure. The structure of both samples shows a different morphology. The microstructure of Ti6Al4V titanium alloy produced by plastic working is visible, equiaxed grains, while the microstructure of Ti6Al4V 3D SLS alloy is acicular.

Phase analysis of the Ti6Al4V alloy produced by various methods was performed. Fig. 2 shows the diffractogram recorded for the Ti6Al4V titanium alloy produced by the plastic working method, and in Fig. 3 for the Ti6Al4V alloy produced by the SLS method. Phase analysis revealed the presence of α and β phases for both materials. Peak positions were read from recorded diffractograms and fitted to individual α and β phases.

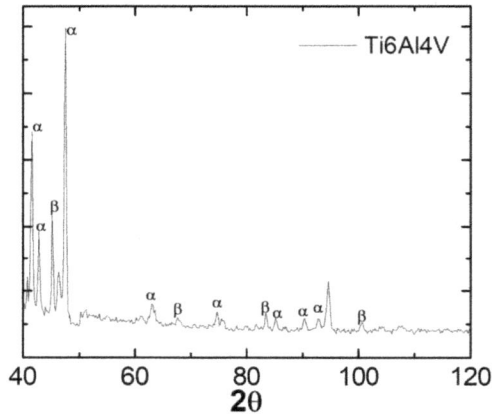

Fig. 2. *The diffractogram registered for the Ti6Al4V titanium alloy produced by the plastic working method.*

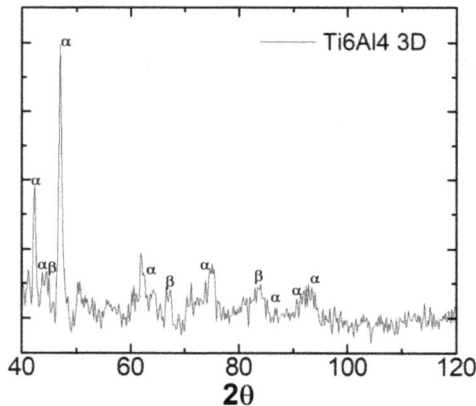

Fig. 3. *Diffractogram registered for Ti6Al4V titanium alloy 3D printed by SLS method.*

The corrosion resistance of the Ti6Al4V titanium alloy was tested. Fig. 4 shows the potentiodynamic polarization curves for the Ti6Al4V titanium alloy produced by plastic working and by selective laser sintering. The corrosive potential of the Ti6Al4V alloy was -0.7 V, while the Ti6Al4V 3D SLS alloy was - 0.63 V. The cathode current densities for both materials were the same. For the Ti6Al4V 3D SLS alloy, a sharp increase in the anode currents density was observed for the potential above +1.9 V.

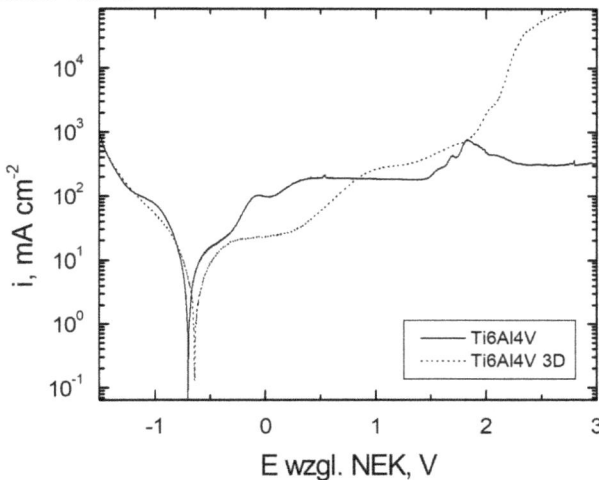

Fig. 4. *Potentiodynamic polarization curves recorded for the Ti6Al4V titanium alloy produced by plastic working and 3D printed with the SLS method in Ringer's fluid.*

Fig. 5 shows the microstructures of the Ti6Al4V titanium alloy produced by various methods after corrosion tests in Ringer's fluid. No signs of pitting corrosion were observed on the surface of these materials, which proves the good corrosion resistance of both the Ti6Al4V titanium alloy produced by plastic working and 3D printed by SLS.

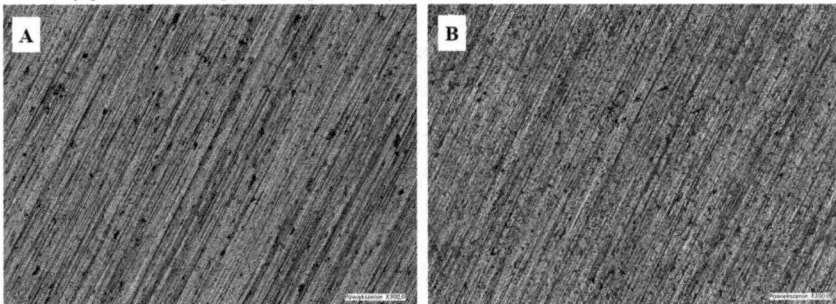

Fig. 5. *Microstructure of (A) Ti6Al4V titanium alloy produced by plastic working and (B) Ti6Al4V titanium alloy produced by SLS method after corrosion tests.*

Conclusions

In this study, an analysis of the physicochemical properties of the Ti6Al4V titanium alloy depending on the production method was carried out. It was revealed that the structure of the tested materials had a different morphology. The Ti6Al4V alloy produced due to plastic working had equiaxial grains, while the Ti6Al4V alloy, 3D printed using the SLS method, had a coniferous structure. The X-ray analysis confirmed the presence of α and β phases in both materials. The corrosion resistance of both materials was analyzed in Ringer's fluid. The conducted analysis revealed slight differences in the corrosion potential values - the corrosion potential of the Ti6Al4V alloy produced by the SLS method was shifted by 0.07 V towards positive values in relation to the corrosion potential of the Ti6Al4V alloy made as a result of plastic working. The cathode current densities were equal in both cases. For the Ti6Al4V 3D SLS alloy, a rapid increase in the anode currents density for the potential above +1.9 V was observed. The microstructural analysis performed after the corrosion tests did not reveal any signs of pitting corrosion.

References

[1] K. Karolewska, B. Ligaj. Comparison analysis of titanium alloy Ti6Al4V produced by metallurgical and 3D printing method, AIP Conference Proceedings 2077 (2019) art. 020025. https://doi.org/10.1063/1.5091886

[2] A.D. Dobrzańska-Danikiewicz, T.G. Gaweł, W. Wolany. Ti6Al4V titanium alloy used as a modern biomimetic material, Archives of Materials Science and Engineering 76 (2015) 150-156.

[3] S. Madhukar, A. Shravan, P. Vidyanand Sai, V.V. Satyanarayana. A critical review on cryogenic machining of titanium alloy (Ti-6Al-4V), International Journal of Mechanical Engineering and Technology (IJMET) 7 (2016) 38-45.

[4] S. Kumar, M. Nehra, D. Kedia, N. Dilbaghi, K. Tankeshwar, K.-H. Kim. Nanotechnology-based biomaterials for orthopaedic applications: Recent advances and future prospects, Materials Science and Engineering: C 106 (2020) art. 110154. https://doi.org/10.1016/j.msec.2019.110154

[5] A. Bandyopadhyay, F. Espana, V.K. Balla, S. Bose, Y. Ohgami, N. M. Davies. Influence of porosity on mechanical properties and in vivo response of Ti6Al4V implants, Acta Biomaterialia 6 (2010) 1640-1648. https://doi.org/10.1016/j.actbio.2009.11.011

[6] J. Fojt, M. Fousova, E. Jablonska, L. Joska, V. Hybasek, E. Pruchova, D. Vojtech, T. Ruml. Corrosion behaviour and cell interaction of Ti-6Al-4V alloy prepared by two techniques of 3D printing, Materials Science & Engineering C 98 (2018) 911-920. https://doi.org/10.1016/j.msec.2018.08.066

[7] Ż.A. Mierzejewska. SLS Technology – characteristics and application of selective laser sintering in biomedical engineering, J. Technol. Exploit. Mech. Eng. 1 (2015) 178-190.

[8] A. Dudek, B. Lisiecka, R. Ulewicz. The effect of alloying method on the structure and properties of sintered stainless steel, Archives of Metallurgy and Materials 62 (2017) 281-287. https://doi.org/10.1515/amm-2017-0042

[9] L. Dąbek, A. Kapjor, Ł.J. Orman. Distilled water and ethyl alcohol boiling heat transfer on selected meshed surfaces, Mechanics and Industry 20 (2019) art. 701. https://doi.org/10.1051/meca/2019068

[10] Ł.J. Orman Ł.J., N. Radek, J. Pietraszek, M. Szczepaniak. Analysis of enhanced pool boiling heat transfer on laser-textured surfaces. Energies 13 (2020) art. 2700. https://doi.org/10.3390/en13112700

[11] A. Dudek. Surface properties in titanium with hydroxyapatite coating, Optica Applicata 39 (2009) 825-831.

[12] A. Dudek, R. Wlodarczyk. Structure and properties of bioceramics layers used for implant coatings, Solid State Phenom. 165 (2010) 31-36. https://doi.org/10.4028/www.scientific.net/SSP.165.31

[13] N. Radek, A. Sladek, J. Broncek, I. Bilska, A. Szczotok. Electrospark alloying of carbon steel with WC-Co-Al2O3: Deposition technique and coating properties, Advanced Materials Research 874 (2014) 101-106. https://doi.org/10.4028/www.scientific.net/AMR.874.101

[14] M. Szala, A. Dudek, A. Maruszczyk, M. Walczak, J. Chmiel, M. Kowal. Effect of atmospheric plasma sprayed TiO2–10% NiAl cermet coating thickness on cavitation erosion, sliding and abrasive wear resistance, Acta Phys. Pol. A 136 (2019) 335-341. https://doi.org/10.12693/APhysPolA.136.335

[15] N. Radek, J. Pietraszek, A. Goroshko. The impact of laser welding parameters on the mechanical properties of the weld, AIP Conf. Proc. 2017 (2018) art.20025. https://doi.org/10.1063/1.5056288

[16] N. Radek, J. Pietraszek, A. Gadek-Moszczak, Ł.J. Orman, A. Szczotok. The morphology and mechanical properties of ESD coatings before and after laser beam machining, Materials 13 (2020) art. 2331. https://doi.org/10.3390/ma13102331

[17] N. Radek, J. Konstanty, J. Pietraszek, Ł.J. Orman, M. Szczepaniak, D. Przestacki. The effect of laser beam processing on the properties of WC-Co coatings deposited on steel. Materials 14 (2021) art. 538. https://doi.org/10.3390/ma14030538

[18] H. Danielewski, A. Skrzypczyk, W. Zowczak, D. Gontarski, L. Płonecki, H. Wiśniewski, D. Soboń, A. Kalinowski, G. Bracha, K. Borkowski. Numerical analysis of laser-welded flange pipe joints in lap and fillet configurations, Technical Transactions 118 (2021) art. e2021030. https://doi.org/10.37705/TechTrans/e2021030

[19] G. Filo, E. Lisowski, M. Domagała, J. Fabiś-Domagała, H. Momeni. Modelling of pressure pulse generator with the use of a flow control valve and a fuzzy logic controller, AIP Conf. Proc. 2029 (2018) art.20015. https://doi.org/10.1063/1.5066477

[20] G. Filo, E. Lisowski, J. Rajda. Flow analysis of a switching valve with innovative poppet head geometry by means of CFD method, Flow Meas. Instrum. 70 (2019) art.101643. https://doi.org/10.1016/j.flowmeasinst.2019.101643

[21] E. Skrzypczak-Pietraszek. Phytochemistry and biotechnology approaches of the genus Exacum. In: The Gentianaceae - Volume 2: Biotechnology and Applications, 2015, 383-401. https://doi.org/10.1007/978-3-642-54102-5_16

[22] J. Pietraszek, E. Skrzypczak-Pietraszek. The uncertainty and robustness of the principal component analysis as a tool for the dimensionality reduction. Solid State Phenom. 235 (2015) 1-8. https://doi.org/10.4028/www.scientific.net/SSP.235.1

[23] E. Skrzypczak-Pietraszek, K. Reiss, P. Żmudzki, J. Pietraszek. Enhanced accumulation of harpagide and 8-O-acetyl-harpagide in Melittis melissophyllum L. agitated shoot cultures analyzed by UPLC-MS/MS. PLoS ONE 13 (2018) art. e0202556. https://doi.org/10.1371/journal.pone.0202556

[24] T. Lipiński, A. Wach. Influence of outside furnace treatment on purity medium carbon steel, METAL 2014 - 23rd Int. Conf. on Metallurgy and Materials (2014), Ostrava, Tanger 738-743.

[25] T. Lipiński. Corrosion resistance of 1.4362 steel in boiling 65% nitric acid, Manufacturing Technology 16 (2016) 1004-1009.

[26] T. Lipiński. Roughness of 1.0721 steel after corrosion tests in 20% NaCl, Production Engineering Archives 15 (2017) 27-30. https://doi.org/10.30657/pea.2017.15.07

[27] A. Dudek. Investigations of microstructure and properties in bioceramic coatings used in medicine, Archives of Metallurgy and Materials 56 (2011) 135-140. https://doi.org/10.2478/v10172-011-0015-y

[28] A. Dudek, L. Adamczyk. Properties of hydroxyapatite layers used for implant coatings, Optica Applicata 43 (2013) 143-151. https://doi.org/10.5277/oa130118

[29] D. Klimecka-Tatar, K. Radomska, G. Pawlowska. Corrosion resistance, roughness and structure of Co64Cr28Mo5(Fe, Si, Al, Be)3 and Co63Cr29Mo6.5(C, Si, Fe, Mn)1.5 biomedical alloys, J. Balkan Tribol. Assoc. 21 (2015) 204-210.

[30] J. Pietraszek, R. Dwornicka, A. Szczotok. The bootstrap approach to the statistical significance of parameters in the fixed effects model. ECCOMAS – Proc. of the 7th European Congress on Computational Methods in Applied Sciences and Engineering 3, 6061-6068. https://doi.org/10.7712/100016.2240.9206

[31] J. Pietraszek, A. Szczotok, N. Radek. The fixed-effects analysis of the relation between SDAS and carbides for the airfoil blade traces. Archives of Metallurgy and Materials 62 (2017) 235-239. https://doi.org/10.1515/amm-2017-0035

[32] J. Pietraszek, N. Radek, A.V. Goroshko. Challenges for the DOE methodology related to the introduction of Industry 4.0. Production Engineering Archives 26 (2020) 190-194. https://doi.org/10.30657/pea.2020.26.33

[33] J. Pietraszek, A. Gadek-Moszczak, N. Radek. The estimation of accuracy for the neural network approximation in the case of sintered metal properties. Studies in Computational Intelligence 513 (2014) 125-134. https://doi.org/10.1007/978-3-319-01787-7_12

[34] J. Fabiś-Domagała, G. Filo, H. Momeni, M. Domagała. Instruments of identification of hydraulic components potential failures, MATEC Web of Conf. 183 (2018) art.03008. https://doi.org/10.1051/matecconf/201818303008

[35] G. Filo, J. Fabiś-Domagała, M. Domagała, E. Lisowski, H. Momeni. The idea of fuzzy logic usage in a sheet-based FMEA analysis of mechanical systems, MATEC Web of Conf. 183 (2018) art.3009. https://doi.org/10.1051/matecconf/201818303009

[36] J. Fabis-Domagala, M. Domagala, H. Momeni. A concept of risk prioritization in fmea analysis for fluid power systems, Energies 14 (2021) art. 6482. https://doi.org/10.3390/en14206482

[37] J. Fabis-Domagala, M. Domagala, H. Momeni. A matrix FMEA analysis of variable delivery vane pumps, Energies 14 (2021) art. 1741. https://doi.org/10.3390/en14061741

[38] M. Ingaldi. Overview of the main methods of service quality analysis, Production Engineering Archives 18 (2018) 54-59. https://doi.org/10.30657/pea.2018.18.10

[39] A. Pacana, R. Ulewicz. Analysis of causes and effects of implementation of the quality management system compliant with iso 9001, Pol. J. Manag. Stud. 21 (2020) 283 296. https://doi.org/10.17512/pjms.2020.21.1.21

[40] D. Siwiec, R. Dwornicka, A. Pacana. Improving the non-destructive test by initiating the quality management techniques on an example of the turbine nozzle outlet, Materials Research Proceedings 17 (2020) 16-22. https://doi.org/10.21741/9781644901038-3

[41] G. Ostasz, K. Czerwińska, A. Pacana. Quality Management of Aluminum Pistons with the Use of Quality Control Points. Management Systems in Production Engineering 28 (2020) 29-33. https://doi.org/10.2478/mspe-2020-0005

[42] D. Nowakowski, A. Gądek-Moszczak, P. Lempa. Application of machine learning in the analysis of surface quality - the detection the surface layer damage of the vehicle body, METAL 2021 – 30th Int. Conf. Metallurgy and Materials (2021), Ostrava, Tanger 864-869. https://doi.org/10.37904/metal.2021.4210

Terotechnology XII
Materials Research Proceedings 24 (2022) 142-147

Materials Research Forum LLC
https://doi.org/10.21741/9781644902059-21

Impact of TNT Storage Time on Its Physicochemical and Explosives Properties

DYLONG Agnieszka[1,a] *, DYSZ Karolina[1,b] and KUBECKI Adam[1,c]

[1]Military Institute of Engineer Technology, 136 Obornicka Str., 50-961 Wroclaw, Poland

[a]dylong@witi.wroc.pl*, [b]dysz@witi.wroc.pl, [c]kubecki@witi.wroc.pl

Keywords: TNT, High Explosives, BAM, Friction Sensitivity, Impact Sensitivity, Melting Point, Decomposition Temperature

Abstract. The Polish Armed Forces have very sizable stocks of explosive ordnance, of which some have exceeded the allowable service life. From the point of view of ageing and acceptable ways of disposal, some high explosives cannot be used if they have been stored for years. That is why studies are performed on the safety of utilizing such kinds of explosive ordnance. During the storage period, high explosives' physical and chemical parameters deteriorate. For example, the sensitivity of such materials increases, resulting in them becoming dangerous. Therefore, diagnostic tests to determine the quality of high explosives for further use (extending the exploitation period or referral for disposal) are conducted. The main goal of this work was to compare how the effect of the ageing process impacts the physical and chemical properties of high explosives and those containing 2,4,6-trinitrotoluene in particular. Many factors effectuate the quality of the stored high explosives, e.g. acidity, melting point, decomposition temperature, friction- and impact sensitivity. The authors investigated high explosives from selected mines produced in different periods and compared these results with those obtained from testing mines of previous years.

Introduction

Considering that Poland has significant stocks of explosive ordnance with some exceeding allowable service life, monitoring the impact of storage time on the physical and chemical properties of 2,4,6-trinitrotoluene (TNT) is a pertinent issue.

TNT's popularity stems from its low melting point (i.e., 80.65°C) and the process occurring with no decomposition. This makes TNT readily cast. TNT is used as an admixture with many other high explosives, mainly due to the low melting point. Comparative studies with other explosives, such as penthrite (PENT), Royal Demolition explosive (RDX) or octogen (HMX), corroborated that TNT is highly stable and has a low sensitivity impact. Studies also reported it being distilled under vacuum (10 – 20 mm) at 210 – 212°C without decomposing. The first significant drawback of the TNT lies in the high environmental impact of the required production process, which caused it no longer be produced in Canada and the USA. Furthermore, the United States record a surplus of this explosive, and research and production are focused on even more powerful compositions. [1]

In Poland, the TNT is still the primary explosive used for explosive ordnance. This is attributable to, first of all, its properties, as it is a durable material, resistant to ageing processes and can retain its performance parameters for an extended time.

From the point of view of ageing and acceptable ways of disposal, some types of high explosives cannot be used if they had been stored for years. That is why safety studies of explosive ordnance determining their safe use are carried out. During the storage period, physical and

Terotechnology XII
Materials Research Proceedings 24 (2022) 142-147

Materials Research Forum LLC
https://doi.org/10.21741/9781644902059-21

chemical parameters of high explosives deteriorate, for example, the sensitivity of such materials rises, resulting in them becoming dangerous.

Moreover, it needs to be remembered that TNT is highly susceptible to photodecomposition [2,3], it reacts easily in an alkaline environment [4], but even though materials that could replace TNT have been sought for over a century, it is still the most commonly used component in the production of high explosives [5] and for almost a hundred years – as a high explosive component of explosive ordnance.

The authors conducted diagnostic tests to determine the quality of high explosives for further use (extending their exploitation time or referring them for disposal) which were enacted under the explosive ordnance quality control and safety system. To evaluate the quality of explosive materials, the authors focused on acidity, melting point, decomposition temperature, friction- and impact sensitivity.

The main goal of the project was to compare, how the effect of the ageing process influences the physical and chemical properties of high explosives, as there are many factors reflecting the quality of the stored high explosives, e.g., acidity, their melting point, decomposition temperature, friction- and impact sensitivity. The same parameters are considered when developing new explosives, especially explosive mixtures, to determine the potential reactions between the components of the mixture and its stability (which is important during long time storage in various conditions). Analysis of high explosives from selected mines from different manufacturing years was performed and the comparison of the results with those obtained in previous years was conducted by the authors.

Experimental Part
To evaluate the quality of explosives, acidity, melting point, decomposition temperature, friction and impact sensitivity testing was performed. The investigation was done according to the following STANAG military standards: "Specification for TNT for deliveries from one NATO nation to another" (STANAG 4025), "Explosive, friction sensitivity tests" (STANAG 4487) and "Explosives, impact sensitivity tests" (STANAG 4489).

Friction sensitivity tests were performed using a friction tester, whereas impact sensitivity tests were carried conducted using a BAM Fallhammer (BFH 10) and thermal analysis involved the use of a TGA/SDTA apparatus.

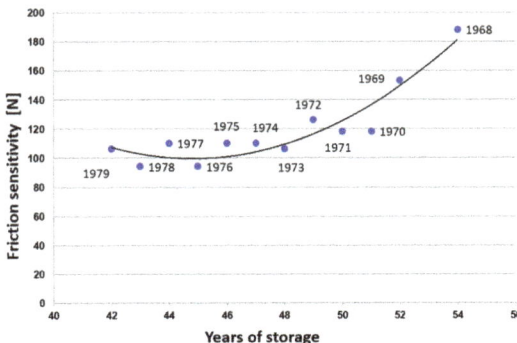

Fig. 1. Friction sensitivity as the function of the years of storage, indicating the TNT ageing process

Terotechnology XII   Materials Research Forum LLC
Materials Research Proceedings **24** (2022) 142-147   https://doi.org/10.21741/9781644902059-21

Acidity testing consists in determining the content of free acids in samples of an explosive by dissolving the tested sample in acetone, adding H_2O and titrating Na_2CO_3 against a mixed indicator. Selected results for TNT friction sensitivity, impact sensitivity and melting temperature as dependent on the storage period are presented in Figs. 1-3.

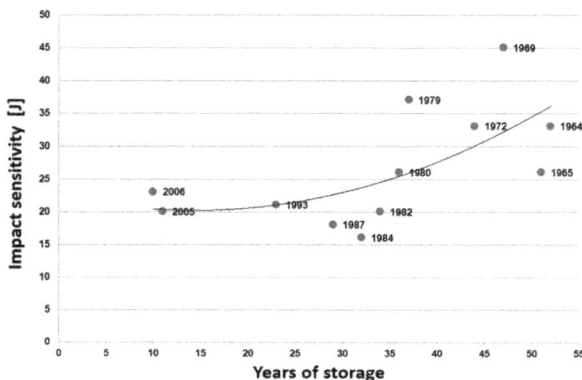

Fig. 2. *Impact sensitivity as the function of the years of storage, indicating the TNT ageing process*

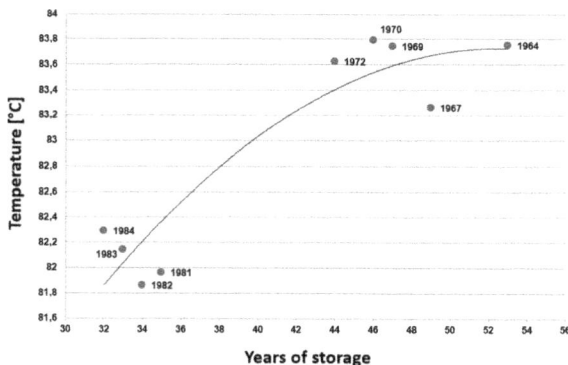

Fig. 3. *Melting point as the function of the years of storage, indicating the TNT aging process*

The chemical and physical parameters of TNT samples display a difference in stability over time between samples from the 1960s-70s and after the 1980s. The diagrams illustrate changes in the quality of the examined TNT samples as dependent on the storage time and they reveal that the quality parameters are superior for older samples. It seems that the difference between the parameters of the samples from the 1960s and early1970s (visible in Fig. 1 - 3) and the values for the samples from the 1980s and 1990s could be a result of the strategy adopted by the Army at that

Terotechnology XII   Materials Research Forum LLC
Materials Research Proceedings 24 (2022) 142-147   https://doi.org/10.21741/9781644902059-21

time. Materials from the earlier periods seem to be more resistant to ageing due to the superior assembly process. These materials were manufactured in a more precise way and with prolonged periods of use and storage in mind. As far as the samples produced after the 1980s, they could be affected by the lower quality of the manufacturing, which translates to lower resistance and increased sensitivity to external stimuli (friction impact and temperature).

Fig. 4 illustrates the condition of TNT as sampled from the mines from the 1960s-70s, and Fig. 5 presents trinitrotoluene which was produced recently.

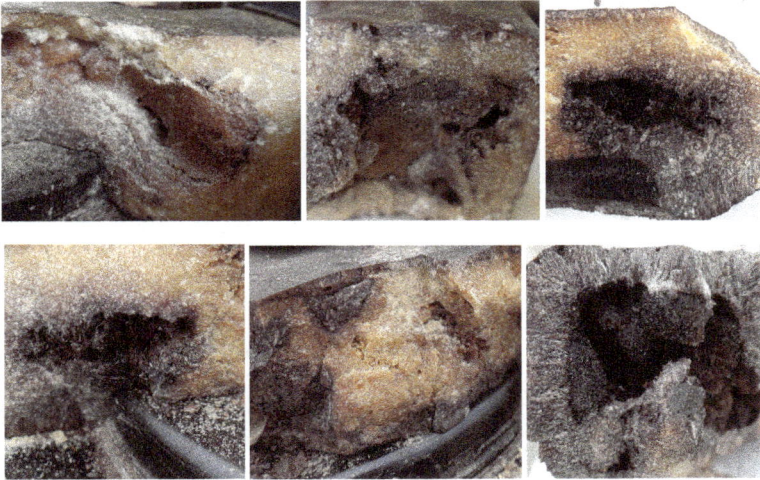

Fig. 4. *TNT fragment sampled from the mines from the 1960s-70s*

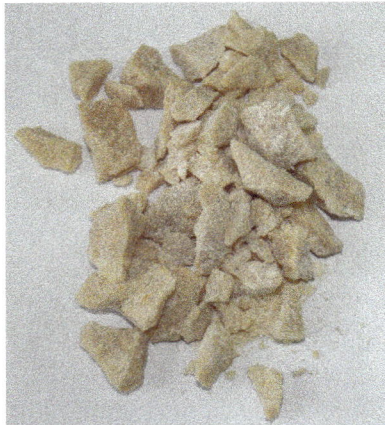

Fig. 5. *TNT from a recent batch*

As a result of the tests of the physicochemical and explosive properties of TNT taken from the mines from the 1960s and 1970s, the authors found that advanced ageing processes of the explosive led to changes in its physicochemical and explosive parameters, which translates into the safety of the product use. TNT recrystallized, it is heterogeneous, contaminated, cracked with local shrinkage in the volume of the mine.

Moreover, the tests of older materials, in comparison with those of "freshly-produced TNT" revealed that their parameters differ notably (Table 1). Low melting points indicate the use of low-quality TNT for elaboration in the case of the former. The external appearance of the tested samples, presented in Figs. 4-5, indicates a lack of homogeneity of the explosive (TNT).

Discolorations, contraction cavities, recrystallized material and its impurities indicate that advanced ageing processes occurred within the material, and point to inferior quality of the explosive itself and the filling process.

Due to the low thermal conductivity of TNT, heat exchange is hindered, and thus the crystallization process is slow. As a result, an undesirable casting is obtained in the form of low-density coarse-crystalline material. Crystallization carried out improperly may result in the so-called contraction cavities. The formation of TNT cracks may have resulted from the improperly carried out crystallization process, yielding an overly dense casting. These are the defects that arise at the stage of explosive's production already.

In order to determine the suitability of explosive ordnance for further storage, a number of additional tests, such as acidity is being conducted.

Table 1. Deterioration of TNT parameters indicated based on acidity testing

TNT year of production	Impact sensitivity	Friction sensitivity	Melting point	Acidity
2019	>50 J	> 356 N	80.2°C	0.005 max. % [6]
1965	88 J	154 N	77.7°C	>0.005
1969	31 J	208 N	77.9°C	>0.005

Summary

Following a long-term (e.g., a few decades) storage was noticed that high explosives are capable of maintaining their physical and chemical properties at an acceptable level. This means that the parameters of the explosives depend mainly on the filling (production) quality, and secondly on the storage conditions.

The explosive stored in the warehouses of the Polish Armed Forces were manufactured 20 to 60 years ago. Furthermore, for the explosive to be suitable for filling, it should meet certain standards at the initial stage of production. For TNT, these are:
- The melting point of 80°C minimum (TNT with a melting point below 80°C was classified into a lower purity category, or even as a TNT waste TNT, when it fell below 76°C;
- Friction sensitivity >360 N;
- Impact sensitivity >50 J;
- Acidity: 0.005 % max [6,7].

The analysis of TNT test results has not yielded a definitive answer, as to the potential response of TNT to the conditions of even more extended storage.

Terotechnology XII   Materials Research Forum LLC
Materials Research Proceedings **24** (2022) 142-147   https://doi.org/10.21741/9781644902059-21

The parameters characterizing the response of explosives and enabling the estimation of further storage period include friction- and impact sensitivity and thermal stability, as well as acidity testing.

Visual assessment of the quality of explosives at the sampling stage (homogeneity of the material, no cracks, crystals) provides the answer to the question, whether verification of all the assumed parameters will be possible. The most common parameter, which lies outside the assumed parameters, is acidity.

Additionally, both the friction- and impact sensitivity of the lower-quality material with which the mine was filled in possible to determine, in the case of a large number of contraction cavities, and high discoloration. These parameters are quite high (e.g., impact sensitivity > 100 J), the melting point is low (< 78.0°C), while the acidity cannot be determined. However, the majority of the explosives filling the mines from the 1980s, 1990s, or at the beginning of the 21st century investigated by the authors reveal stable properties, as their parameters are characteristic for the commonly used TNT, and such mines can be successfully stored in warehouses for many years. In case of the explosive from the 1960s and 1970s it is the quality of the final product (the type of mine casing and its airtightness), the quality of the filling process and the storage conditions (stable temperature) which guarantees that the acceptable parameters for the explosive itself will be maintained.

References

[1] L. Davison, Y. Horie, L. Davison. High-Pressure Shock Compression of Condensed Matter, Chapter 9, J. C. Oxley, Safe Handling of Explosives (1997) 341–350. https://doi.org/10.1007/978-1-4612-2292-7

[2] W. R. Mabey, D. Tse, A. Baraze, T. Mill. Photolysis of nitroaromatics in aquatic systems: 2,4,6-trinitrotoluene, Chemosphere 12 (1983) 3–16. https://doi.org/10.1016/0045-6535(83)90174-1

[3] D. J. L. Prak, J. E. T. Breuer, E. A. Rios, E. E. Jedlicka, D. W. O'Sullivan. Photolysis of 2,4,6-trinitrotoluene in seawater and estuary water: Impact of pH, temperature, salinity, and dissolved organic matter, Marine Pollution Biulletin 114 (2017) 977–986. https://doi.org/10.1016/j.marpolbul.2016.10.073

[4] Ch. A. Latendresse, S. C. Fernandes, S. You, W. B. Euler. Speciation of the Products of and Establishing the Role of Water in the Reaction of TNT with Hydroxide and Amines: Structure, Kinetics, and Computational Results, J. Phys. Chem. A 117 (2013) 11167–11182.

[5] M. Szala, T. Sałaciński, 2,4,6-trinitrotoluene as a source of modern explosives, a review. High-Energetic Materials 7 (2015) 125–143. https://doi.org/10.1021/jp408992n

[6] STANAG 4025 MMS (Edition 3) – Specification for TNT for deliveries from one NATO nation to another (1991).

[7] T. Urbański. Chemia i technologia materiałów wybuchowych. Tom 1. Wyd. MON, Warszawa, 1954, 120–125.

Terotechnology XII
Materials Research Proceedings 24 (2022) 148-153

Materials Research Forum LLC
https://doi.org/10.21741/9781644902059-22

Microstructure and Mechanical Properties of Powder Metallurgy Fe-Cu-Sn-Ni Alloys

BOROWIECKA-JAMROZEK Joanna [1,a] and LACHOWSKI Jan [1,b*]

[1]Kielce University of Technology, al. Tysiclecia P.P. 7, 25-314 Kielce, Poland

[a]jamrozek@tu.kielce.pl, [b]jlach@tu.kielce.pl

Keywords: Metallic Matrix, Milling Process, Powder Metallurgy, Diamond Tools

Abstract. The main purpose of the work was to determine the powder composition on the microstructure and properties of iron-based sinters used as a matrix in metallic diamond tools. Fe-Cu-Sn-Ni sinters obtained from a mixture of elemental powders were used for research. Sintering was performed using the hot-pressing technique in a graphite mold. The microstructure and mechanical properties of the sinters were investigated. The investigations of obtained sinters included density, hardness, static tensile test, X-ray analysis, microstructure, and fracture surface observations. The results obtained indicate that the sinters produced have relatively high hardness and good mechanical properties.

Introduction

Diamond blades are commonly used to cut natural stone and ceramics. The cutting section of the tools consists of synthetic diamonds embedded in a metallic matrix by powder metallurgy technology [1,2]. There are many metals that have been used as matrices of diamond-impregnated tools. In general examples of such metals are cobalt, aluminum, titanium, copper, iron, zinc, tin, and nickel. For a few years, the use of inexpensive pre-mixed and milled powders in ball mills, which may replace cobalt powders, has been observed [3,4]. Diamond-metallic compounds based on metal matrices containing iron, copper, nickel and tin possess favorable mechanical properties, relatively low sintering temperature, and low cost of materials [5,6].

Based on previous research [7], the iron-based material was designed for the matrix of metallic diamond tools. The material was obtained from elemental powders subjected to grinding for 30 h. The material was selected for the tests, with the mass fractions of individual powders as follows: 60% Fe, 23.8% Cu, 4.2% Sn, and 12% Ni. The new matrix material had a porosity not exceeding 3% and mechanical properties that would allow replacement of cobalt-based sinters [8]. The very important role of a matrix in diamond impregnated segments is to hold diamond particles as long as possible. The effective way to improve diamond retention in the matrix is through the use of diamond particles with a thin film of strong carbide obtained with coating technology [9,10].

Material and Its Mechanical Properties

The experimental powder mixtures were made from the following elemental powders (Fig. 1):
a) Höganäs NC100.24 grade, carbon-reduced iron powder (20-180 μm),
b) Vale T255 grade, carbonyl nickel powder (average size = 2.4 μm),
c) water-sprayed bronze tin powder containing 15% by mass. tin (25GR85 / 15.325 from ECKA), with a particle size below 45 μm.

The mass fractions of individual powders in the tested material FeCuSnNi were: 60% Fe, 28% bronze and 12% Ni, which corresponds to the chemical composition: 60% Fe, 23.8% Cu, 4.2% Sn, 12% Ni. Before the consolidation process, the powders were premixed in the right proportions in a Turbula mixer for 30 minutes and then milled in an EnviSense RJM-102 laboratory ball mill in ethyl alcohol with a small amount of glycerol. The milling vial was filled to half its volume with 100Cr6 steel balls of 12 mm diameter.

The initial mixtures of FeCuSnNi powders and ground powders of both materials were subjected to consolidation by hot pressing, in a nitrogen atmosphere, using the Unidiamond press furnace Idea (Italy), within 30 hours. The prepared mixtures were pressed in a graphite matrix, allowing the simultaneous production of 4 samples with nominal dimensions of 7 x 6 x 40 mm. A temperature of 900°C and a pressure of 25 MPa were used for sintering the material.

a)

b)

c)

Fig. 1. *Powders used for research, a) iron powder NC100.24, b) nickel powder T255,
c) tin bronze powder 25GR85 / 15.325*

Table 1. *Physical Properties of the Material*

Material	Density, [g/cm^3]	Porosity, [%]	HV10	oxygen content, [% mas.]
FeCuSnNi, without milling	8.000 ± 0.028	2.20 ± 0.20	241.7 ± 15.4	0.34 ± 0.03
FeCuSnNi, milling 30 h	8.008 ± 0.037	2.20 ± 0.40	272.7 ± 8.4	0.59 ± 0.01

Table 2. *Mechanical Properties of the Material*

Material	Elastic modulus E [GPa]	Poisson ratio v	Yield Strength $R_{p0,2}$ [MPa]	Tensile strength R_m [MPa]	Relative elongation ε [%]
without milling			169+/-10	481.1+/-15.6	7.5+/-0.2
milling 30h	165 +/- 3	0.32	273+/-12	739.4+/-20.5	11.0+/-0.5

Terotechnology XII | Materials Research Forum LLC
Materials Research Proceedings **24** (2022) 148-153 | https://doi.org/10.21741/9781644902059-22

The static tensile test was carried out using a Universal Testing Machine UTS-100 with an automatic control and data recording system from Zwick. Based on the recorded data, the following were calculated: yield strength $R_{p0.2}$, tensile strength Rm, and relative elongation ε. The parameters mentioned above were determined as the arithmetic mean of three tensile tests. The elastic modulus E and the Poisson ratio v were also measured by ultrasound using a Panametrics Epoch III flaw detector. The sinters obtained from powders after milling show a higher hardness and a clearer tensile strength compared to sinters made from powders without milling.

Fractography of Fracture
Fractographic tests were performed on the fracture obtained in the tensile test. Research was carried out using the JSM-7100F scanning electron microscope, integrated with the OINA-AZtec X-ray microanalysis system.

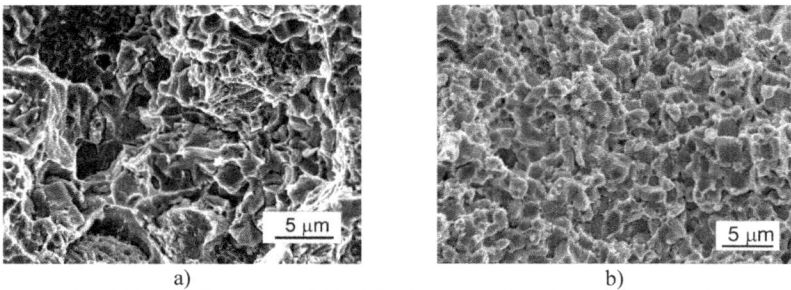

a)   b)
Fig. 2. *Surface fracture of FeCuSnNi sinter samples obtained from powders*
a) without milling, b) after milling for 30 h

Observations of the fracture surfaces lead to the conclusion that, for all sinters of the tested materials without milling, the fracture is brittle and transcrystalline. Cracking occurs at the grain boundaries. The mechanism of fracture is mixed. On the other hand, for sinters made from powder mixtures, after a grinding time of 30 h, there is a fracture surface developed. The fractures take a ductile form; only areas showing the features of an intercrystalline fracture are visible in places.

Microstructure Studies
Observations of the microstructure of the produced sinters were carried out using the JSM-7100F scanning electron microscope, integrated with the X-Max-AZtec series EDS microanalysis system from OXFORD INSTRUMENTS. The tests were carried out using a backscattered electron detector. The microstructure of the FeCuSnNi sinter, obtained from powder mixtures without milling and after milling 30 h (Fig. 3).

The FeCuSnNi sinter revealed a complex phase structure (Fig. 4). Point chemical analysis showed the presence of Fe solution,, Cu solution and iron oxides (Fig. 5). Nickel atoms were distributed throughout the entire sinter volume with a distinct advantage in the white phase (Fig. 6). The oxides were concentrated in the region of the iron solution. For FeCuSnNi sinters obtained from powders without milling and after 30 h of milling, the chemical composition of the point and the surface distribution of the elements are shown in Fig. 4…Fig.7.

In the area rich in iron, there are also Ni atoms, which easily integrate into the network of Fe atoms because nitrogen has similar sizes of atomic radii.

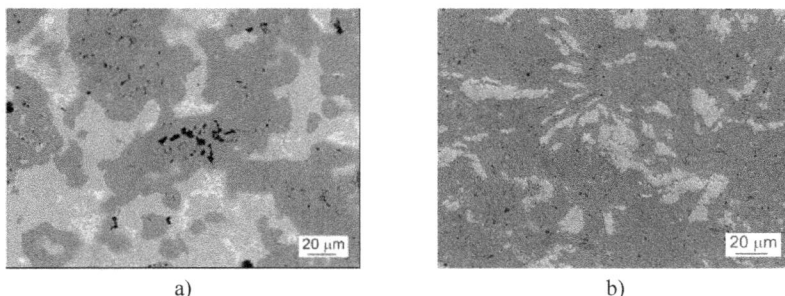

a)

b)

Fig. 3. *Microstructure of sinters made of FeCuSnNi powder,*
a) not subjected to the milling process, b) subjected to milling for 30 h,

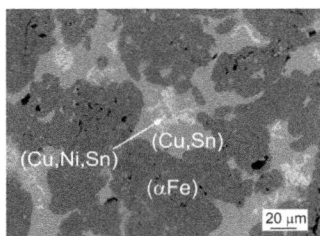

	O [%]	Fe [%]	Cu [%]	Sn [%]	Ni [%]
(Fe)		85.73	4.56	0.34	9.37
(Cu)		3.26	85.55	5.65	5.54
Oxides	15.12	72.64	3.33	2.05	6.86

Fig. 4. *Microstructure of the FeCuSnNi sinter (without milling) and the results*
of the microanalysis (wt%) (from the X-ray spectra) collected from the areas:
Fe solution, Cu solution and the area of iron oxides

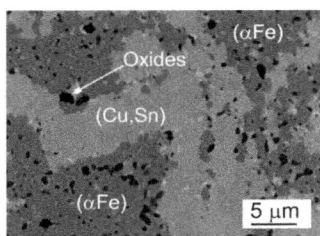

	O [%]	Fe [%]	Cu [%]	Sn [%]	Ni [%]
(Fe)	-	72.59	10.54	0.93	15.94
(Cu)	-	5.24	78.04	11.32	5.40
Oxides	17.18	71.64	1.97	0.75	8.46

Fig. 5. *Microstructure of FeCuSnNi sinter (milling 30 h) and the results*
of the microanalysis (wt%) (from the spectra of X-rays) collected from the areas:
Fe solution,, Cu solution and the area of iron oxides.

Terotechnology XII
Materials Research Proceedings **24** (2022) 148-153

Materials Research Forum LLC
https://doi.org/10.21741/9781644902059-22

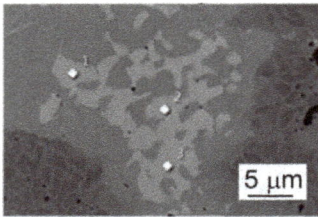

	Fe [%]	Cu [%]	Sn [%]	Ni [%]
1	1.59	28.20	38.47	31.74
2	2.04	29.28	36.03	32.65
3	2.30	26.17	38.09	33.44

*Fig. 6. Microstructure of the FeCuSnNi sinter (without milling)
with the chemical composition of the point.*

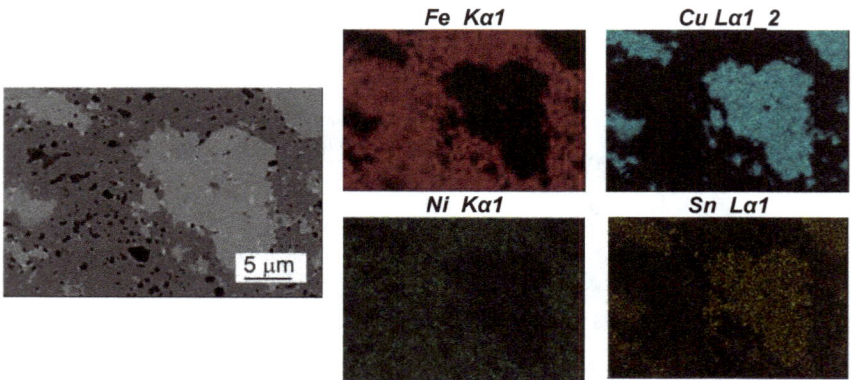

*Fig. 7. Surface distribution of elements in FeCuSnNi sinter, obtained
from powder mixtures after milling for 30 h*

Additionally, a detailed point analysis of the white chemical composition of the area in white was performed. The conducted analysis showed the presence of an intermetallic relationship of copper, nickel, and tin.

Conclusions
Research shows that the milling process causes greater fragmentation of the microstructure of the tested sinters (Figs. 3 and 7). The microstructure of the sinter obtained from mixtures without milling, also after 30 hours of milling, is heterogeneous. The large areas of iron solution are noticeable in microstructure, in which greater amounts of nickel were found (Fig. 7). Hence, it can be concluded that nickel diffusion proceeds into the iron solution.

The microstructures of the milled sinters obtained from the mixes milled through 30 h show distinctive features. There is a banding and lamellar structure, which is due to the flake shape of the powder particles. In the microstructure of the FeCuSnNi alloy made of the unmilled powder mixture, there is a white phase that clearly decreases after 30 h of milling. A point analysis of the chemical composition of this phase showed a high concentration of copper, tin and nickel. Analysis of chemical composition shows that it is the intermetallic phase $(CuNi)_3Sn$. In the milling process, nickel is dissolved in bronze, and tin diffuses into the iron solution (Fig. 7). The introduction of a

Terotechnology XII   Materials Research Forum LLC
Materials Research Proceedings **24** (2022) 148-153   https://doi.org/10.21741/9781644902059-22

tin bronze addition, instead of copper, to Fe and Ni powders resulted in the obtaining of a liquid phase (> 798°C) during the hot pressing process, which helped to consolidate and intensify the diffusion of the phase components.

References

[1] J. Konstanty, X.P. Xu. Production of Diamond Sawblades for Stone Sawing Applications, Key Engineering Materials 250 (2003) 1-12.
https://doi.org/10.4028/www.scientific.net/KEM.250.1

[2] X.P. Xu, Y.B. Hong, S. Chen. Performance of diamond segments in different machining processes. Materials Science Forum 471-472 (2004) 77-81.
https://doi.org/10.4028/www.scientific.net/MSF.471-472.77

[3] F.A.C. Oliveira, C.A. Anjinho. PM materials selection: The Key for Improved Performance of Diamond Tools, Metal Powder Report 72 (2017) 339-344.
https://doi.org/10.1016/j.mprp.2016.04.002

[4] A.P. Barbosa, G.S. Bobrovnitchii, A.L.D. Skury, R.S. Guimaraes, M. Filgueira. Structure, microstructure and mechanical properties of PM Fe–Cu–Co alloys, Mater. Des. 31 (2010) 522–526. https://doi.org/10.1016/j.matdes.2009.07.027

[5] V.G. Mechnik, M.O. Bondarenko, V.M. Kolodnitskyi, V.I. Zakiev, I.M. Zakiev, M. Kuzin, E.S. Gevorkyan. Influence of diamond–matrix transition zone structure on mechanical properties and wear of sintered diamond-containing composites based on Fe–Cu–Ni–Sn matrix with varying CrB2 content, Int. J. Refract. Hard Met. 100 (2021) art. 105655.
https://doi.org/10.1016/j.ijrmhm.2021.105655

[6] V.G Mechnik, N.A. Bondarenko, S.N. Duba, V.M. Kolodnitskyi, Yu V. Nesterenko, N.O. Kuzin, I.M. Zakiev, E.S. Gevorkyan. A study of the microstructure of Fe-Cu-Ni-Sn and Fe-Cu-Ni-Sn-VN metal matrix for diamond-containing composites, Mater. Charact. 146 (2018) 209–216. https://doi.org/10.1016/j.matchar.2018.10.002

[7] J. Borowiecka-Jamrozek, J. Lachowski. Microstructure and Mechanical Properties of Fe-Cu-Ni Sinters Prepared by Ball Milling and Hot Pressing, Defect Diffus. Forum 405 (2020) 379-384.
10.4028/www.scientific.net/DDF.405.379

[8] J. Borowiecka-Jamrozek. Engineering structure and properties of materials used as a matrix in diamond impregnated tools, Arch. Metall. Mater. 58 (2013) 5-8.
https://doi.org/10.2478/v10172-012-0142-0

[9] N. Radek, A. Szczotok, A. Gądek-Moszczak, R. Dwornicka, J. Broncek, J. Pietraszek. The impact of laser processing parameters on the properties of electro-spark deposited coatings, Arch. Metall. Mater. 63 (2018) 809-816. https://doi.org/10.24425/122407

[10] N. Radek, A. Sladek, J. Broncek, I. Bilska, A. Szczotok. Electrospark alloying of carbon steel with WC-Co-Al2O3: deposition technique and coating properties, Adv. Mater. Res. 874 (2014) 101-106. https://doi.org/10.4028/www.scientific.net/AMR.874.101

Terotechnology XII
Materials Research Proceedings **24** (2022) 154-158

Materials Research Forum LLC
https://doi.org/10.21741/9781644902059-23

Study of the Impact of –NH₂ Modification on Adsorptive Properties of Graphene Oxide

KWAK Anna[1,a], DYLONG Agnieszka[1,b*] and MALISZEWSKI Waldemar[1,c]

[1]Military Institute of Engineer Technology, Obornicka 136 Str., 50-961 Wroclaw, Poland

[a]kwak@witi.wroc.pl, [b]dylong@witi.wroc.pl*, [c]maliszewski@witi.wroc.pl

> *In memoriam of Kazimierz Szyszka*

Keywords: Graphene Oxide, Modified Graphene Oxide, Amine Compounds, Adsorbent, Water Treatment, Metal Ions, Arsenic

Abstract. Graphene oxide is a nanomaterial of very high adsorption capabilities due to its vast surface area. Moreover, numerous oxygen functional groups present on the surface of graphene oxide enable its modifications to be performed. The authors aimed to create adsorbents based on activated carbon impregnated with amine-modified graphene oxide. The study showed that the amino group functionalisation, both with the use of ethylenediamine (GO-EDA) and using polyaniline (GO-PANI), causes the adsorbent to remove lead, mercury, copper, and iron ions from aqueous solutions very efficiently. Both adsorbents also reduce the cadmium, nickel, zinc, and arsenic ion content, however to a lesser extent but nevertheless still significantly. The two sorbents can be applied in field water treatment to remove specific contaminants.

Introduction

Graphene oxide (GO) indicated increased adsorption, as far as the removal of heavy metal compounds dissolved in water is concerned, when compared to active carbons, widely used in water purification processes. The addition of graphene oxide to active carbon increases the adsorption capacity of the adsorbent obtained in this manner relative to selected heavy metals. The surface of graphene oxide contains hydroxyl-, carboxyl-, ketone- and epoxy functional groups, as well as others which enable the formation of hydrogen or polarized bonds not only with organic compounds but also with inorganic compounds of heavy metals [1]. It was also proven [2÷9] that a further increase in adsorption properties can be achieved via modification of graphene oxide with –NH2 and –SH groups or with some metal oxides (including iron-, titanium-, manganese- and tin oxides). The authors developed detailed technologies for the fabrication of carbon adsorbents containing GO modified with compounds with –NH₂ groups. The obtained adsorbents were evaluated for aqueous solution heavy metal adsorption, and this allowed for the assessment of their possible suitability for use in field water treatment processes to be performed.

Experimental Procedure

Synthesis of Graphene Oxide Flakes. Graphene oxide was obtained via the modified Hummers method [10, 11]. 20 g of flake graphite was added to a 1-litre beaker and the acid mixture (432 cm³ of H₂SO₄ and 42 cm³ of H₃PO₄) was slowly poured into the same beaker. After the mixture was stirred for 30 minutes, the graphite oxidation process was initiated using KMnO₄ (~60 g), which was being added to the beaker for 2 hours and stirred continuously, keeping the temperature of the mixture at below 60°C. After the oxidant was dosed into the mixture, the contents of the beaker were stirred for another 2 hours and then set aside sealed for 24 hours. While stirring continuously, the oxidized mixture was slowly added to the beaker containing water. It was topped up with water up to 5 dm^3 and stirred for 6 hours. The brown liquid obtained after stirring was first set aside for 24 hours and then centrifuged, while the sediment was topped up with water up to 5 dm^3 and mixed. Following another centrifugation, the liquid was sonicated for 6 hours. A thick, brown, greasy substance containing graphene oxide was obtained, and derivative thermogravimetry was used for its analysis.

Compounds Used to Modify Graphene Oxide with Amino Groups. To modify graphene oxide with amino groups, the following compounds containing such groups were selected: ethylenediamine (EDA) and polyaniline (PANI). In its molecule, ethylenediamine contains two -NH2 amine groups, which makes it easy for it to bond the metal and form heterocyclic structures. Moreover, EDA coupled with graphene oxide (GO) in the presence of N,N'-dicyclohexylcarbodiimide (N-DHC) [6] which is an organic chemical compound mainly used as a condensing agent in organic synthesis, e.g. for the coupling of amino acids in peptide synthesis.

Polyaniline (PANI) is an aniline polymer synthesized directly on the surface of graphene oxide. Aniline polymerization is conducted in a hydrochloric acid environment. Ammonium persulfate [9] is used as the oxidant. Depending on the degree of oxidation, some segments of the polyaniline molecule can transform into a quinone form, which allows polyaniline to exist in various forms that differ in colour and their properties.

Active Carbon Impregnation with Graphene Oxide with Ethylenediamine. The initial stage in the preparation of a new adsorbent was functionalizing graphene oxide with ethylenediamine. A concentrated aqueous slurry of graphene oxide containing 2 g of GO was mixed with 80 g of ethylenediamine and 7 g of N,N'-dicyclohexylcarbodiimide. The mixture was stirred for 48 hours under reflux at 60°C, then filtered through a funnel with sintered disc G4 and washed first with water, and then with ethanol. The obtained greasy substance was dried for 8 hours in a dryer at 80°C. Graphene oxide functionalised with ethylenediamine (GO-EDA) was synthesised.

100 g of active carbon was weighed out and dried in a drier for 2 hours at a temperature of 120°C, then cooled to room temperature. Distilled water, in the amount equal to the mass of prepared active carbon, was poured onto 2 g of the weighted out GO-EDA and stirred for 1 hour using a magnetic stirrer, followed by sonication for 3 hours in an ultrasonic cleaner.

The dried active carbon was mixed with the entirety of the obtained composition of graphene oxide and ethylenediamine. It was stirred for 30 minutes and set aside for 2 hours at room temperature. The impregnated carbon was dried for 12 hours at a temperature of 120°C.

The cooled GO-EDA modified sorbent containing 2% of the impregnant was ready to be used in the water purification process for the removal of heavy metal compounds.

Active Carbon Impregnation with Graphene Oxide with Polyaniline. To obtain graphene oxide with polyaniline, 2 g of GO were mixed with 200 cm^3 of distilled water. 10 g of aniline was added to the homogeneous suspension while stirring, and then 30 cm^3 of concentrated hydrochloric acid was poured into the suspension. A magnetic stirrer was used to stir the obtained suspension for 1 hour. Then 20 g of ammonium persulfate dissolved in 1 molar hydrochloric acid was added to the mixture. The mixture was first stirred for another 2 hours in a cooling bath with ice and sodium chloride, maintaining the temperature at $5 \div 7$°C, and then transferred to a refrigerator and set aside for 16 hours. The cooled reactor was transferred to the bath with a cooling mixture, and it was kept at a temperature of $5 \div 7$°C for another 3 hours while stirring.

The resulting greasy, homogeneous suspension was diluted with 4 litres of distilled water and precipitated through a funnel with sintered disc G4. The precipitate was washed using 1 litre of distilled water on the filter, and 62 g of wet sorbent containing 15% GO-PANI was obtained. 100 g of active carbon was prepared for impregnation as described above. 13.3 g of wet GO-PANI was mixed with 90 cm^3 of distilled water for 1 hour using a magnetic stirrer and then sonicated for 3 hours in an ultrasonic cleaner. The obtained product was added to a dried active carbon sample, seasoned for 2 hours at room temperature, and then dried for another 2 hours at 120°C and cooled.

The adsorbent obtained in this manner contained containing 2% of the impregnant and was ready to be used in the water purification process for removal of heavy metal compounds

Method for verifying the effectiveness of modified adsorbents. The following cations were selected for the study of the adsorption of metals from water: Hg^{2+}, Pb^{2+}, Cd^{2+}, As^{3+}, Ni^{2+}, Cu^{2+}, Zn^{2+}, Fe^{3+}. Selected metals salt solutions with cation concentrations equal to 1mg/dm3 and 5mg/dm3 were prepared. The study utilized the static method, which consisted in pouring 100 cm3 of individual solutions onto 1 g weights of the adsorbent (containing GO-EDA or GO-PANI) and stirring the mixture for 30 minutes using a magnetic stirrer. The contents of the flask were then filtered, and the metal content in the filtrate was determined by means of inductively coupled plasma optical emission spectrometry (ICP-OES), and then the adsorption efficiency of the studied element was calculated according to the formula:

$$A\% = [(c_i-c_f)/c_i] \cdot 100 \qquad (1)$$

where:

A – adsorption efficiency, in percent;

ci – initial metal ion concentration in the solution, in milligrams per cubic decimetre;

cf – final metal ion concentration in the solution, in milligrams per cubic decimetre.

Table 1. Heavy metal adsorption efficiency for GO-PANI- and GO-EDA-modified activated carbon.

p	C_i	$C_{fGO\text{-}PANI}$	$A_{GO\text{-}PANI}$	$C_{fGO\text{-}EDA}$	$A_{GO\text{-}EDA}$
	[mg/dm^3]	[mg/dm^3]	[%]	[mg/dm^3]	[%]
Pb^{2+}	1	0.000	100	0.000	100
Cd^{2+}	1	0.079	92.1	0.061	94.0
Ni^{2+}	1	0.225	77.5	0.156	84.4
Hg^{2+}	1	0.000	100	0.000	100
Zn^{2+}	1	0.222	77.8	0.072	92.8
Cu^{2+}	1	0.022	97.8	0.010	99.0
Fe^{2+}	1	0.023	97.7	0.024	97.6
As^{3+}	1	0.251	74.9	0.242	75.8
Pb^{2+}	5	0.007	99.9	0.000	100
Cd^{2+}	5	1.720	64.8	1.322	73.6

Ni^{2+}	5	2.108	57.8	2.019	59.6
Hg^{2+}	5	0.041	99.2	0.062	98.8
Zn^{2+}	5	2.806	43.9	2.340	53.2
Cu^{2+}	5	0.481	90.4	0.048	99.0
Fe^{2+}	5	0.556	88.9	0.265	94.7
As^{3+}	5	2.325	53.5	2.068	58.64

As shown in Table 1, both the GO-PANI and GO-EDA adsorbents are characterized by high adsorption efficiency with relation to the lead and mercury ions. Both at the concentration of 1 mg/dm3 and 5 mg/dm3, water contaminated with these metal compounds was completely purified. The Pb and Hg adsorption efficiency approximated 100%.

As far as water containing copper ions is concerned, the absorption efficiency in case of both concentrations (1 mg/dm3 and 5 mg/dm3) equalled 99% for the GO-EDA adsorbent, whereas for the GO-PANI adsorbent it is lower and amounts to 97.8% at the Cu^{2+} concentration of 1mg/dm3 and 90.4% for 5 mg/dm3 copper concentration in the initial solution. Such a high adsorption efficiency for copper ions from water solutions proves that the filter beds prepared from these adsorbents will significantly reduce the copper content in water.

The adsorption of iron ions from solutions with a metal content of 1 mg/dm3 was at the level of 97.6%, whilst from solutions when the metal amounts to 5 mg/dm3, it was lower and equalled 88.9% for GO-PANI and 94.7% for GO-EDA. Both adsorbents are characterised by the capability to significantly reduce iron content in the water.

GO-PANI and GO-EDA are weak adsorbents with regard to cadmium, zinc, nickel, and arsenic salts in particular. When examining the cadmium sorption, the adsorption efficiency recorded was above 90% for the initial solution of 1 mg/ dm3 Cd, but in the case of 5 mg/ dm3 Cd, the value was under 74%. Adsorbents of such kind do not ensure sufficient removal of cadmium, zinc, nickel, and arsenic cations from the water to be purified. If the content of one of these metal cations is at 5 mg/dm3, more than half of its mass remains unadsorbed.

Summary

The GO-EDA adsorbent is characterized by a very high (i.e. close to 100%) absorption efficiency towards lead and mercury ions, as well as for copper (99%) and iron (approx. 95%). The GO-EDA adsorbent sorbs cadmium to a lesser degree (about 74%). The same could be seen for nickel (ca. 60%), zinc (about 53%) and arsenic (about 58%). The disadvantage of this adsorbent is the long preparation process, lasting even several days which includes 48-hour heating under a reflux condenser.

The GO-PANI sorbent is characteristic in that it exhibits a remarkably high (close to 100%) lead and mercury ion adsorption efficiency, while the same for copper and iron was approx. 90%. The remaining investigated metals are adsorbed much less efficiently: cadmium at about 65%, nickel at 58%, zinc at about 44% and arsenic (ca. 53%). GO-PANI is therefore an adsorbent with a narrowed metal ion absorption spectrum, capable of removing selected ions from contaminated water to a high degree. The preparation process requires a number of hours of stirring at a low temperature (approx. 5°C).

Both sorbents seem to be useful in removing specific contaminants, such as metal ions or arsenic from water.

References

[1] P. Janik, K. Komorowska, E. Turek, B. Zawisza, R. Sitko. Graphene oxide as a new adsorbent in analytical chemistry, 6th Scientific Seminar: Current Issues in Analytical Chemistry, Katowice, May 18th, 2012.

[2] B. Zawisza, A. Baranik, E. Malicka, E. Talik, R. Sitko. Preconcentration of Fe(III), Co(II), Ni(II), Cu(II), Zn(II) and Pb(II) with ethylenediamine-modified graphene oxide. Microchimica Acta 183 (2016) 231-240. https://doi.org/10.1007/s00604-015-1629-y

[3] R. Sitko, P. Janik, B. Zawisza, E. Talik, E. Margui, I. Queralt. Green approach for ultratrace determination of divalent metal ions and arsenic species using total-reflection X-ray fluorescence spectrometry and mercapto-functionalised graphene oxide nanosheets as a novel adsorbent, Analytical Chemistry 87 (2015) 3535-3542. https://doi.org/10.1021/acs.analchem.5b00283

[4] P. Janik, B. Zawisza, E. Talik, E. Margui, I. Queralt, R. Sitko. Application of graphene oxide functionalised with thiol groups in the determination of heavy metals and arsenic speciation using the total reflection x-ray fluorescence (TXRF) technique, 9th Scientific Seminar: Current Issues in Analytical Chemistry, Katowice, May 15th, 2015.

[5] P. Janik, B. Feist, E. Talik, A. Gagor, R. Sitko. Graphene oxide functionalised with aminosilanes as a selective adsorbent in the determination of Pb^{2+} ions using electrothermal atomic absorption spectrometry (ET-AAS), Winter Congress of SSPTChem, Wroclaw, December 13th, 2014.

[6] P. Janik, A. Baranik, U. Porada. Testing the adsorption capacity of graphene oxide and carbon nanotubes modified with amine groups relative to lead ions, 8th Scientific Seminar: Current Issues in Analytical Chemistry, Katowice, May 16th, 2014.

[7] P. Janik, U. Porada, A. Baranik, R. Sitko. Sorption properties of graphene oxide functionalised with sulfonic groups relative to Co(II), Ni(II), Cu(II), Zn(II), Cd(II), and Pb(II) metal ions, 23rd Poznan Analytical Seminar: Modern Methods of Sample Preparation and Determination of Trace Quantities of Elements, Poznan, May 8th-9th, 2014.

[8] X. Luo, C. Wang, S. Luo, R. Dong, X. Tu, G. Zeng. Adsorption of As(III) and As(V) from water using magnetite Fe3O4-reduced graphite oxide-MnO2 nanocomposites, Chemical Engineering Journal 187 (2012) 45-52. https://doi.org/10.1016/j.cej.2012.01.073

[9] R. Li, L. Liu, F. Yang. Preparation of polyaniline/reduced graphene oxide nanocomposite and its application in adsorption of aqueous Hg(II), Chemical Engineering Journal 229 (2013) 460-468. https://doi.org/10.1016/j.cej.2013.05.089

[10] W. S. Hummers, Jr., R.E. Offeman. Preparation of Graphitic Oxide, J. Am. Chem. Soc. 80 (1958) 1339-1339. https://doi.org/10.1021/ja01539a017

[11] D.C. Marcano, D.V. Kosynkin, J.M. Berlin, A. Sinitskii, Z. Zun, A. Slesarev, L.B.A. Lemany, W. Lu, J.M. Tou. Improved synthesis of graphene oxide, ACS Nano 4 (2010) 4806-4814. https://doi.org/10.1021/nn1006368

Terotechnology XII

Materials Research Forum LLC

Materials Research Proceedings 24 (2022) 159-165

https://doi.org/10.21741/9781644902059-24

Empirical Model of the Rail Head Operational Crack on the Example of a Head Check Defect

ANTOLIK Łukasz [1,a]

[1] Instytut Kolejnictwa (The Railway Research Institute), Materials & Structure Laboratory, 50 Chłopicki Street, 04-275 Warsaw, Poland

[a]LANTOLIK@IKOLEJ.PL

Keywords: Rail Defects, Ultrasonic Testing

Abstract. The paper presents a method and the results of the CT test of the head check defect, which commonly occurs in rails. It is one of the most dangerous defects which cause rail discontinuity. The methods of simulation of the ultrasonic wave and projection possibilities of the real defect have been taken under consideration. The paper outlines the ultrasonic wave simulation method and describes the ultrasonic probe optimization results for the considered discontinuities.

Introduction

The defects of the rails arise at several stages of production and use. The internal defects of the rails appear in the metallurgical process, i.e. at the stage of smelting and casting. The defects that arise can subsequently be included in all internal discontinuities of the material, mainly the remains of the shrinkage cavity and large clusters of non-metallic inclusions. The rolling process of the rails may cause surface defects of the rails in the form of scratches, scale indentations on the rail running surface, or the formation of a coarse-grained structure of the material [1, 2]. On the other hand, straightening the rails after rolling promotes the formation of internal stress in the rails, which may lead to the curvature of the rails and cracking of the rails in the tracks. Due to the common use of non-contact termite welded and electrofusion welded rails in the tracks, the number of damage in the connection zone increased. This was mainly due to the failure to meet the parameters recommended by the welding technology. Joining the rails by electrofusion welding significantly improved the quality of the joints. Therefore the number of defects caused by this technology significantly decreased due to the higher repeatability and control of the welding process. The increase of the train velocity and traffic intensity causes the rails to be subjected to increasing contact stresses and hence the overall increase in their damage. In particular, the number of visible defects as cracks in the railhead surface has increased. They constitute a huge percentage of recorded rail defects, approximately 10% [1]. All head check defects occurring in the tracks in operation are found based on visual inspection. Thus, it is impossible to determine their approximate depth [1]. The analysis of many practical cases and literature shows that the approximate depth of the defect may be determined, for example, by the eddy current method [2]. Unfortunately, it is not used successfully in practice. In addition to this, the estimation of the depth of the defect is more accurate for rails not mechanically processed during operation e.g. grinding. The length of the defect on the surface may correlate less with a crack depth after multiple grinding operations. It was considered to develop an optimized system of ultrasonic transducers dedicated to detecting head check defects. The different degrees of the head of the railhead were taken into account. For this purpose, computed tomography was used as the only available method that allows getting to know the whole geometry of the defect with a required resolution in the entire railhead area.

Terotechnology XII   Materials Research Forum LLC
Materials Research Proceedings **24** (2022) 159-165   https://doi.org/10.21741/9781644902059-24

Determination of the Crack Geometry
In the presented part of the experiment, the relatively advanced defects were considered, referring to the defect catalog with code 2223. They are characteristic of a straight section of the track. These cyclically repeated defects pose a significant risk due to the possibility of multiple rail breakages over several meters. Due to the thickness of the railhead area and taking into account the limited power of the available CT stations, the prepared samples were made to scan at the available lamp power. The real samples and the final results of the 3D geometry of the head check crack are presented in Fig. 1 and Fig. 2.

Fig. 1. Head check defect prepared to CT test

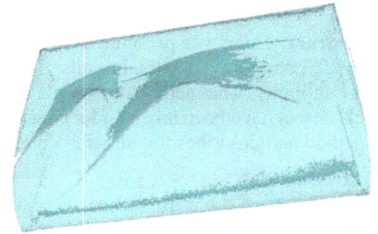

Fig. 2. Visualization of the fracture using editor „myVGL"

First of all, it was noticed that the angle of propagation (α) in relation to the longitudinal axis of the rail was in the range of 35° to 52°. This is a characteristic value for all head check defects. Different sources specify an even wider angular range of crack propagation on the rail surface. It means (35-70)°, depending on the prevailing geometry of the wheel-running surface contact area. In the analyzed cases, the distances between the cracks were usually 3 to 20 mm, which depends on the used material and local operating conditions. The angle of the defect penetration into the material was also characteristic. It was measured as the crack's angular deviation from the normal. Based on several measurements, it was estimated that the value of this angle (β) for about 50% of the cases is in the range of approximately (60-70)°. Due to local conditions and the change in the propagation angle in the depth of the material, the measured angle was from 52° to 105°. The last value indicates that the defect may develop deep into the rail material and also may propagate parallel under the running surface. This may result in a risk of detachment of the running surface.

If the crack penetration angle were measured at the point of surface penetration, this angle would usually fluctuate in the range from 0 to 20°. It would not be a valuable parameter referenced to the rail running surface on which the ultrasonic test heads are placed.

The parameters listed in Table 1 present linear dimensions concerning the depth of the crack under the running surface, measured from the highest point of the railhead. A correlation between the crack length and depth, in this case, was not found. On the other hand, it was proved that the depth of crack penetration ranged from 2 to 10 mm. It impeded ultrasonic testing with typical single transducers and a pulse-echo mode.

The Model of the Head Check Defect
Depending on the required accuracy of the simulation, the expected effects, and the capabilities of the simulation environment, the equivalent defect may have a different nature. The most precise

solution is to convert the radiograph layers into a point cloud readable by the simulation environment. This solution enables to implementation of a geometry of the natural defect, taking into account the imperfections of the tested object. Such an environment includes the FEM method, but the application for modeling the propagation of ultrasonic waves in specific objects encounters significant methodological difficulties and practical limitations. To reach an acceptable simulation accuracy, the distance between nodes should be 1/10 of the wavelength [6].

Table 1. The results of the fracture geometry measurements of the head check defect.

Sample No.	The length of the crack on the running surface [mm]	The depth of the crack under the running surface [mm]	The angle of crack propagation (angle of attack) on the horizontal plane α [deg]	The angle of crack propagation on the vertical plane β [deg]
5_19	35	10	43	69
	53	5.5	43	67-100
5_19_a	27	9.5	48	72
	30	6.5	46	68-105
	17	8	44	65-85
8_19	36	11	46	65-100
	28	6	52	52
19_19_II	10	3	35	88
	20	3	35	85
	27	2.5	35	88

Optimization analyses were performed based on the system environment based on the ray-tracing method to simulate the ultrasonic beam propagation in the rail. The predefined defect patterns were prepared based on the empirical models. The two solutions were applied there. Firstly, the size, simplified geometry, and location were defined after collecting data from CT scans (Fig. 2). Secondly, the representative equivalent flat bottom hole (*EFBH*) was located in the internal space of the rail model (Fig. 3).

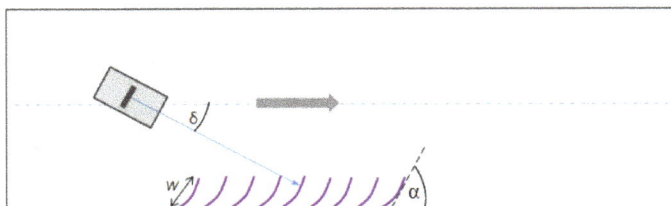

Fig. 2. Schema of detecting the multiple head check defect using surface wave [6].

The industry regulations were established and should be remembered during planning ultrasonic simulations and designing material discontinuities. The sensitivity during ultrasonic testing of rail rails was set as EFBH Ø3mm. Graph samples of received sound pressure are presented in the next chapter.

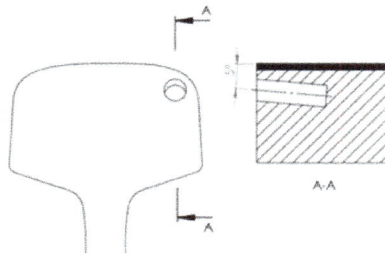

Fig. 3. *Schema of the applied flat bottom hole (FBH).*

Simulation of the Ultrasonic Wave Propagation

The ray-tracing modeling technique allowed the visualization of the ultrasonic wave propagation in whole and cross-section geometry [6]. An example 3D model of testing a rail with the angular head of transverse waves is presented in Fig. 4a. Due to the test object's complicated geometry, the utilization of 3D simulation software is crucial. It allows the presentation of beam propagation after reflections and transformations from non-parallel or curved boundary surfaces. The axis of the beam is visible as a blue line of the transverse wave. It falls to the bottom of the railhead, then it reflects and transforms into a wave of the longitudinal type. Its axis is visible as the red line (Fig. 4b). Both types of waves propagate towards the upper surface of the railhead. Then they are reflected multiple times, and longitudinal wave transformations occur into a transverse wave again. The ultrasonic wave modeling ends after elapsing the assumed time interval, corresponding with the range observation.

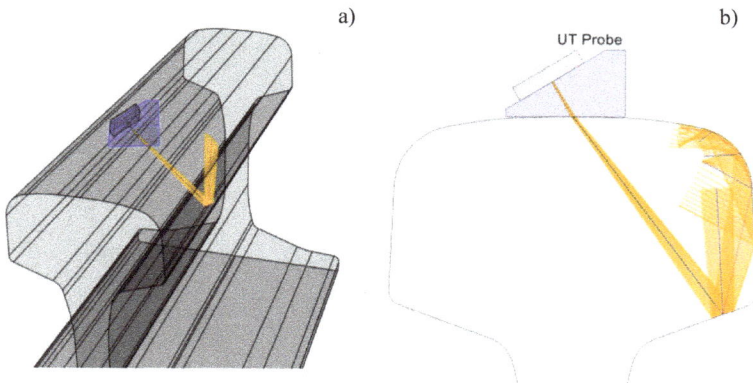

Fig.4. *3D model of the ultrasonic rail testing with a prototype of UT probe.*

The basic limitation of the used simulation environment is its form of the physical foundations based on the ray-tracing model [4]. That means that the ultrasound beam is modeled geometrically as a bundle of rays coming from the center of the transducer. According to the laws of geometry and optics, they are subject to subsequent refractions and reflections from defects or boundary shapes of the tested element. This approach is very beneficial and effective from the point of view of spatial modeling of the wave propagation in the complex shape models. But it does not include the effects of beam diffraction for rays reflected from discontinuity edges [3]. The defect will produce its reflected beam with specific diffraction, and part of that energy will go back to the head of the transducer.

The calculations were based on the ray-tracing model to complete the more advanced and accurate simulation results. The calculation algorithm follows the Huygens principle, performs calculations for all elementary point sources on the transducer surface, and sums the calculated partial pressures at the field point. In this way, we obtain the value of the sound pressure generated at each field point by the entire transducer. Additionally, the used program calculates the defect reflections using the wave solution with the Green's function with Kirchhoff's approximation [5]. It allows taking into account the diffraction divergence of the reflected beams from the model defect. Apart from the finite element method, it is the most accurate theoretical model currently available to calculate the reflection/scattering of ultrasound waves on any model defects.

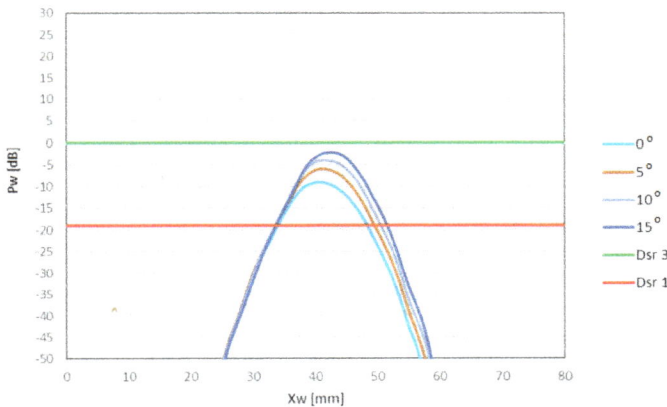

Fig 5. *Echo course of the model defect (DSR 8mm) situated 10mm under the running surface in relation to the deviation from the horizontal plane [6].*

The configuration of the prototype double probe with the longitudinal wave was subjected to numerical tests on the characteristic model defects of advanced head check discontinuities. The model defects DSR 8 mm located at a depth of 10 mm were considered. The variable values in the simulations were the orientation of defects and their distance from the rail axis. Fig. 5 shows the course of the echo envelope of model defects depending on the deviation of their orientation from the horizontal orientation. The Xw value is the distance between the transducer and the crack. The centers of all model defects were at the same distance from the rail axis, equal to 20 mm. The reference reflector, in this case, was a DSR 3 mm, located 50 mm from the head. All model defects gave echo envelopes with a maximum height exceeding the detection threshold by at least 10 dB. As shown in Fig. 6, only defects located less than 20 mm from the rail axis provide useful echoes to detect them.

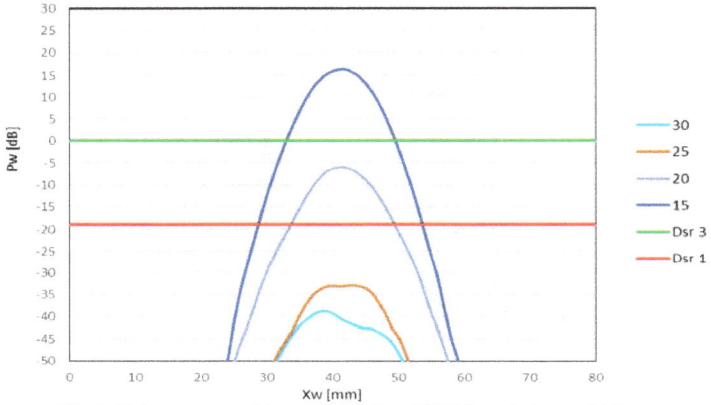

Fig 6. *Echo course of the model defect (DSR 8mm) situated 10mm under the running surface in relation to the horizontal distance between the axis of the defect and the longitudinal axis of the rail [6].*

Summary

Knowing the nature of the fracture and the way of propagation is very helpful in designing model defects. The used software is an essential tool for computer-aided ultrasonic testing of rails. The commercial program (ray tracing) was valuable in projecting the head guiding. It was also helpful to observe the shape of the beam and the wave transformations. In the aftermath, it is well prepared for the initial analysis of various test configurations, especially where different reflections from the rail's boundary surfaces may be important.

The proprietary program used there allows for a much more accurate quantitative modeling of the amplitude distribution of the ultrasonic beam and the waves' reflections from model defects. The diffraction phenomena are considered in the pencil tracing model implemented in the program.

Both programs used together and compared the developed results were a very effective set of software for computer modeling ultrasonic rails tests.

References

[1] D. Kowalczyk, Ł. Antolik, I. Mikłaszewicz, M. Chalimoniuk. Investigations of the head check defects in rails. Materials Research Proceedings 17 (2020) 246-250. https://doi.org/10.21741/9781644901038-36

[2] P. Clayton, M. B. P. Allery, Metallurgical Aspects of Surface Damage Problems in Rails. Canadian Metallurgical Quarterly 21 (1982) 31-46. https://doi.org/10.1179/cmq.1982.21.1.31.

[3] P. Calmon, A. Lhemery, I. Lecoeur-Taıbi, R. Raillon, L. Paradis. Models for the computation of ultrasonic fields and their interaction with defects in realistic NDT

Terotechnology XII
Materials Research Proceedings **24** (2022) 159-165

Materials Research Forum LLC
https://doi.org/10.21741/9781644902059-24

configurations, Nuclear Engineering and Design 180 (1998) 271–283. https://doi.org/10.1016/S0029- 5493(97)00299-9.

[4] N. Gengembre, Pencil method for ultrasonic beam computation, 5th World Congress on Ultrasonics, Paris 2003. [viewed: 2022-01-14]. Available from: http://www.conforg.fr/wcu2003/procs/cd1/articles/000538.pdf

[5] L. W. Schmerr Jr, S-J. Song, Ultrasonic Nondestructive Evaluation Systems. Models and Measurements 1st Ed. Springer, Boston, 2007. https://doi.org/10.1007/978-0-387-49063-2

[6] Z. Ranachowski, S. Mackiewicz, T. Katz. The optimization of the transducers system focused on internal defects detection according to the defect catalog. Research project No. POIR.04.01.01.-00-0011/17. Chapter 3. 2020. Railway Research Institute–Institute of Fundamental Technological Research–Polish Academy of Sciences.

Terotechnology XII
Materials Research Proceedings 24 (2022) 166-173

Materials Research Forum LLC
https://doi.org/10.21741/9781644902059-25

The Model for Analyzing and Improving Aluminum Castings

CZERWIŃSKA Karolina[1,a] *, DWORNICKA Renata [2,b] and PACANA Andrzej[3,c]

[1] Rzeszow University of Technology, The Faculty of Mechanical Engineering and Aeronautics al. Powstańców Warszawy 8, 35-959 Rzeszów,Poland, ORCID: 000-0003-2150-0963

[2] Cracow University of Technology, Faculty of Mechanical Engineering al. Jana Pawła II 37, 31-864, Kraków, Poland, ORCID: 0000-0002-2979-1614

[3] Rzeszow University of Technology, The Faculty of Mechanical Engineering and Aeronautics al. Powstańców Warszawy 8, 35-959 Rzeszów, Poland, ORCID: 0000-0003-1121-6352

[a] k.czerwinska@prz.edu.pl, [b] renata.dwornicka@pk.edu.pl, [c] app@prz.edu.pl

Keywords: Quality Control, Quality Management Tool, Casting Defects

Abstract. The purpose of this study is to develop a model for the analysis and improvement of aluminum castings and to perform its test in the framework of verifying the quality of industrial robot arm parts, and then, determine the causes of potential and root causes of nonconformities, including proposing improvement actions in the production area. Aluminum castings have become the focus of research due to significant problems in maintaining their expected level of quality. The use of the model (consisting of eddy current testing and consecutive brainstorming sessions, Ishikawa diagram, and ABCD - Suzuki method) would contribute to the detection of the causes of nonconformities and consequently to the elimination of non-conforming products. The study demonstrates the desirability of using an integrated approach to solving quality problems. The actions taken were a new solution for the company, as no in-depth analysis of quality problems using a sequence of quality management techniques had been carried out before.

Introduction

The economic progress of recent years has caused both the quality of offered products and services, and ISO standardization have become a priority criterion that determines the success of manufacturing companies [1, 2]. These aspects can support the achievement of key economic, environmental, and social objectives, which in turn helps to facilitate the implications of the concept of sustainability [3]. Equally important is the knowledge and experience of the management regarding the possibilities of increasing the efficiency of work and implemented processes [4].

Quality management in manufacturing enterprises requires control of processes to identify non-conformities appearing in these processes, and their effective improvement, so as to completely eliminate the occurrence of both non-conformities and their causes. In general, quality control consists of checking the conformity of the manufacture of a specific product in terms of meeting the requirements intended for it [5]. The manufacturing process represents the cost of production that is borne by the organization. A key aspect of business operations is controlling these costs. At the moment when there are inconsistencies in the manufactured product, it ceases to meet the requirements of buyers, which generates excessive costs and affects negatively the evaluation of the customer. In such a situation, it is crucial to identify the causes of the quality problem [6]. To this end, the subject literature recommends the use of quality management tools as an effective way to improve quality. Increasing attention has begun to be paid to the use of more quality management instruments in the in-depth analysis of quality problems, thus indicating their universal nature [7, 8]. However, comprehensive and effective quality control methods are still being sought to realize the in-depth causal analysis of production problems.

The objective of this research was to develop an integrally configured model for the analysis and improvement of aluminum castings consisting of the integration of nondestructive testing (eddy current method) and consecutive quality management tools (brainstorming, Ishikawa diagram, ABCD-Suzuki method) for use in post-implementation control of the casting process or inter-operational quality control. The proposed model makes it possible to perform an in-depth analysis of a manufacturing problem and to propose effective improvement actions.

Model for Analyzing and Improving Aluminum Castings

The proposed analysis and improvement model consisted of a sequence of appropriately selected quality management tools that are used to sequentially determine the cause of a problem identified using nondestructive testing (NDT). The tools were brainstorming (BM), Ishikawa diagram, and ABCD - Suzuki method. On the other hand, eddy current (ET) testing was used to diagnose the castings.

When a casting nonconformity is detected and the root cause of the nonconformity is identified, appropriate improvement actions can be taken. Eliminating the cause of the problem will eliminate or reduce product nonconformance. A schematic of the model is shown in Fig. 1.

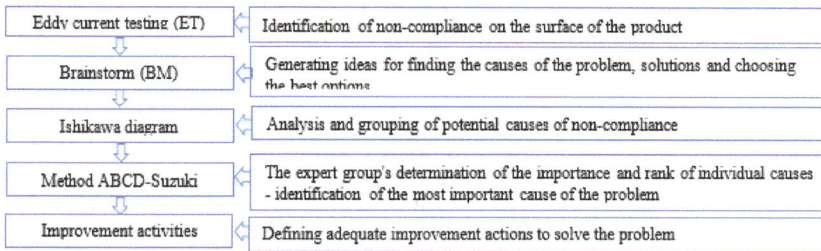

Eddy current testing (ET)	Identification of non-compliance on the surface of the product
Brainstorm (BM)	Generating ideas for finding the causes of the problem, solutions and choosing the best options
Ishikawa diagram	Analysis and grouping of potential causes of non-compliance
Method ABCD-Suzuki	The expert group's determination of the importance and rank of individual causes - identification of the most important cause of the problem
Improvement activities	Defining adequate improvement actions to solve the problem

Fig. 1. *Schematic of a model for analysis and diagnosis of aluminum castings including quality management tools and NDT testing*

Limitations of the model include the subjective approach of the combined techniques. This is a result of the specificity of the techniques used, which are based on the knowledge and experience of the experts appointed to solve the qualitative problem [9]. The issue of the applicability of the eddy current method may also be a limitation. The main limitations of the method include [10, 11]:

– difficulty in testing castings with rough surfaces, non-uniform and irregular shapes,
– difficulty in testing ferromagnetic components due to the so-called skin effect (concentration of eddy current field near the surface),
– limiting the depth of verification of material within the audit,
– difficulty in detecting discontinuities lying parallel to the probe winding course and test direction. (for example, material delamination, cracks),
– vibrations and impacts make it difficult to find defects.

A brief description of the proposed model was developed in six main steps, as presented in the following section.

Step 1. Research subject selection. Considering the applicability of the selected NDT method in the model, the diagnosed test object should have axisymmetric elements or be an axisymmetric product made of conductive material [12]. The object of study may be an aluminum casting, which should be diagnosed for the presence of the most dangerous discontinuities: flat, narrow-slot, and other object discontinuities.

Step 2. Eddy current testing - product quality control. Eddy current defectoscopy was used to inspect the castings for the presence of surface flat discontinuities, narrow-slit discontinuities, and relatively large near-surface surface discontinuities. It is possible to detect cracks with a depth of about 0. 1 mm, a width of about 0. 0005 mm and a length of about 0. 4 mm [13]. Once a product nonconformity has been detected, a characterization of that nonconformity must be made, such as identifying the type (type) of nonconformity and the product material on which the nonconformity occurred.

Step 3. Brainstorm - identify potential causes of non-compliance and step 4. Ishikawa diagram - categorize potential causes of non-compliance. In the study, causes are identified by a panel of experts in a brainstorming (BM) session. A detailed implementation of brainstorming is presented in the literature e. g. [14]. You need to group all the generated causes according to appropriately selected categories, thus taking into account the next step of the method - the Ishikawa diagram. It is effective to use a participatory factor system Ishikawa diagram, which includes the following range of causal categories, i. e. , human, method, machine, material, measurement, management, and environment [15].

Step 4. ABCD method - Suzuki - selection of causes of major inconsistencies. The selection of root causes is done by a panel of experts in a brainstorming (BM) session. All causes placed on the Ishikawa diagram (step 4) are considered. The root causes should be considered those that are as likely as possible to result in noncompliance. The expert team uses the ABCD-Suzuki method [16] to determine the importance and rank of each potential cause. The ABCD method is easy to use and analyze the results because the final result is obtained from simple mathematical operations.

Stage 6. Suggest improvement activities. Once the root causes of a quality problem have been identified, appropriate improvement actions should be determined. That is actions that, when properly implemented, will contribute to the elimination or significant reduction of existing nonconformities. These activities are determined by a panel of experts.

After implementation of the proposed improvement actions, the obtained results should be analyzed and, if necessary, the proposed model should be applied to further nonconformities.

Model Verification and Results
Verification of the proposed model for the analysis and improvement of aluminum castings was carried out against a nonconformity that is relatively often identified in one of the enterprises located in the southern part of Poland.

A part of an industrial robot arm weighing 92 kilograms, gravity cast from AlSi7Mg0. 3 alloy, was tested. It is a product whose specificity depends on the type of purpose of the industrial robot. The selection of the product was due to the declining quality level and the lack of clearly defined reasons for critical nonconformities. Therefore, the purpose of the study was to verify the quality

Terotechnology XII

Materials Research Proceedings **24** (2022) 166-173

Materials Research Forum LLC

https://doi.org/10.21741/9781644902059-25

of the industrial robot arm part, and then, identify the causes of potential and root causes of nonconformities, including proposing improvement actions.

In the next step, quality control of the axially symmetric part of the industrial robot arm was performed. The inspection was performed using non-destructive testing - eddy current testing. The choice of the method of product diagnostics resulted from the specifics of the tested object (material, geometry) and the production process. The presence of oxides in the casting was identified after non-destructive testing was realized. An example of the indications is shown in Fig. 2.

As part of the implementation of the following research steps, potential causes for the presence of oxides in the castings of the robot arm component were determined. To this end, a team of experts conducted brainstorming sessions. The generated potential causes were grouped and visualized in an Ishikawa diagram. The result of the analyses is shown in Fig. 3.

According to the fifth step of the analysis, the expert team applied the ABCD-Suzuki method to determine the importance and rank of each potential cause. All causes placed on the Ishikawa diagram are considered. The analysis result sheet is shown in Table 1.

As a result of the conducted analyses it turned out that the three most significant reasons for the occurrence of nonconformities (the presence of oxides) in the casting (in order of importance) were: inappropriate temperature of pouring the mold (too low temperature), impurities in the liquid melt and improper ventilation of the mold.

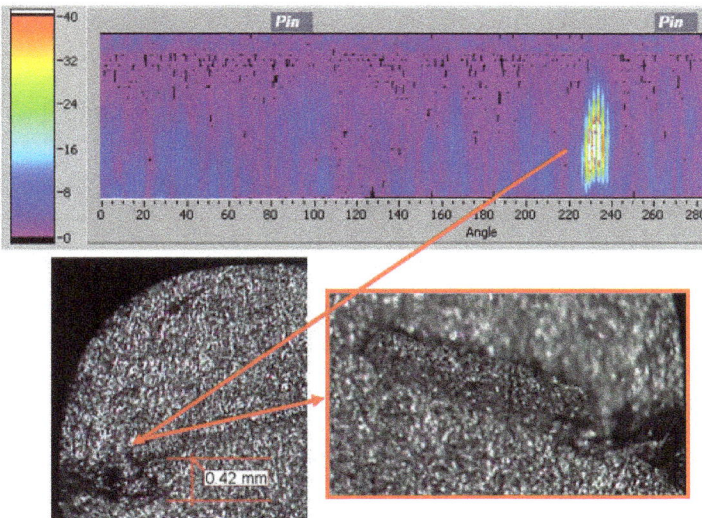

Fig. 2. *Example of oxide presence in a casting*

Materials Research Forum LLC

https://doi.org/10.21741/9781644902059-25

METHOD   **MACHINE**   **MAN**

low-intensity feeding of individual casting zones — inadequate filling system — failure to take account of material specificity in delivery control

pouring molds — furnace renovation — rush

quality of foundry molds

order of pouring the moulds — too low a pouring temperature in the mould — poor venting of the form — inadequate supervisio — non-compliance with the position instructions

failure to remove impurities from the liquid metal mirror — a large number of tasks to be performedn

inadequate melting of the metallurgical charge — moisture of the modifier — outdated measurement methods

poor batch quality — lack of adequate instructions at the place of use — atmospheric factors

impurities in the liquid melt — lack of compliance with the specifications — amount of additives used — lack of an adequate motivation system — layout of the production hall

METERIAL   **MANAGEMENT**   **SURROUNDINGS**

Presence of oxide inclusions

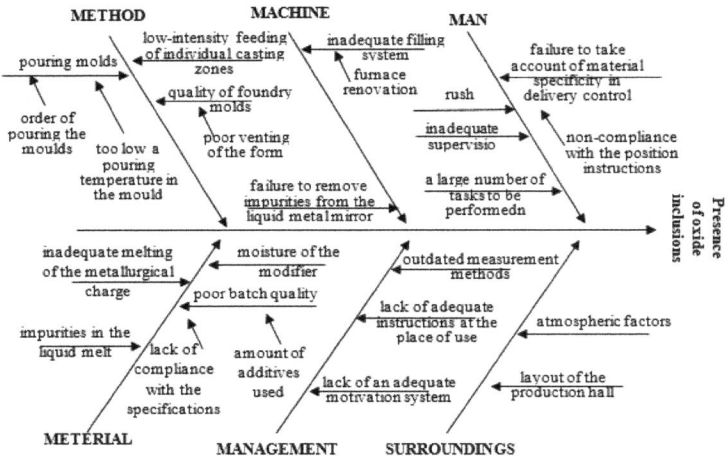

Fig. 3. Ishikawa diagram for the problem of the presence of oxides in a casting

Table 1. *Summary of the results of the ABCD-Suzuki method for the problem of the presence of oxides in a casting*

Cause of non-compliance	The rank of the criteria										The adjusted sum of meanings	The number of undeleted answers	Rank indicator	Rank
	1	2	3	4	5	6	7	8	9	10				
Bad feed quality	3	3	6	2	1						31		2.38	4
Low-intensity supply of individual zones of the casting	1	5	5	2		1	0				34		2.62	5
Bad mold bleeding	0	5	2	6	0						40		3.08	6
The temperature of pouring the mold is too low	5	5	1	1		1	0				28		2.15	1
Impurities in the liquid melt	4	4	3	3	2						29		2.23	2
Moisture in the modifier	3	2	4	1		3	0				41		3.15	8
Inadequate melt of the metallurgical charge	0	4	4	4	2	1					41		3.15	10
Bad mold bleeding	2	6	4	2	1						30	13	2,31	3
Not removing the scum from the liquid metal mirror	1	2	5	6	5						40		3,08	7
Failure to comply with the workplace instructions	0	3	6	3	2	1					41		3,15	9
Inadequate supervision	0	1	7	1	3	2		0			52		4,0	14
Obsolete measurement methods		3	6	2	1	1	0				43		3,31	11
Lack of proper instructions at the place of use	0	3	6	3		1			0		42		3,32	12
Lack of an adequate motivation system	0	2	5	4	2		0				45		3,46	13

According to step six, improvement actions were proposed according to the indicated root cause of the verified problem. It was concluded that up-to-date and adequate job instructions should be developed and that additional training directed at foundry department employees is recommended. Additionally, it is recommended that the effectiveness of these actions be periodically reviewed. Subsequently, improvement actions can be taken for further nonconformities.

Conclusion

In this paper, a model for casting analysis and refinement is proposed and its test is performed. The purpose of the study was to verify the quality of the industrial robot arm parts, and then, to identify the causes of potential and root causes of nonconformities, including proposing improvement actions in the manufacturing area.

Using eddy current testing, the most serious type of nonconformity – the presence of oxides in the casting – was detected. The presence of an identified nonconformity disqualifies the product. To characterize the problem, the expert team completed a brainstorming session to isolate potential causes of noncompliance. An Ishikawa diagram was drawn to classify the potential causes of nonconformance. To indicate the importance and rank of each potential cause, the ABCD – Suzuki method was used, according to which the main cause of nonconformity was the inadequate temperature of pouring the mold (too low temperature), impurities in the liquid melt, and improper venting of the mold.

The applied model for the analysis and improvement of aluminum castings combines the diagnostic testing method in combination with quality management methods, which are largely complementary. The model presented can be useful in terms of methods to support quality management processes.

The model presented in this paper may be useful in many areas of research (tribological tests [17], biotechnology [18], waste treatment [19-21]) and industrial production (special coatings [22], laser surface treatment [23, 24], machine regeneration [25], power hydraulics [26]), where castings may appear as part of devices. Regardless, it is a development impulse both for the methods of experimental data analysis [27-29] and for the analysis of possible failure scenarios [40, 41].

References

[1] U. Mentel, M. Hajduk-Stelmachowicz. Does standardization have an impact on innovation activity in different countries? Problems and Perspectives in Management, 18(4) 2020. https://doi.org/10.21511/ppm.18(4).2020.39

[2] G. Ostasz, K. Czerwińska, A. Pacana, A. Quality management of aluminum pistons with the use of quality control points, Management Systems In Production Engineering 28 (2020) 29-33. https://doi.org/10.2478/mspe-2020-0005

[3] M. Hajduk-Stelmachowicz, External Benefits of Organisational Eco-Innovations Functioning as Results of Strategic Choices, Handel Wewnętrzny 368 (2017) 132-141.

[4] R, Wolniak, B. Skotnicka. Metody i narzędzi zarządzania jakością. Teoria i praktyka. Politechnika Śląska, Gliwice, 2005.

[5] A. Pacana, K. Czerwińska. Improving the quality level in the automotive industry, Production Engineering Archives 26 (2020) 162-166. https://doi.org/10.30657/pea.2020.26.29

[6] A. Ahsen. Cost oriented failure mode and effects analysis, International Journal of Quality & Reliability Management 25 (2008) 466-476. https://doi.org/10.1108/02656710810873871

[7] R. Bris. Cost oriented statistical decision
problem in acceptance sampling and quality control, Applied Mathematics in Engineering and
Reliability – Proc. 1st Int. Conf. Applied Mathematics in Engineering and Reliability, ICAMER
2016, 19-26. https://doi.org/10.1201/b21348-5

[8] A. Pacana, A. Gazda, D. Malindzak, R. Stefko. Study on improving the quality of stretch
film by Shainin method, Przemysł Chemiczny 93 (2014) 243-246.
https://doi.org/10.12916/przemchem.2014.243

[9] A. Pacana, K. Czerwinska, L. Bednárová. Comprehensive improvement of the surface
quality of the diesel engine piston, Metalurgija 58(3-4) 2019, pp. 329-332.

[10] A.N. AbdAlla, M.A. Faraj, F. Samsuri, D. Rifai, K. Ali, Y. Al-Douri. Challenges in
improving the performance of eddy current testing: Review. Measurement & Control 52 (2019)
52-64. https://doi.org/10.1177/0020294018801382

[11] O. Beer, J. Fössel. Experience in the Eddy Current Testing of Rolling Element Bearing
Components. HTM-Journal of Heat Treatment and Materials 75 (2020) 236-247.
https://doi.org/10.3139/105.110414

[12] M. Pelkner, R. Casperson, R. Pohl, D. Münzke, B. Becker. Eddy current testing of
composite pressure vessels. Int. J. Appl. Electromagn. Mech. 59 (2019) 1221-1226.
https://doi.org/10.3233/JAE-171044

[13] Z-C. Wang, J-W. Lu. Pulse Eddy Current Testing Thin Metal Thickness. Proc. 2018 5th Int.
Conf. Information Science and Control Engineering, ICISCE 2018, 1176-1179.
https://doi.org/10.1109/ICISCE.2018.00241

[14] V. Putman, P. Paulus. Brainstorming, Brainstorming Rules and Decision Making, Journal of
Creative Behavior 43 (2009) 29-40. https://doi.org/10.1002/j.2162-6057.2009.tb01304.x

[15] K. Czerwińska, P. Bełch, M. Hajduk-Stelmachowicz, D. Siwiec, A. Pacana. Doskonalenie
procesu grafitowania wyrobów aluminiowych, Przemysł Chemiczny 100 (2021) 1191-1193.
https://doi.org/10.15199/62.2021.12.8

[16] M. Orlik, K. Knop. The use of the ABCD-Suzuki method to rank the needs and
requirements of customers in relation to the selected electronic product, Archiwum Wiedzy
Inżynierskiej 4 (2019) 22-25.

[17] J. Bronček, P. Fabian, N. Radek. Tribological research of properties of heat-treated cast
irons with globular graphite, Materials Science Forum 818 (2015) 209-212.
https://doi.org/10.4028/www.scientific.net/MSF.818.209

[18] E. Skrzypczak-Pietraszek. Phytochemistry and biotechnology approaches of the genus
Exacum. In: The Gentianaceae - Volume 2: Biotechnology and Applications, 2015, 383-401.
https://doi.org/10.1007/978-3-642-54102-5_16

[19] M. Zenkiewicz, T. Zuk, J. Pietraszek, P. Rytlewski, K. Moraczewski, M. Stepczyńska.
Electrostatic separation of binary mixtures of some biodegradable polymers and poly(vinyl
chloride) or poly(ethylene terephthalate), Polimery/Polymers 61 (2016) 835-843.
https://doi.org/10.14314/polimery.2016.835

[20] E. Radzyminska-Lenarcik, R. Ulewicz, M. Ulewicz. Zinc recovery from model and waste solutions using polymer inclusion membranes (PIMs) with 1-octyl-4-methylimidazole, Desalination and Water Treatment 102 (2018) 211-219. https://doi.org/10.5004/dwt.2018.21826

[21] M. Ulewicz, A. Pietrzak. Properties and structure of concretes doped with production waste of thermoplastic elastomers from the production of car floor mats. Materials 14 (2021) art. 872. https://doi.org/10.3390/ma14040872

[22] N. Radek, J. Pietraszek, A. Gadek-Moszczak, Ł.J. Orman, A. Szczotok. The morphology and mechanical properties of ESD coatings before and after laser beam machining, Materials 13 (2020) art. 2331. https://doi.org/10.3390/ma13102331

[23] Ł.J. Orman Ł.J., N. Radek, J. Pietraszek, M. Szczepaniak. Analysis of enhanced pool boiling heat transfer on laser-textured surfaces. Energies 13 (2020) art. 2700. https://doi.org/10.3390/en13112700

[24] N. Radek, J. Konstanty, J. Pietraszek, Ł.J. Orman, M. Szczepaniak, D. Przestacki. The effect of laser beam processing on the properties of WC-Co coatings deposited on steel. Materials 14 (2021) art. 538. https://doi.org/10.3390/ma14030538

[25] S. Marković, D. Arsić, R.R. Nikolić, V. Lazić, B. Hadzima, V.P. Milovanović, R. Dwornicka, R. Ulewicz. Exploitation characteristics of teeth flanks of gears regenerated by three hard-facing procedures, Materials 14 (2021) art. 4203. https://doi.org/10.3390/ma14154203

[26] Barucca G. et al. PANDA Phase One: PANDA collaboration. European Physical Journal A 57 (2021) art. 184. https://doi.org/10.1140/epja/s10050-021-00475-y

[27] T. Styrylska, J. Pietraszek. Numerical modeling of non-steady-state temperature-fields with supplementary data. Zeitschrift für Angewandte Mathematik und Mechanik 72 (1992) T537-T539.

[28] J. Pietraszek. Response surface methodology at irregular grids based on Voronoi scheme with neural network approximator. 6th Int. Conf. on Neural Networks and Soft Computing JUN 11-15, 2002, Springer, 250-255. https://doi.org/10.1007/978-3-7908-1902-1_35

[29] J. Pietraszek, R. Dwornicka, A. Szczotok. The bootstrap approach to the statistical significance of parameters in the fixed effects model. ECCOMAS 2016 – Proc. 7th European Congress on Computational Methods in Applied Sciences and Engineering 3, 6061-6068. https://doi.org/10.7712/100016.2240.9206

[30] J. Jura, M. Ulewicz. Assessment of the Possibility of Using Fly Ash from Biomass Combustion for Concrete. Materials 14 (2021) art. 6708. https://doi.org/10.3390/ma14216708

[31] A. Kubecki, C. Śliwiński, J. Śliwiński, I. Lubach, L. Bogdan, W. Maliszewski. Assessment of the technical condition of mines with mechanical fuses, Technical Transactions 118 (2021) art. e2021025. https://doi.org/10.37705/TechTrans/e2021025

Terotechnology XII
Materials Research Proceedings **24** (2022) 174-180

Materials Research Forum LLC
https://doi.org/10.21741/9781644902059-26

Improving the Non-Destructive Testing Process of the Outer Bearing Ring

SIWIEC Dominika[1,a] , DWORNICKA Renata [2,b] and PACANA Andrzej [1,c]

[1]Rzeszow University of Technology, The Faculty of Mechanical Engineering and Aeronautics, al. Powstancow Warszawy 12, 35-959, Rzeszow, Poland

[2]Cracow University of Technology, Faculty of Mechanical Engineering, al. Jana Pawła II 37, 31-864 Krakow, Poland

[a]d.siwiec@prz.edu.pl, [b]renata.dwornicka@pk.edu.pl, [c]app@prz.edu.pl

Keywords: Mechanical Engineering, Bearing Housing, Non-Destructive Test, Quality Management

Abstract. The aim was to identify the unconformities in the product with the magnetic-powder method and extension of the analysis process about selected quality management techniques in order to identify the root of unconformities. The product subject to magnetic-powder testing was the outer ring of the four-point ball bearing. On the product, the cracks were identified, so in order to point to the root of unconformities, the techniques like Ishikawa diagram and the 5Why method were implemented. Analyses showed that the source of the cracks on the outer bearing ring was defective material from the supplier. The proposed process of using quality management techniques together with non-destructive testing can be used in any enterprise to detect the unconformities of the products and the reasons for their creation.

Introduction

Performing quality analysis of the products is a key stage in the product creation and improvement process. An important part of the quality analysis has non-destructive tests (NDT), which allow identifying the nonconformities of the product without its destructive [1]. It is very important in case of the mechanical products; which production is expensive. Although the NDT is effective in identifying nonconformities of the product, they don't indicate the root cause of nonconformities [2, 3]. So, improving the process of the NDT about the techniques which allow identifying the root of unconformities is a challenge for enterprises. A review of the literature on the subject indicates that the improving process of NDT research used other methods of research, simulations, and industrial robots [4-8] and moderated the substances, equipment, and test parameters used in NDT research [9-11]. Also, a few NDT methods in one research were used [12-14]. However, in this research, after detecting unconformities, no activities were performed to identify the source of the unconformities. Only focused on the effective identify the unconformities by NDT methods, but not a complex analysis of the problem. Therefore, it was purposeful to improve the non-destructive testing process by using quality management techniques in order to identify the source of unconformities. The problem lack of identifying the root of unconformities detected by NDT methods efforts was made in an enterprise located in south-eastern Poland.

In the enterprise, the unit control of the product with NDT methods (magnetic-powder and fluorescent) was made. In view of unit control of research after identifying the unconformities the additional analyses in order to identify the root of the problem were not made. This was due to the fear of time-consuming analyzes for a large number of different types of unit products. The results of NTD research showed repeated types of unconformities (including cracks), the

Terotechnology XII   Materials Research Forum LLC
Materials Research Proceedings 24 (2022) 174-180   https://doi.org/10.21741/9781644902059-26

number of which increased. However, in the enterprise, the quality analyses in terms of the type of unconformities not taken into account, because only the unit character of testing products was focused. Therefore, it was considered appropriate to extend the product analysis process by implementing quality management techniques for the types of unconformities identified. The aim was to identify the unconformities on the product with used the magnetic-powder method and extension of the analysis process about selected quality management techniques in order to identify the root of unconformities. The tested product was the outer ring of the four-point ball bearing. At the request of the client, the product was tested using the magnetic-powder method, after which the unconformities (cracks) were identified. In order to identify the root of cracks, the process of analysis of the product was expanded to the Ishikawa diagram and 5Why method. The main causes of the problem were selected by the Ishikawa diagram. These causes were analyzed by using the 5Why method and pointed out the root of the problem which was defective material from the supplier.

Material and Method
The outer ring of the four-point ball bearing was analyzed. Choice of product to analysis was conditioned by individual preferences of quality control manager, who claimed that earlier research made on these types of products showed different unconformities, mainly cracks. The outer ring of the four-point ball bearing was made from 418 alloy (i.e. 5616). It is a chrome-tungsten-nickel stainless steel alloy that is used for products subjected to heavy loads up to 649 °C. The outer ring of the four-point ball bearing as well as ball bearings in general is applicable in virtually every technical field, for example automotive, machinery, and light industry [15]. So, improving the NDT process on the basis of the outer bearing ring has considered important.

In the enterprise, the research was made with used NDT methods – fluorescent and magnetic-powder methods. The selection of methods for the tested products depended on individual customer requirements and the type of material. In the case of the analyzed product (outer bearing ring) the outer customer ordered to carry out analysis with the magnetic-powder method, which was adequate because this method has been applied to ferromagnetic products. This method has been applied to ferromagnetic material [16]. It consists in applying magnetic powder and inducing magnetic flux in material discontinuities [17]. The magnetic-powder method is considered the most sensitive, reliable, and efficient NDT method. In this method, color and fluorescent techniques are used. This method has been applied for example in the aviation, automotive, and foundry industries [18, 19].

The research was carried out with a MAG 50 magnetic flaw detector. This detector was powered by rectified 3-phase alternating current. The cleanliness of the product was checked (after washing operation and eventually magnetic residue). On the MAG 50 the destination of current flow and magnetic field force lines were set at 17 kJ. After fixing the product in the detector, the surface of the product was poured with magnetic suspension (Chemetall – MPI Diluent (HF)). Three magnetic pulses turned on (current 1600 A, time 0.5 s). The first and second magnetic pulses were turned on at a time when on the product the magnetic suspension was poured and the third without poured. These actions were repeated three times (every 120 s). Contact field value in magnetic time was measure and it was ≥ 2.4 kA/m. The product was demagnetized after inspection and the results recorded. On the outer bearing ring, the cracks were identified. Because, the root cause of the problem was not known, further analysis was made. In order to identify the potential causes of the problem, the Ishikawa diagram was prepared.

Techniques implemented in NDT process were Ishikawa diagram and *5Why* method. The choice these techniques to improving the process of NDT was conditioned their effectiveness in identifying the root of the problem. Added, these methods are simple in realization and not expensive. Application these methods in a sequence way allow finding the potential, main and root causes of the problem [2, 3, 20].

Ishikawa diagram is called a fishbone diagram or causes and effects diagram. Allows to graphic presentation the problem and causes which has influence on its uprising. The division of causes into categories helps in the analysis of the problem. Basic categories are rule 5M+E, i.e. method, machine, material, man, management and environment. Using this diagram to identify potential causes of the problem, among which the root causes can be selected [2, 21-23]. In the central part of the diagram, the problem (cracks) was noted. From main categories (5M+E) were selected: man, method, material, management and environment. Category like method was omitted, because it did not apply. To these categories the potential causes of cracks were noted. The main causes of the problem were selected and analyzed by *5Why* method.

The *5Why* method otherwise called *Why-Why* diagram has applies to identify the root of the problem. Analyzing the problem using the 5Why method is to ask the question "why?". The end of method is when the answer is exhausted and the source of the problem is found. The *5Why* method allows taking actions adequate to the root of problem in order to eliminate or minimalize the problem [3, 24]. It was started from the problem and main causes (pointed by Ishikawa diagram). The "Why?" question was asked sequentially until the root cause of the problem was identified. After identified the root cause of the problem (defective material from supplier) the improvement actions were proposed.

Results
After the magnetic-powder testing on the outer bearing ring the unconformities were identified (cracks), an example of which is shown in Fig. 1.

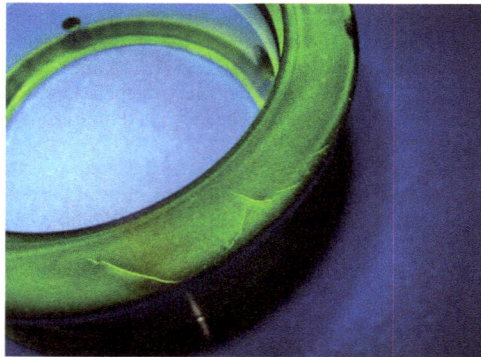

Fig. 1. Cracks on the outer bearing ring.

Next, in order to show the potential causes of the cracks problem on the outer bearing ring the Ishikawa diagram was prepared, which is shown in figure 2. The main causes of the problem were selected, i.e. defective material and lack of knowledge about material supplier. In order to identify the root cause of cracks the 5Why method was made, which is shown in figure 3. The root cause of the problem was defective material from the supplier. The improvement action was informed the customer about the root cause of the problem.

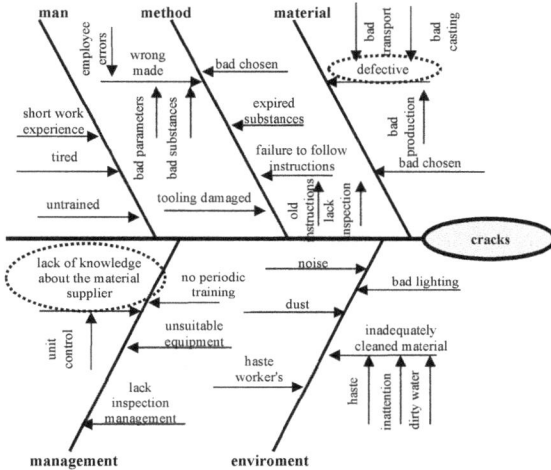

Fig. 2. *The Ishikawa diagram for the cracks problem on the outer bearing ring.*

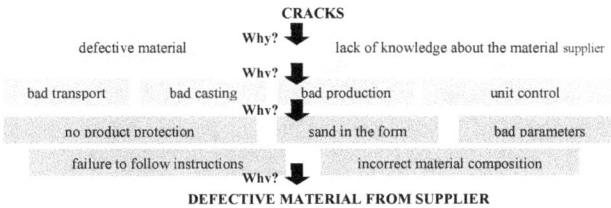

Fig. 3. *The 5Why method for the cracks problem on the outer bearing ring.*

Summary and Conclusion

Making the effective quality control of the product allows developing the product, enterprise and satisfaction of the customer. The desire to improve the process of NDT research has been demonstrated by an enterprise located in south-eastern Poland. In the enterprise the quality management techniques i.e. Ishikawa diagram and *5Why* method were implemented after NDT research. The aim was to identify the unconformities on the product with used the magnetic-powder method and extension of the analysis process about selected quality management techniques in order to identify the root of unconformities. The outer ring of the four-point ball bearing was analyzed. By the magnetic-powder method, the cracks on the product were identified. Expanding the process of NDT analyze the Ishikawa diagram was used, by which the main causes of the cracks were selected, i.e. defective material and lack of knowledge about the material supplier. Next, used the *5Why* method, which showed the root cause of unconformity i.e. defective material from the supplier. The proposed process of using quality management techniques together

Terotechnology XII   Materials Research Forum LLC
Materials Research Proceedings **24** (2022) 174-180   https://doi.org/10.21741/9781644902059-26

with non-destructive testing can be used in any enterprise to detect unconformities of the products and the reasons for their creation.

Non-destructive testing methods [25] are very important in many areas of research (special alloys [26], implants [27], tribological tests [28], welding [29] and biotechnology [30]) and industrial production (special coatings [31, 32], laser surface treatment [33, 34], machine regeneration [35], power hydraulics [36, 37], steel industry [38]), so the presented method may be inspiring for many recipients. Regardless, it is a development impulse both for the methods of experimental data analysis [39-41] and for the analysis of possible failure scenarios [42].

Reference

[1] M. Zielińska, M. Rucka. Non-destructive assessment of masonry pillars using ultrasonic tomography. Materials 11 (2018) art. 2543. https://doi.org/10.3390/ma11122543

[2] A. Pacana, D. Siwiec, L. Bednárová. Analysis of the incompatibility of the product with fluorescent method. Metalurgija 58 (2019) 337-340.

[3] A. Pacana, A. Radon-Cholewa, J. Pacana J. The study of stickiness of packaging film by Shainin method. Przemysl Chemiczny 94 (2015) 1334-1336.

[4] W. Swiderski. Non-destructive testing of light armours of CFRP after ballistic impacts by IR thermography methods. Composite Structures 224 (2019) art. 111086.

[5] C. Guo, C. Xu, J. Hao, D. Xiao, W. Yang. Ultrasonic Non-Destructive Testing System of Semi-Enclosed Workpiece with Dual-Robot Testing System. Sensors 19 (2019) art. 3359. https://doi.org/10.3390/s19153359

[6] I. Kryukov, S. Boehm. Prospects and limitations of eddy current shearography for non-destructive testing of adhesively bonded structural joints. Journal of Adhesion 95 (2019) 874-886.

[7] R. Mulaveesala, V. Arora, A. Rani. Coded thermal wave imaging technique for infrared non-destructive testing and evaluation. Nondestructive Testing and Evaluation 34 (2019) 243-253.

[8] A. Imperiale, N. Leymarie, T. Fortuna. Coupling Strategies Between Asymptotic and Numerical Models with Application to Ultrasonic Non-Destructive Testing of Surface Flaws. Journal Of Theoretical And Computational Acoustics 27 (2019) art. 1850052. https://doi.org/10.1142/S2591728518500524

[9] K. Trieb, J. Glinz, M. Reiter. Non-Destructive Testing of Ceramic Knee Implants Using Micro-Computed Tomography. Journal of Arthroplasty 34 (2019) 2111-2117.

[10] M. Sofi, Y. Oktavianus, E. Lumantarna. Condition assessment of concrete by hybrid non-destructive tests. Journal of Civil Structural Health Monitoring 9 (2019) 339-351.

[11] E. Jasiuniene, L. Mazeika, V. Samaitis. Ultrasonic non-destructive testing of complex titanium/carbon fibre composite joints. Ultrasonics 95 (2019) 13-21.

[12] H. Rathod, R. Gupta. Sub-surface simulated damage detection using Non-Destructive Testing Techniques in reinforced-concrete slabs. Construction and Building Materials 215 (2019) 754-764.

[13] A.N. Hoshyar, M. Rashidi, R. Liyanapathirana. Algorithm Development for the Non-Destructive Testing of Structural Damage. Applied Sciences 9 (2019) art. 2810. https://doi.org/10.3390/app9142810

[14] S. Farhangdoust, A. Mehrabi. Health Monitoring of Closure Joints in Accelerated Bridge Construction: A Review of Non-Destructive Testing Application. Journal of Advanced Concrete Technology 17 (2019) 381-404.

[15] Alloy 418. https://www.neonickel.com/pl/alloys/stale-nierdzewne/greek-ascoloy-alloy-418/ (access: 16.09.2019).

[16] L. Sozański. Welded joints magnetic-particle inspection standards. Welding Review 10 (2012) 58-60.

[17] L. Kloskowski, S. Pałubicki, K. Kukiełka. Non-destructive testing of welded joints on the exemplary selected structural elements of the wind turbines. Coaches 6 (2015) 113-122.

[18] P. Zientek. Non-destructive testing methods for selected elements of small power turbogenerators. Electrical Machines - Problem Notebooks 3 (2016) 115-120.

[19] J. Krysztofik, W. Manaj. Non-destructive testing technology application in aviation. Prace Instytutu Lotnictwa 211 (2011) 120-129.

[20] R. Ulewicz. Quality control system in production of the castings from spheroid cast iron. Metalurgija 42 (2003) 61-63.

[21] R. Wolniak. Application methods for analysis car accident in industry on the example of power. Support Systems In Production Engineering 6 (2017) 34-40.

[22] A. Pacana, D. Siwiec, L. Bednárová. Method of Choice: A Fluorescent Penetrant Taking into Account Sustainability Criteria. Sustainability 12 (2020) art. 5854. https://doi.org/10.3390/su12145854

[23] B. Skotnicka-Zasadzien, R. Wolniak, M. Zasadzien. Use of quality engineering tools and methods for the analysis of production processes – case study. Proc. 2nd Int. Conf. on Economic and Business Management "AEBMR-Advances in Economics Business and Management Research". Shanghai – 33 (2017) 240-245.

[24] A. Pacana, L. Bednarova, I. Liberko. Effect of selected production factors of the stretch film on its extensibility. Przemysl Chemiczny 93 (2014) 1139-1140.

[25] A. Pacana, D. Siwiec, L. Bednárová. Method of choice: A fluorescent penetrant taking into account sustainability criteria, Sustainability 12 (2020) art. 5854. https://doi.org/10.3390/su12145854

[26] A. Dudek, B. Lisiecka, R. Ulewicz. The effect of alloying method on the structure and properties of sintered stainless steel, Archives of Metallurgy and Materials 62 (2017) 281-287. https://doi.org/10.1515/amm-2017-0042

[27] A. Dudek, M. Klimas. Composites based on titanium alloy Ti-6Al-4V with an addition of inert ceramics and bioactive ceramics for medical applications fabricated by spark plasma sintering (SPS method), Materialwissenschaft und Werkstofftechnik 46 (2015) 237-247. https://doi.org/10.1002/mawe.201500334

[28] J. Bronček, P. Fabian, N. Radek. Tribological research of properties of heat-treated cast irons with globular graphite, Materials Science Forum 818 (2015) 209-212. https://doi.org/10.4028/www.scientific.net/MSF.818.209

[29] I. Miletić, A. Ilić, R.R. Nikolić, R. Ulewicz, L. Ivanović, N. Sczygiol. Analysis of selected properties of welded joints of the HSLA Steels, Materials 13 (2020) art.1301. https://doi.org/10.3390/ma13061301

[30] E. Skrzypczak-Pietraszek. Phytochemistry and biotechnology approaches of the genus *Exacum*. In: The Gentianaceae - Volume 2: Biotechnology and Applications, 2015, 383-401. https://doi.org/10.1007/978-3-642-54102-5_16

[31] A. Szczotok, N. Radek, R. Dwornicka. Effect of the induction hardening on microstructures of the selected steels. METAL 2018 - 27th Int. Conf. Metall. Mater. (2018), Ostrava, Tanger 1264-1269.

[32] N. Radek, J. Pietraszek, A. Gadek-Moszczak, Ł.J. Orman, A. Szczotok. The morphology and mechanical properties of ESD coatings before and after laser beam machining, Materials 13 (2020) art. 2331. https://doi.org/10.3390/ma13102331

[33] Ł.J. Orman Ł.J., N. Radek, J. Pietraszek, M. Szczepaniak. Analysis of enhanced pool boiling heat transfer on laser-textured surfaces. Energies 13 (2020) art. 2700. https://doi.org/10.3390/en13112700

[34] N. Radek, J. Konstanty, J. Pietraszek, Ł.J. Orman, M. Szczepaniak, D. Przestacki. The effect of laser beam processing on the properties of WC-Co coatings deposited on steel. Materials 14 (2021) art. 538. https://doi.org/10.3390/ma14030538

[35] S. Marković, D. Arsić, R.R. Nikolić, V. Lazić, B. Hadzima, V.P. Milovanović, R. Dwornicka, R. Ulewicz. Exploitation characteristics of teeth flanks of gears regenerated by three hard-facing procedures, Materials 14 (2021) art. 4203. https://doi.org/10.3390/ma14154203

[36] E. Lisowski, J. Rajda, G. Filo, P. Lempa. Flow Analysis of a 2URED6C Cartridge Valve, Lecture Notes in Mechanical Engineering 24 (2021) 40-49. https://doi.org/10.1007/978-3-030-59509-8_4

[37] Barucca G. et al. PANDA Phase One: PANDA collaboration. European Physical Journal A 57 (2021) art. 184. https://doi.org/10.1140/epja/s10050-021-00475-y

[38] A. Maszke, R. Dwornicka, R. Ulewicz. Problems in the implementation of the lean concept at a steel works – Case study, MATEC Web of Conf. 183 (2018) art.01014. https://doi.org/10.1051/matecconf/201818301014

[39] T. Styrylska, J. Pietraszek. Numerical modeling of non-steady-state temperature-fields with supplementary data. Zeitschrift für Angewandte Mathematik und Mechanik 72 (1992) T537-T539.

[40] J. Pietraszek. Response surface methodology at irregular grids based on Voronoi scheme with neural network approximator. 6th Int. Conf. on Neural Networks and Soft Computing JUN 11-15, 2002, Springer, 250-255. https://doi.org/10.1007/978-3-7908-1902-1_35

[41] J. Pietraszek, R. Dwornicka, A. Szczotok. The bootstrap approach to the statistical significance of parameters in the fixed effects model. ECCOMAS 2016 – Proc. 7th European Congress on Computational Methods in Applied Sciences and Engineering 3, 6061-6068. https://doi.org/10.7712/100016.2240.9206

[42] G. Filo, J. Fabiś-Domagała, M. Domagała, E. Lisowski, H. Momeni. The idea of fuzzy logic usage in a sheet-based FMEA analysis of mechanical systems, MATEC Web of Conf. 183 (2018) art.3009. https://doi.org/10.1051/matecconf/201818303009

Terotechnology XII
Materials Research Proceedings 24 (2022) 181-188

Materials Research Forum LLC
https://doi.org/10.21741/9781644902059-27

Solving Critical Quality Problems by Detecting and Eliminating their Root Causes – Case-Study from the Automotive Industry

KNOP Krzysztof[1,a] * and ULEWICZ Robert[1,b]

[1] Department of Production Engineering and Safety, Faculty of Management, Czestochowa University of Technology, Al. Armii Krajowej 19b, 42-218 Czestochowa, Poland

[a] krzysztof.knop@wz.pcz.pl, [b] robert.ulewicz@wz.pcz.pl

Keywords: Quality Problems, Root Cause Analysis, Quality Tools, Automotive Industry

Abstract. The article presents the results of using selected quality tools to detect and analyze critical quality problems of low-pressure hoses for car radiators and their root causes for their elimination. The research aims to analyze the nonconformities of the tested product, identify critical nonconformities in terms of frequency of occurrence, and analyze potential causes of their creation to propose effective corrective and preventive actions. The paper presents the use of such quality tools as the Pareto-Lorenz diagram, Ishikawa diagram, and 5 WHY analysis. It was identified that the critical nonconformities of low-pressure hose that should be addressed first are nonconforming insert mounting depth and crooked cut detail. It has been shown that most of the factors causing critical nonconformances are in the "man" and "machine" type of problem areas. They identified the root causes of the quality problems, which were the incorrect setting of the machine by the operator and a lack of frequent inspections of the machine's technical condition. It was proposed corrective actions to eliminate the possibility of critical nonconformities due to their root causes.

Introduction

Effective satisfying customers' needs and requirements is one of the most critical objectives in any manufacturing enterprise, conditioning its success in the market [1]. The goal of any enterprise should be to make money by their meeting and even exceeding [2]. The processes of continuous globalization and the dynamics of change have established new challenges for enterprises. Changes in the perspective of quality perception are a consequence of its development. Quality is interpreted as perfection or excellence [3]. However, in reality, quality measurement is complex and focuses on the process of reaching it [4]. Quality is defined as a set of product features that determine how the product will meet customer expectations [4]. Product features, in turn, are its functionality, which carries a specific utility for the customer. One of the essential product features is its durability and reliability over time [5]. Companies strive to gain the highest level of quality, understood as offering products of higher quality at a cost-effective price for the consumer, focusing attention on not making mistakes and errors during production, producing products of higher quality than the competing company, and preventing the emergence of non-compliant products, which are associated with higher costs for the organization, introducing the process of improving the way of performing the assigned tasks and the product for all employees to involve them in the process of continuous improvement, planning and organizing the work of employees for comprehensive quality management [5, 6]. The basic tool for achieving these goals has become a quality management system, whose elements are methods, tools, rules, procedures, job descriptions, people, and the relationship between these elements. The effective implementation

Terotechnology XII   Materials Research Forum LLC
Materials Research Proceedings 24 (2022) 181-188   https://doi.org/10.21741/9781644902059-27

of these system elements increases the company's ability to satisfy customer needs and requirements [6].

Methodology

The study analyzes quality problems in manufacturing low-pressure hoses for car radiators. The analysis was conducted using selected quality tools such as the Pareto-Lorenz diagram, Ishikawa diagram, and 5WHYs.

The research object is a plant of a leading international manufacturer of fluid flow systems, seals for car bodies as well as anti-vibration systems located in the Silesia voivodeship in Poland. The analyzed plant is engaged in the production of hoses: cooling, turbocharger, vacuum, and air transfer. The subject of the research is a low-pressure hose used for the cooling system in automobiles.

A Pareto-Lorenz diagram was used to identify the most critical quality problems in terms of the frequency of occurrence in relation to low-pressure cooling system hoses. A Pareto-Lorenz diagram is a tool that allows data on emerging problems to be recorded and analyzed so that the most significant problem areas are highlighted [7]. In order to significantly reduce the likelihood of quality problems, the most frequently occurring nonconformances should be addressed first [8]. The remaining nonconformities should not be underestimated; they are less important but not invalid [7].

For the most frequently occurring nonconformity of the analyzed product, an analysis of the causes of its occurrence was carried out using the Ishikawa diagram. The analysis is based on five main areas of the problem, namely: man, machine, tools, materials, and method, within which there are looking for probable causes of the problem [9, 10].

For the second, most frequent occurring nonconformity of the analyzed product, an analysis of its root cause was carried out using the 5 WHY method. Its primary goal of using is to find the exact, fundamental reason that causes a given problem by asking a sequence of "why" questions [9, 10]. It is one of the most effective tools for root cause analysis in the Lean management concept [11].

Results

10000 pieces of the low-pressure hoses were inspected, and certain nonconformities were found, such as scratches on the hose (N1), too short hose (N2), too long hose (N3), illegible print (N4), crooked cut hose (N5), deformations (N6), under-bottomed overmoulding (N7), hose leakage (N8), incompatible insert mounting depth (N9), broken hose end (N10). The Pareto-Lorenz diagram (Fig. 1) was built based on the data on the frequency of occurrence of these nonconformities.

The Pareto-Lorenz diagram showed that 20% of the nonconformities (2 out of 10) were responsible for 52.4% of all hoses quality problems. The remaining 8 nonconformities are responsible for only 47.6% of the quality problems. The critical nonconformities in terms of frequency of occurrence were, therefore: nonconforming insertion depth (N9) and crooked cut hose (N10).

The most frequent nonconformity, i.e. incompatible insert mounting depth (N9), was addressed first. All possible causes of this nonconformity were identified using brainstorming and plotted on the Ishikawa diagram (Fig. 2).

Terotechnology XII

Materials Research Forum LLC

Materials Research Proceedings **24** (2022) 181-188

https://doi.org/10.21741/9781644902059-27

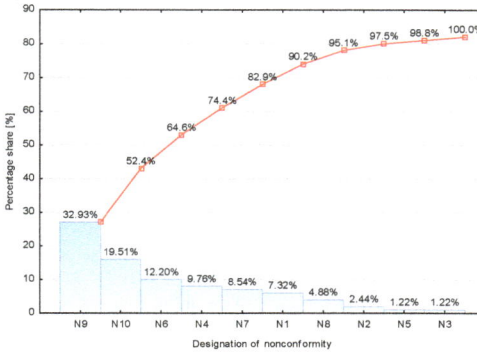

Fig. 1. *Pareto-Lorenz diagram for the analysis of the frequency of low-pressure hose nonconformities [own study with the use of Statistica 13.3 software]*

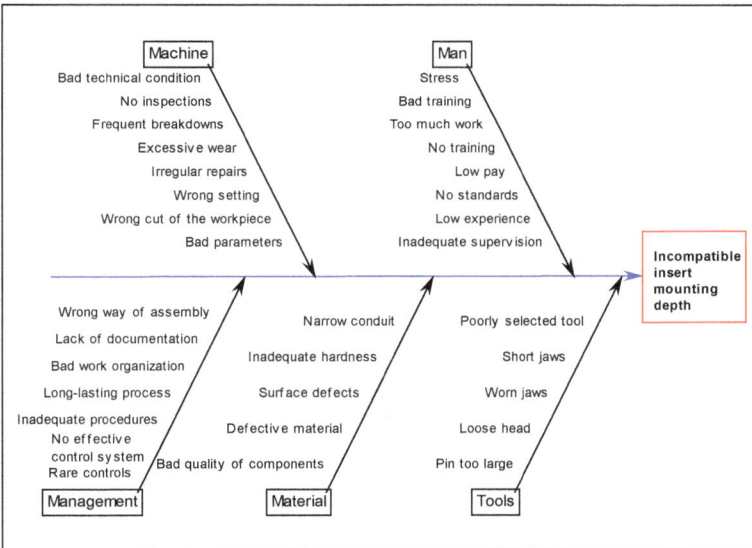

Fig. 2. *Ishikawa diagram for root cause analysis of critical low pressure hose nonconformance [own study with the use of Statistica 13.3 software]*

It can be seen from the diagram that most of the factors causing nonconforming insert mounting depths appear in the "human" and "machine" type areas. The poor condition of machine parts resulted in lower qualitative capability and turned into incomplete or improper insert mounting. Many "hard" and "soft" factors related to humans affected the formation of nonconforming products, such as lack of appropriate knowledge and experience, lack of adequate motivation, low wages, and lack of training. Also, improper management so bad organization of work, lack of work standards, not-effective control systems, rare inspections, or lack of current documentation at the workplace affected the critical nonconformance occurrence. In the effect of this last situation, the customer after making corrections on the workpiece still receives the hose before the corrections, because the employees are unaware of the corrections made and continue to produce a workpiece that does not meet the current requirements of the customer. Poor quality of materials and tools led to the production of poor-quality details or rapid wear of parts such as jaws, which are to compress the hose. Worn jaws posed a risk of producing a non-conforming product or perforation of the conduit during assembly.

A 5 WHY analysis was used to identify the root cause of the second most common low-pressure hose nonconformance, namely, a crooked cut workpiece (Fig. 3).

The root cause of the problem was the lack of frequent inspections of the machine's technical condition, and its parts, which was caused by too long a defined time between such inspections in the maintenance work schedule. In the analyzed case, the machine work became out of control because of a broken washer. Its absence resulted in backlash at the washer location, as well as the slow loosening of the nut, which increased disc movement. To solve this problem in a permanent way machine inspection intervals should be increased. Thus, the maintenance inspection schedule should be updated. Mechanics should inspect the machine at least once a month (the best twice a month), while operators should inspect the machine each time before starting and after ending work on the machine. This corrective action will minimize the possibility of the machine malfunctioning due to its poor technical condition.

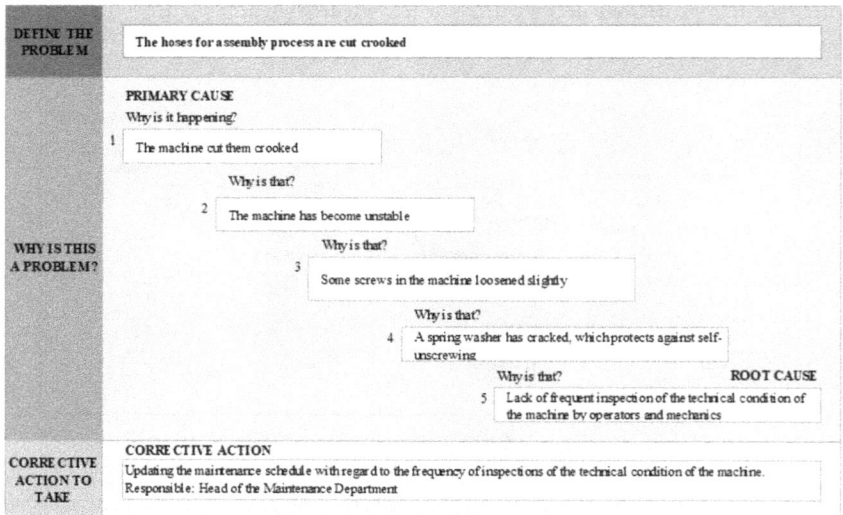

Fig. 3. 5WHY analysis for a crooked cut detail [own study]

Terotechnology XII   Materials Research Forum LLC
Materials Research Proceedings 24 (2022) 181-188   https://doi.org/10.21741/9781644902059-27

In order to reduce errors caused by human work (operators, mechanics), it is necessary to introduce further remedial measures aimed at raises for employees that will motivate them and give them a desire to develop; development of visual work standards in workplaces (visual controls & management tools [12]), which will allow the production of uniform, standardized parts and also allow rapid assessment of their quality by a glance. It is important to provide job training for employees to increase their awareness about the impact of their work on quality, and increase knowledge, and skills. The greatest attention should be paid to poor machine alignment. In order to eliminate this cause, it is necessary, first of all, to train people who will set up and changeover machines to a new workpiece; pay attention to excessive loading of the machines because it leads to rapid wear or overheating of its parts; more often diagnose the technical condition of machine parts so that the process carried out by them is not disrupted; conduct training for employees on machine diagnostics, so that they can react immediately if they notice a problem on a machine; develop documentation with pictures of correct and incorrect functioning of a machine (One Point Lessons - OPL cards in a version of "basic knowledge" and "problem" [13], boards and KAMISHIBAI cards [14]). In the event of detected non-compliance with requirements, the person using these types of visual control tools will have to report the problem immediately in order to implement necessary corrective actions.

Summary

The article presents the method of identifying the most important quality problems of low-pressure hoses for car radiators, the results of their analysis in order to discover the root causes, and the methods of eliminating the possibility of their occurrence - preventing the critical nonconformities. For this purpose, selected, popular quality tools were used.

Using data that was collected during the manufacturing process of a low-pressure hose for automotive radiators, 10 major nonconformities were listed that contributed to nonconforming products. The most common nonconformities were a nonconforming insert depth and a crooked cut hose. In order to eliminate the first main nonconformity, the Ishikawa diagram was made, which allowed identifying the basic cause of its occurrence, which is poor alignment of the machine by the operator. In order to eliminate this nonconformity, it was proposed to conduct periodic training for employees responsible for setting up and changeover of machines, to introduce more frequent inspections of the technical condition of machines, and to develop visual documentation of the correct operation of the machine. Using the 5 WHY method, the root cause for the crooked cut hose was identified, which was the lack of frequent machine maintenance. This caused the machine to become out of adjustment, which resulted in this nonconformance. It was suggested that the maintenance schedule should be updated and that in addition, a machine job card should be developed for employees to fill out before and after work, allowing for quicker detection of worn parts by which failures and quality problems can arise.

In conclusion, the application of quality tools to analyze low-pressure hose nonconformities was intended to prevent the possibility of their occurrence in the future by paying attention to their root, fundamental causes. In order to have a lasting effect of eliminating analyzed critical nonconformities of the tested product, defined and listed corrective and preventive actions must be implemented in an effective way and quality analyses focused on investigating potential causes of nonconformities must be generally conducted in an anticipatory manner. Success in the permanent elimination of nonconformities is a combination of employee involvement (a necessary condition), the use of appropriate methods and tools for detecting and analyzing their root causes, and the implementation of effective ways of preventing them (good first time).

The presented analysis may inspiring in many branches of the industry interested in high quality e.g. service [15], automotive [16-19], BIM [20], biotechnology [21, 22] and quality general management [23]. Such approach leads to desirable sustainable production [24-26], increases the trend for automation [27, 28] and gives new ideas for analysis methods [29, 30], both parametric [31, 32] and non-parametric [33-35].

References

[1] G. Kyriakopoulos. The role of quality management for effective implementation of customer satisfaction, customer consultation and self-assessment, within service quality schemes: A review, African Journal of Business Management 5 (2011) 4901–4915. https://doi.org/10.5897/AJBM10.1584

[2] Y.T. Chong, Ch.-H. Chen. Customer needs as moving targets of product development: A review, The International Journal of Advanced Manufacturing Technology 48 (2009) 395–406. https://doi.org/10.1007/s00170-009-2282-6

[3] G.F. Smith. The meaning of quality, Total Quality Management 4 (1993) 235–244. https://doi.org/10.1080/09544129300000038

[4] J. Martin, M. Elg, I. Gremyr. The Many Meanings of Quality: Towards a Definition in Support of Sustainable Operations, Total Quality Management & Business Excellence (2020) 1–14. https://doi.org/10.1080/14783363.2020.1844564

[5] R. Ulewicz, M. Mazur, K. Knop, R. Dwornicka. Logistic Controlling Processes and Quality Issues in a Cast Iron Foundry. Materials Research Proceedings 17 (2020) 65–71. https://doi.org/10.21741/9781644901038-10

[6] I. Gremyr, J. Lenning, M. Elg, J. Martin. Increasing the value of quality management systems, International Journal of Quality and Service Sciences 13 (2021) 381–394. https://doi.org/10.1108/IJQSS-10-2020-0170

[7] D.R. Bamford, R.W. Greatbanks. The use of quality management tools and techniques: a study of application in everyday situations, Int. J. Quality & Reliability Management 22 (2005) 376–392. https://doi.org/10.1108/02656710510591219

[8] K. Knop. Analysis and Quality Improvement of the UV Printing Process on Glass Packagings, Quality Production Improvement. QPI 2021, (Eds.) Ulewicz R., Hadzima B., De Gruyter, Warszawa (2021) 314–325.

[9] D. Siwiec, A. Pacana. The use of quality management techniques to analyse the cluster of porosities on the turbine outlet nozzle, Production Engineering Archives 24 (2019) 33–36. https://doi.org/10.30657/pea.2019.24.08

[10] D. Pavletic, M. Sokovic, G. Paliska. Practical Application of Quality Tools, International Journal for Quality Research 2 (2008) 1-6.

[11] R. Ulewicz, D. Kleszcz, M. Ulewicz. Implementation of Lean Instruments in Ceramics Industries, Management Systems in Production Engineering 29 (2021) 203–207. https://doi.org/10.2478/mspe-2021-0025

[12] K. Knop. Indicating and Analysis the Interrelation Between Terms – Visual: Management, Control, Inspection and Testing, Production Engineering Archives 26 (2020) 110–120. https://doi.org/10.30657/pea.2020.26.22

Terotechnology XII
Materials Research Proceedings **24** (2022) 181-188

Materials Research Forum LLC
https://doi.org/10.21741/9781644902059-27

[13] M. Ebrahim, A. Baboli, E. Rother. The evolution of world class manufacturing toward Industry 4.0: A case study in the automotive industry, IFAC-PapersOnLine 52 (2019) 188–194. https://doi.org/10.1016/j.ifacol.2019.10.021

[14] K. Knop, R. Ulewicz. Analysis of the Possibility of Using the Kamishibai Audit in the Area of Quality Inspection Process Implementation, Organization & Management: Scientific Quarterly 3 (2018) 31-49. https://doi.org/10.29119/1899-6116.2018.43.3

[15] M. Ingaldi. Overview of the main methods of service quality analysis, Production Engineering Archives 18 (2018) 54-59. https://doi.org/10.30657/pea.2018.18.10

[16] A. Pacana, K. Czerwińska, L. Bednárová. Comprehensive improvement of the surface quality of the diesel engine piston, Metalurgija 58 (2019) 329-332.

[17] D. Siwiec, R. Dwornicka, A. Pacana. Improving the non-destructive test by initiating the quality management techniques on an example of the turbine nozzle outlet, Materials Research Proceedings 17 (2020) 16-22. https://doi.org/10.21741/9781644901038-3

[18] G. Ostasz, K. Czerwińska, A. Pacana. Quality Management of Aluminum Pistons with the Use of Quality Control Points. Management Systems in Production Engineering 28 (2020) 29-33. https://doi.org/10.2478/mspe-2020-0005

[19] D. Nowakowski, A. Gądek-Moszczak, P. Lempa. Application of machine learning in the analysis of surface quality - the detection the surface layer damage of the vehicle body, METAL 2021 - 30th Int. Conf. Metallurgy and Materials (2021), Ostrava, Tanger 864-869. https://doi.org/10.37904/metal.2021.4210

[20] G. Majewski, Ł.J. Orman, M. Telejko, N. Radek, J. Pietraszek, A. Dudek. Assessment of thermal comfort in the intelligent buildings in view of providing high quality indoor environment, Energies 13 (2020) art. 1973. https://doi.org/10.3390/en13081973

[21] E. Skrzypczak-Pietraszek, A. Szewczyk, A. Piekoszewska, H. Ekiert. Biotransformation of hydroquinone to arbutin in plant in vitro cultures – Preliminary results. Acta Physiol. Plant 27 (2005) 79-87. https://doi.org/10.1007/s11738-005-0039-x

[22] E. Skrzypczak-Pietraszek. Phytochemistry and biotechnology approaches of the genus exacum. In: The Gentianaceae - Volume 2: Biotechnology and Applications, 2015, 383-401. https://doi.org/10.1007/978-3-642-54102-5_16

[23] A. Pacana, R. Ulewicz. Analysis of causes and effects of implementation of the quality management system compliant with iso 9001, Pol. J. Manag. Stud. 21 (2020) 283-296. https://doi.org/10.17512/pjms.2020.21.1.21

[24] S. Lazar, D. Klimecka-Tatar, M. Obrecht. Sustainability orientation and focus in logistics and supply chains. Sustainability 13 (2021) art. 3280. https://doi.org/10.3390/su13063280

[25] R. Ulewicz, D. Jelonek, M. Mazur. Implementation of logic flow in planning and production control, Management and Production Engineering Review 7 (2016) 89-94. https://doi.org/10.1515/mper-2016-0010

[26] M. Ulewicz, A. Pietrzak. Properties and structure of concretes doped with production waste of thermoplastic elastomers from the production of car floor mats. Materials 14 (2021) art. 872. https://doi.org/10.3390/ma14040872

Terotechnology XII
Materials Research Proceedings 24 (2022) 181-188

Materials Research Forum LLC
https://doi.org/10.21741/9781644902059-27

[27] K. Antosz, A. Pacana. Comparative analysis of the implementation of the SMED method on selected production stands, Tehnicki Vjesnik 25 (2018) 276-282. https://doi.org/10.17559/TV-20160411095705

[28] R. Ulewicz, M. Mazur. Economic aspects of robotization of production processes by example of a car semi-trailers manufacturer, Manufacturing Technology 19 (2019) 1054-1059. https://doi.org/10.21062/ujep/408.2019/a/1213-2489/MT/19/6/1054

[29] J. Pietraszek, E. Skrzypczak-Pietraszek. The uncertainty and robustness of the principal component analysis as a tool for the dimensionality reduction. Solid State Phenom. 235 (2015) 1-8. https://doi.org/10.4028/www.scientific.net/SSP.235.1

[30] Ł.J. Orman Ł.J., N. Radek, J. Pietraszek, M. Szczepaniak. Analysis of enhanced pool boiling heat transfer on laser-textured surfaces. Energies 13 (2020) art. 2700. https://doi.org/10.3390/en13112700

[31] J. Pietraszek, R. Dwornicka, A. Szczotok. The bootstrap approach to the statistical significance of parameters in the fixed effects model. Proc. ECCOMAS Congress (2016) 3, 6061-6068. https://doi.org/10.7712/100016.2240.9206

[32] J. Pietraszek, A. Szczotok, N. Radek. The fixed-effects analysis of the relation between SDAS and carbides for the airfoil blade traces. Archives of Metallurgy and Materials 62 (2017) 235-239. https://doi.org/10.1515/amm-2017-0035

[33] J. Pietraszek, A. Gadek-Moszczak, N. Radek. The estimation of accuracy for the neural network approximation in the case of sintered metal properties. Studies Comp. Intell. 513 (2014) 125-134. https://doi.org/10.1007/978-3-319-01787-7_12

[34] J. Pietraszek, N. Radek, A.V. Goroshko. Challenges for the DOE methodology related to the introduction of Industry 4.0. Production Engineering Archives 26 (2020) 190-194. https://doi.org/10.30657/pea.2020.26.33

[35] L. Radziszewski, M. Kekez. Application of a genetic-fuzzy system to diesel engine pressure modeling, International Journal of Advanced Manufacturing Technology 46 (2010) 1-9. https://doi.org/10.1007/s00170-009-2080-1

Terotechnology XII
Materials Research Proceedings 24 (2022) 189-195

Materials Research Forum LLC
https://doi.org/10.21741/9781644902059-28

Vickers Method Application for Quality Assessment the First Patterns of a Selected Product of the Automotive Industry – a Case Study

JAGUSIAK-KOCIK Marta[1,a] and IDZIKOWSKI Adam[1,b*]

[1] Department of Production Engineering and Safety, Faculty of Management, Czestochowa University of Technology, Al. Armii Krajowej 19b, 42-218 Czestochowa, Poland

[a] m.jagusiak-kocik@pcz.pl, [b] adam.idzikowski@pcz.pl

Keywords: Hardness Measurement, Vickers Method, Product Quality Assessment

Abstract. Production is one of the basic branches of the economy. Along with the development of techniques and technology, it expanded the scope of its activities, including a wide flow of information concerning, inter alia, the quality of manufactured products in order to increase competitiveness in meeting customer needs. The article assesses the quality of the first patterns of the selected product (detail X) by measuring the hardness using the Vickers method. The measurements were carried out for four samples marked with the letters A1, B1, C1 and D1, which for the Vickers test had to be properly prepared, which was associated with the need to meet both the requirements relating to surface roughness, the size of the detail and getting rid of contamination. This process follows the procedure prepared by the quality control section, specified in the detailed execution manual for detail X. The measurement of the samples confirmed the research carried out in the scope of quality assessment and the validity of the selection of the Vickers method for hardness measurement.

Introduction

Production is an activity that is spread over time and is mainly based on the physical production of a specific product [1, 2]. It is also considered to combine various types of inputs and production resources in order to obtain a specific effect in the form of a semi-finished or finished product [3-5].

The variety of materials used in production requires defining, inter alia, their mechanical properties, which determine the load-bearing capacity of the material [x6]. Mechanical properties affect the quality and strength of materials, and thus – the properties of the finished product [7, 8]. The study of mechanical properties is extremely important, e.g. during the production of large structures, where every smallest element is responsible for the safety and quality of the entire structure. One of the mechanical properties is hardness [9], and hardness is assessed by static hardness tests, dynamic hardness tests and scratch tests. Hardness measurement is a simple, quick and low-cost test. The obtained result can be converted into material strength by applying an appropriate formula, taking into account the linear relationship between the material hardness and tensile strength. Hardness is the resistance of a material that occurs when a specific object (indenter) is pressed into it. If the indenter goes deeper into the material with the same load, it means that the indenter is softer. To perform the measurement, a ball-shaped element is used, which is made of tungsten carbide or hardened steel. Sometimes an indenter is also used, which is in the form of a diamond cone or diamond pyramid. The element is pressed into the material in a static or dynamic manner.

Terotechnology XII   Materials Research Forum LLC
Materials Research Proceedings **24** (2022) 192-198   https://doi.org/10.21741/9781644902059-28

The methods of static hardness measurement include the Brinell, Vickers and Rockwell methods. The Brinell method is one of the oldest methods of hardness measurement [10] and consists in perpendicularly pressing a hardened ball (or carbide ball) with a diameter of D and a force of F value into the tested material. The Rockwell method consists in pressing an indenter in the form of a diamond cone or steel ball into the surface material and consists of 2 stages: loading with the initial force F0 (in order to reduce the impact on the measurement of non-uniformity of the sample surface), and then applying the main force F1 and unloading to the force F0. This determines the permanent increase in the depth of the impression resulting from the initial force after unloading [11]. The Vickers method [12] consists in measuring the hardness (Fig. 1) using an indenter in the form of a simple pyramid, characterized by a square base and an apex angle equal to the Brinell ball indentation angle, i.e. 136° with an accuracy of 0.5". The hardness is calculated as the ratio of the load F to the side surface of the permanently imprinted part of the pyramid.

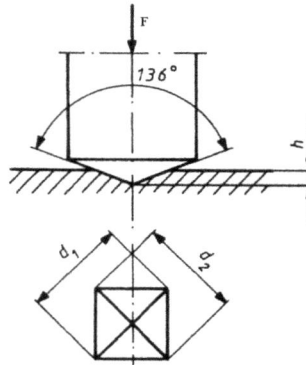

Fig. 1. *Scheme of hardness measurement by the Vickers method [12]*

The aim of the study is to use the Vickers hardness method to assess the quality of the first patterns of a selected product in the automotive industry [13].

The obtained results and methodology should be of interest to researchers in the field of material sciences [14-17], energy [18, 19], machining [20-22] and coating [23-25] but also quality engineers of the steel industry [26, 27], food [28, 29] and the automotive industry [30-33]. Increasing the quality of the materials produced will have a positive effect on the production of highly responsible parts and machines, where there is a risk of contamination [34], costly breakdowns [35, 36] or injury [37, 38]. The results can also inspire the further development of data analysis methods [39-41] and the development of risks in failure scenarios [42, 43] as a valuable source of experimental information.

Methodology

The tests used a hardness test with a low loading force, symbol HV1, i.e. with a nominal force value of 9.807 N. In order to carry out a hardness test using the Vickers method, it is necessary to properly prepare a sample of the detail. The surface of the sample should be free of impurities and characterized by a roughness of not more than 2.5 μm. The duration of the load for most of the tested materials should be in the range of 10 to 15 seconds, assuming an accuracy of 2 seconds. The sample thickness should not be less than 1.5d.

Terotechnology XII
Materials Research Proceedings 24 (2022) 192-198

Materials Research Forum LLC
https://doi.org/10.21741/9781644902059-28

This is done according to an appropriate procedure, specified in the X detail execution manual, prepared by the quality control department. First, the upper part of the X detail is cut. It is necessary to properly set the handle before the cut, in order to properly set the laser line of the cutter, which determines the direction of the cut. Then the detail is fixed in the holder. Before the first cut, it is necessary to adjust the table setting by a blade thickness of 2 mm. The cuts are made along the angles determined for the X detail. In the selected detail for analysis, the cut was made along an angle of 88° and 110° along the cross-section of the entire detail. After each cut, the target area is marked. A transverse cut is then made below the X detail at a distance of 2 mm to avoid damaging the measuring surfaces. Fragments of the detail are then placed on the plunger of the metallographic press with the measuring side facing down, and then covered with thermosetting resin. Then the samples are included. After this process, the samples are polished and placed on the hardness tester pad, ready for measurement. In the case of the lower part of X detail, the first cut is made 2 mm above the upper part in order to obtain a fragment for further examination. Then the detail is placed in the holder in such a way that the laser line of the cutter passes through the center of the cam. The last cut is made 2 mm above in order not to damage the measuring surfaces. The obtained fragments are placed on the plunger of the metallographic press, and then, after the completed process, they are placed on the pad of the polishing machine. After its completion, the samples can be measured.

Results

The measurements were carried out for four samples marked with the letters A1, B1, C1 and D1. Basic measurement parameters are presented below (Tab. 1)

Table 1. Measurement results for four samples by the Vickers method

MEASUREMENT RESULTS FOR SAMPLE A1		
POINT	HARDNESS	POSITION
1	716	0,10
2	279	0,50
3	257	1,00
4	271	1,25
MEASUREMENT RESULTS FOR SAMPLE B1		
1	683	0,10
2	277	0,50
3	261	1,00
4	265	1,25
MEASUREMENT RESULTS FOR SAMPLE C1		
1	714	0,10
2	276	0,50
3	250	1,00
4	272	1,25
MEASUREMENT RESULTS FOR SAMPLE D1		
1	715	0,10
2	293	0,50
3	267	1,00
4	269	1,25

Obtaining similar hardness values for all analyzed samples proves about the good effectiveness of the Vickers method. Therefore, it can be used to assess the quality of the first standards, with special care for the proper preparation of the sample.

Summary

Modern enterprises operate in difficult market conditions. The demand for specific products depends on many factors, including meeting customers' requirements. Due to the variety of available offers, users have more and more requirements, including the high quality of manufactured products.

Checking the first patterns ensures early detection of abnormalities in manufactured items. It is very useful in simulating production processes, which are then translated into production. However, the effectiveness of this control depends on the methods used. The solution analyzed at work, used in the surveyed company of the automotive industry, is based on the application of the Vickers method, which based on the PN-EN ISO 6507-1: 2018-05 standard, it consists in indenting a diamond indenter into the sample surface with a predetermined force. The advantage of the Vickers method over the Brinell method is the universality of the diamond indenter. In addition, considering the advantages and disadvantages, it can be stated that the Vickers method, having the advantages, has the widest scale and high measurement accuracy, and the measurement result does not depend on the load applied during the test. On the other hand, the disadvantages of the Vickers method include: the difficulty of applying the method on heterogeneous or coarse-grained materials.

References

[1] T. Karpiński. Inżynieria produkcji, WNT, Warszawa, 2011.

[2] B. Liwowski, R. Kozłowski. Podstawowe zagadnienia zarządzania produkcją, Oficyna Ekonomiczna, Kraków 2007.

[3] A. Burduk. Modelowanie systemów narzędziem oceny stabilności procesów produkcyjnych, Politechnika Wrocławska, Wrocław, 2013

[4] B. Liwowski, R. Kozłowski. Podstawowe zagadnienia zarządzania produkcją, Oficyna Ekonomiczna, Kraków, 2006

[5] E. Masłyk-Musiał, A. Rakowska, E. Krajewska-Bińczyk. Zarządzanie dla inżynierów, PWE, Warszawa, 2012.

[6] R. Ulewicz, F. Nový, M. Mazur, P. Szataniak. Fatigue properties of the hsla steel in high and ultra-high cycle region, Production Engineering Archive 4 (2014) 18-21.

[7] N. Kopiika, J. Selejdak, Y. Blikharskyy. Specifics of physico-mechanical characteristics of thermally-hardened rebar, Production Engineering Archive 28 (2022) 73-81. https://doi.org/10.30657/pea.2022.28.09

[8] M. Szary, K. Knop. Ocena technologii i możliwości technologicznych przedsiębiorstwa z branży metalowej. Archiwum Wiedzy Inżynierskiej 3 (2018) 31-34.

[9] W. J. Klimasara. Badanie materiałów konstrukcyjnych 315[01].O2.02. Poradnik dla ucznia, Instytut Technologii Eksploatacji – Państwowy Instytut Badawczy, Radom, 2007

[10] Z. Zatorski, R. Biernat. Energy-based determination of Brinell hardness, Scientific Journal of Polish Naval Academy 4 (2014) 103-116.

Terotechnology XII   Materials Research Forum LLC
Materials Research Proceedings 24 (2022) 192-198   https://doi.org/10.21741/9781644902059-28

[11] W.J. Klimasara. Badanie materiałów konstrukcyjnych, Instytut Technologii Eksploatacji – Państwowy Instytut Badawczy, Radom, 2007.

[12] PN-EN ISO 6507-1:2018-05. Metale. Pomiar twardości sposobem Vickersa. Część 1: Metoda badań.

[13] E. Staniszewska, D. Klimecka-Tatar, M. Obrecht. Eco-design processes in the automotive industry, Production Engineering Archive 26 (2020) 131-137. https://doi.org/10.30657/pea.2020.26.25

[14] A. Dudek, B. Lisiecka, R. Ulewicz. The effect of alloying method on the structure and properties of sintered stainless steel, Archives of Metallurgy and Materials 62 (2017) 281-287. https://doi.org/10.1515/amm-2017-0042

[15] A. Szczotok, N. Radek, R. Dwornicka. Effect of the induction hardening on microstructures of the selected steels. METAL 2018 - 27th Int. Conf. Metall. Mater. (2018), Ostrava, Tanger 1264-1269.

[16] N. Radek, J. Pietraszek, A. Goroshko. The impact of laser welding parameters on the mechanical properties of the weld, AIP Conf. Proc. 2017 (2018) art.20025. https://doi.org/10.1063/1.5056288

[17] T. Lipiński. Corrosion resistance of 1.4362 steel in boiling 65% nitric acid, Manufacturing Technology 16 (2016) 1004-1009.

[18] Ł.J. Orman. Boiling heat transfer on single phosphor bronze and copper mesh microstructures, EPJ Web of Conf. 67 (2014) art. 2087. https://doi.org/10.1051/epjconf/20146702087

[19] Ł.J. Orman Ł.J., N. Radek, J. Pietraszek, M. Szczepaniak. Analysis of enhanced pool boiling heat transfer on laser-textured surfaces. Energies 13 (2020) art. 2700. https://doi.org/10.3390/en13112700

[20] P. Szataniak, F. Novy, R. Ulewicz. HSLA steels - Comparison of cutting techniques, METAL 2014 – 23rd Int. Conf. Metallurgy and Materials (2014), Ostrava, Tanger, 778-783.

[21] A. Szczotok, J. Pietraszek, N. Radek. Metallographic Study and Repeatability Analysis of γ' Phase Precipitates in Cored, Thin-Walled Castings Made from IN713C Superalloy. Archives of Metallurgy and Materials 62 (2017) 595-601. https://doi.org/10.1515/amm-2017-0088

[22] I. Miletić, A. Ilić, R.R. Nikolić, R. Ulewicz, L. Ivanović, N. Sczygiol. Analysis of selected properties of welded joints of the HSLA Steels, Materials 13 (2020) art.1301. https://doi.org/10.3390/ma13061301

[23] J. Bronček, P. Fabian, N. Radek. Tribological research of properties of heat-treated cast irons with globular graphite, Materials Science Forum 818 (2015) 209-212. https://doi.org/10.4028/www.scientific.net/MSF.818.209

[24] N. Radek, J. Pietraszek, A. Gadek-Moszczak, Ł.J. Orman, A. Szczotok. The morphology and mechanical properties of ESD coatings before and after laser beam machining, Materials 13 (2020) art. 2331. https://doi.org/10.3390/ma13102331

[25] N. Radek, J. Konstanty, J. Pietraszek, Ł.J. Orman, M. Szczepaniak, D. Przestacki. The effect of laser beam processing on the properties of WC-Co coatings deposited on steel. Materials 14 (2021) art. 538. https://doi.org/10.3390/ma14030538

[26] A. Maszke, R. Dwornicka, R. Ulewicz. Problems in the implementation of the lean concept at a steel works - Case study, MATEC Web of Conf. 183 (2018) art.01014. https://doi.org/10.1051/matecconf/201818301014

[27] H. Danielewski, A. Skrzypczyk, W. Zowczak, D. Gontarski, L. Płonecki, H. Wiśniewski, D. Soboń, A. Kalinowski, G. Bracha, K. Borkowski. Numerical analysis of laser-welded flange pipe joints in lap and fillet configurations, Technical Transactions 118 (2021) art. e2021030. https://doi.org/10.37705/TechTrans/e2021030

[28] N. Baryshnikova, O. Kiriliuk, D. Klimecka-Tatar. Management approach on food export expansion in the conditions of limited internal demand, Polish Journal of Management Studies 21 (20200 101-114. https://doi.org/10.17512/pjms.2020.21.2.08

[29] N. Baryshnikova, O. Kiriliuk, D. Klimecka-Tatar. Enterprises' strategies transformation in the real sector of the economy in the context of the COVID-19 pandemic, Production Engineering Archives 27 (2021) 8-15. https://doi.org/10.30657/pea.2021.27.2

[30] K. Antosz, A. Pacana. Comparative analysis of the implementation of the SMED method on selected production stands, Tehnicki Vjesnik 25 (2018) 276-282. https://doi.org/10.17559/TV-20160411095705

[31] A. Pacana, R. Ulewicz. Analysis of causes and effects of implementation of the quality management system compliant with iso 9001, Pol. J. Manag. Stud. 21 (2020) 283 296. https://doi.org/10.17512/pjms.2020.21.1.21

[32] A. Pacana, D. Siwiec, L. Bednárová. Method of choice: A fluorescent penetrant taking into account sustainability criteria, Sustainability 12 (2020) art. 5854. https://doi.org/10.3390/su12145854

[33] G. Ostasz, K. Czerwińska, A. Pacana. Quality Management of Aluminum Pistons with the Use of Quality Control Points. Management Systems in Production Engineering 28 (2020) 29-33. https://doi.org/10.2478/mspe-2020-0005

[34] E. Skrzypczak-Pietraszek. Phytochemistry and biotechnology approaches of the genus Exacum. In: The Gentianaceae - Volume 2: Biotechnology and Applications, 2015, 383-401.https://doi.org/10.1007/978-3-642-54102-5_16

[35] S. Marković, D. Arsić, R.R. Nikolić, V. Lazić, B. Hadzima, V.P. Milovanović, R. Dwornicka, R. Ulewicz. Exploitation characteristics of teeth flanks of gears regenerated by three hard-facing procedures, Materials 14 (20210 art. 4203. https://doi.org/10.3390/ma14154203

[36] E. Lisowski, J. Rajda, G. Filo, P. Lempa. Flow Analysis of a 2URED6C Cartridge Valve, Lecture Notes in Mechanical Engineering 24 (2021) 40-49. https://doi.org/10.1007/978-3-030-59509-8_4

[37] A. Dudek, M. Klimas. Composites based on titanium alloy Ti-6Al-4V with an addition of inert ceramics and bioactive ceramics for medical applications fabricated by spark plasma

Terotechnology XII   Materials Research Forum LLC
Materials Research Proceedings **24** (2022) 192-198   https://doi.org/10.21741/9781644902059-28

sintering (SPS method), Materialwissenschaft und Werkstofftechnik 46 (2015) 237-247. https://doi.org/10.1002/mawe.201500334

[38] A. Kubecki, C. Śliwiński, J. Śliwiński, I. Lubach, L. Bogdan, W. Maliszewski. Assessment of the technical condition of mines with mechanical fuses, Technical Transactions 118 (2021) art. e2021025. https://doi.org/10.37705/TechTrans/e2021025

[39] J. Pietraszek, E. Skrzypczak-Pietraszek. The uncertainty and robustness of the principal component analysis as a tool for the dimensionality reduction. Solid State Phenom. 235 (2015) 1-8. https://doi.org/10.4028/www.scientific.net/SSP.235.1

[40] J. Pietraszek, R. Dwornicka, A. Szczotok. The bootstrap approach to the statistical significance of parameters in the fixed effects model. ECCOMAS Congress 2016 - Proceedings of the 7th European Congress on Computational Methods in Applied Sciences and Engineering 3, 6061-6068. https://doi.org/10.7712/100016.2240.9206

[41] J. Pietraszek, A. Szczotok, N. Radek. The fixed-effects analysis of the relation between SDAS and carbides for the airfoil blade traces. Archives of Metallurgy and Materials 62 (2017) 235-239. https://doi.org/10.1515/amm-2017-0035

[42] J. Pietraszek, N. Radek, A.V. Goroshko. Challenges for the DOE methodology related to the introduction of Industry 4.0. Production Engineering Archives 26 (2020) 190-194. https://doi.org/10.30657/pea.2020.26.33

[43] J. Fabis-Domagala, M. Domagala, H. Momeni. A matrix FMEA analysis of variable delivery vane pumps, Energies 14 (2021) art. 1741. https://doi.org/10.3390/en14061741

Terotechnology XII
Materials Research Proceedings **24** (2022) 196-203

Materials Research Forum LLC
https://doi.org/10.21741/9781644902059-29

Analysis of the Causes of the Non-Conformity of the Bearing Shell Casting Used on Rail Vehicles

PACANA Andrzej[1, a *], CZERWIŃSKA Karolina[2,b] and DWORNICKA Renata[3,c]

[1] Rzeszow University of Technology, Faculty of Mechanical Engineering and Aeronautics
al. Powstańców Warszawy 8, 35-959 Rzeszów, Poland, ORCID: 0000-0003-1121-6352

[2] Rzeszow University of Technology, Faculty of Mechanical Engineering and Aeronautics
al. Powstańców Warszawy 8, 35-959 Rzeszów,Poland, ORCID: 000-0003-2150-0963

[3] Cracow University of Technology, Faculty of Mechanical Engineering
al. Jana Pawła II 37, 31-864, Kraków, Poland, ORCID: 0000-0002-2979-1614

[a] app@prz.edu.pl, [b] k.czerwinska@prz.edu.pl, [c] renata.dwornicka@pk.edu.pl*

Keywords: Quality Control, Quality Management Tools, Pareto-Lorenza Diagram, Casting Defects

Abstract. Continuous improvement of the quality of manufactured products and monitoring of the production process is the key to the success of every company. Skillful use of available technologies and quality management instruments makes it possible to eliminate casting incompatibilities and prevent their recurrence in the future. The aim of the article was to analyze the types of defects occurring in castings, locate the areas with the most frequent occurrence of defects and identify the reasons for the presence of defects in castings of bearing housings used in railway vehicles. The paper presents the usefulness of a combination of quality management instruments for diagnosing material discontinuities in the analyzed castings.

Introduction

Castings are an essential part of any industry, and their quality is determined by the technical conditions of their acceptance [1-2]. The production of casting, and thus maintaining the high quality of the finished product, is related to several technological parameters that may affect the quality of the finished product. The problem occurring during the casting production process is the inability to control all the factors of the technological process simultaneously. One and the most crucial option affecting the quality and competitiveness of casting is to confirm that the casting is free from defects. [3-5].

Inconveniences in aluminum castings are frequent problems causing a decrease in the strength of the casting and an increase in the costs of the technological process, thus influencing further mechanical processing and operation of the casting. [6]. Quick determination of the type of defect, its cause, and place of occurrence has a significant impact on the course of the casting process and reduction of the number of defects. Currently used diagnostic methods in the casting control process are characterized by effectiveness depending on the type of the tested product, the sensitivity of the method depending on the type of object, and the place of defect occurrence [7]. The aspect that hinders effective diagnostics is the great variety of shapes, often with a significant degree of complication, and the complexity of casting processes with a large number of parameters that can affect their course. This makes it difficult to clearly identify a diagnostic method that would be fast, simple and efficient, and applicable at every stage of casting production. Hence, it is justified and necessary to select appropriate technological parameters not only to stabilize the process but also to select appropriate quality management instruments contributing to the achievement of the desired level of quality and measurable organizational and financial benefits [8-10].

Analysis

The aim of the conducted research was to diagnose, at individual stages between operational quality control, the state of castings of rolling bearing housings used in railway vehicles and to indicate the areas of shielding in which, most often, there are discrepancies. In addition, the objective was to identify the causes of non-compliance in castings where appropriate corrective and preventive action could significantly reduce the occurrence of non-compliant castings.

The subject of research was the casting of rolling bearing housings used in railway vehicles. The finished castings are worth 110 kg, and their dimensions are 504 x 480 x 356. The 3D model of the product is shown in Fig. 1.

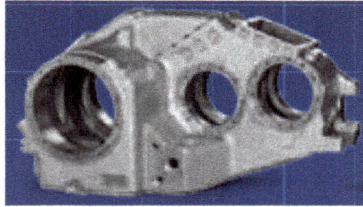

Fig. 1. *Test subject – sink rolling bearing housing used in rail vehicles [11]*

The study was carried out on a batch of products cast in the 2nd quarter of 2021 in a Polish production company located in the southern part of the country.

The construction material used for gravity casting of the housing is ENAC-AlSi7Mg0. 6 (ENAC-42200) alloy. The chemical composition of the alloy and its mechanical properties is shown in Table 1.

Table 1. *Chemical composition and mechanical properties of AlSi7Mg0.6 alloy*

Chemical composition of AlSi7Mg0. 6 alloy										
Element		Fe	Si	Mn	Ti	Cu	Mg	Zn	Others	Al
Value [%]	Min	-	6.50	-	-	-	0.45	-	each: 0.03; total: 0.01	remainder
	Max	0.19	7.50	0.10	0.25	0.05	0.70	0.07		
Mechanical properties of alloy AlSi7Mg0,6										
Property name		Tensile strength [Rm]		Yield strength [R02]		Elongation at break [A]		Brinell hardness		
Value [%]	Min	300	320	240	240	4	6	100	115	
	Max	350		280		6		115		
Unit of measure		[N/mm^2]	[Mpa]	[N/mm^2]	[Mpa]	[%]	[%]	[HB]	[HB]	

Source: Own elaboration based on: [12]

AlSi7Mg0.6 alloy combines silicon and magnesium as alloying elements, which give very good mechanical properties [13]. The distinguishing feature of the alloy is its exceptional corrosion resistance, good weldability, and excellent machining properties [14-16]. For this reason, the alloy

Terotechnology XII   Materials Research Forum LLC
Materials Research Proceedings 24 (2022) 199-206   https://doi.org/10.21741/9781644902059-29

has been used in architecture, aviation, automotive [17-19], food and chemical industry, mechanical engineering, shipbuilding, models, and forms. [20-23].

In order to analyze the quality of the products, tests were conducted, the scope of which included verification of shape defects (on the internal surface of the casting), defects of the raw surface, continuity breaks and internal defects in the casting, and additionally marking the place of occurrence of non-compliance together with a precise determination of the type of identified non-compliance. The realized control of castings was performed in accordance with the internal procedure of the company according to each production order.

The check was also subject to a visual check of the casting designation.

Results of Analysis

Currently, in the analyzed company, each casting of rolling bearing housings used in railway vehicles is subject to a visual quality control carried out after the completion of each production operation. In order to minimize the number of castings verified as non-compliant, an analysis of the reasons for discards has been undertaken. The first step of the analysis was to determine the areas in the housing casting (Fig. 2), in which the most frequent irregularities occur and to determine the type of these defects. The preliminary analysis allowed us to identify 9 types of incompatibilities in the housing casting. The proposed instrument for an in-depth analysis of the defectiveness of castings was the Pareto-Lorenz analysis combined with the ABC method, whose aim is to specify the most significant inconsistencies in terms of the number of their occurrence and the severity of their effects (Fig. 3). In the Pareto Lorenzo diagram, the nonconformities have been marked as shown in Fig. 2.

Fig. 2. Rolling bearing housing casting used in railway vehicles to identify the most common areas of non-compliance (1 - shrinkage cavity; 2 - sanding; 3 - gas bladders; 4 - under casting; 5 - surface roughness incompatibility). 6 - mechanical damage; 7 - foreign material inclusions; 8 - peeling; 9 - incorrect or illegible marking of the casting)

The analysis of casting showed that the most important incompatibilities were systolic cavities (52, 5%) and sanding cavities (27, 3%). These defects contribute to 79. 8% of all defects after the casting process. In accordance with the ABC method, area A, to which the listed non-conformities have been qualified, is determined as critical.

Terotechnology XII   Materials Research Forum LLC
Materials Research Proceedings 24 (2022) 199-206   https://doi.org/10.21741/9781644902059-29

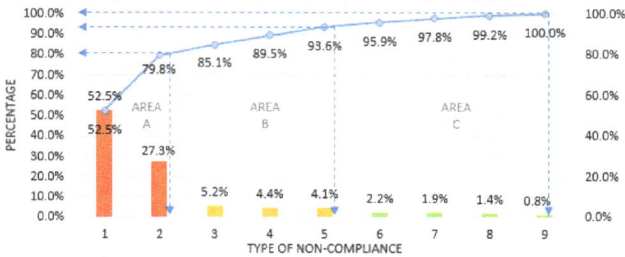

Fig. 3. *Lorenzo's Pareto Chart and ABC method for non-conformity of castings of rolling bearing housings used in railway vehicles*

Due to the number of skin cavities and crusts and the fact that the presence of specified nonconformities eliminates the product, the metallographic examination was performed to observe the nonconformities and analyze them. Fig. 4 shows one of the observed inconsistencies, which most often occurs in the casting of the casing - the dermal cavity.

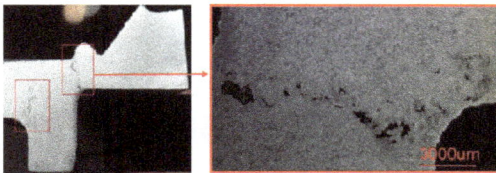

Fig. 4. *Survey results from the area of discontinuities - systolic cavity*

In order to reduce costs and make the right decision regarding the handling of non-compliant products, defects should be disclosed at the earliest possible stage of the production process, because then they generate lower costs than their detection at the time of final inspection. Additionally, the disclosure of a defect in the place of its occurrence enables the application of effective corrective and preventive actions, which in the future may eliminate it.

Data on the type of quality control in which the most inconsistencies with a given type of defect are detected and decisions on handling the non-compliant product are presented in Table 2. Table 2 uses the type of defect markings used in Fig. 2.

Table 2. *Percentage of decisions on handling non-compliant products*

The casting of bearing housing used in railway vehicles									
Type of defect	1	2	3	4	5	6	7	8	9
Quality control most often identifies non-compliance	X-ray method, UT	X-ray method	X-ray, PT, ET	Visual inspection	Using a contact profiler	Visual inspection	Visual inspection	Visual inspection	Visual inspection
Decision — Disposal of casting	76%	63%	71%	66%	8%	39%	64%	31%	0%
Decision — Repair	6%	15%	11%	17%	81%	42%	29%	53%	100%
Decision — Release for casting	18%	22%	18%	17%	11%	19%	7%	16%	0%

Materials Research Forum LLC

https://doi.org/10.21741/9781644902059-29

The largest number of the most serious discrepancies in castings, i. e. systolic cavities, and porosity, is detected during the initial control using the NDT X-ray method.

Proposal for Improvement

As part of the in-depth analysis, it was decided to use a combination of quality management instruments, i. e. the brainstorming method and the Ishikawa diagram, to identify the causes of the most severe quantitative inconsistencies (systolic cavity and sanding) in the analyzed casting.

The brainstorming method was used to isolate all possible causes of incompatibilities in products and their hierarchy. The causes identified during the brainstorming session have been arranged in an Ishikawa diagram showing the interrelationship of causes causing a specific problem. Due to volume limitations, the article contains a part of the Ishikawa diagram containing the key causes of the most acute inconsistency - the systolic cavity, within the material (Fig. 5a) and the method (Fig. 5b).

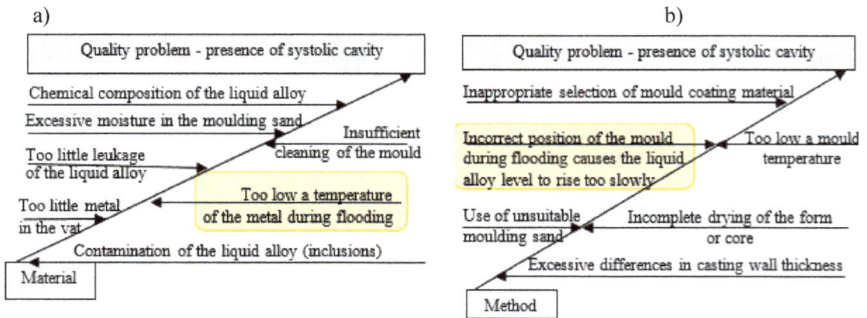

Fig. 5. *A fragment of the Ishikawa diagram showing the reasons f or the occurrence of the systolic cavity in the category "material" and "method".*

After analyzing the reasons for the presence of shrinkage cavities in castings, it was found that the main reason for the presence of shrinkage cavities in the material was too low a temperature of metal during flooding, while in the area of human beings the use of unsuitable molding sand was mentioned.

Conclusion

Continuous monitoring of the production process and improvement of the quality of manufactured products is the key to the success of every company. The presented proposal for a detailed analysis of the types of incompatibilities in castings, locating the areas where the most common defects are located, and identifying the reasons for the presence of defects in castings contributes to their elimination and implementation of effective measures to prevent the occurrence of incompatibilities in castings. The key reason for the most important defect in the casting (shrinkage cavity) was too low a temperature of the metal during flooding and incorrect setting of the mold during flooding causes too slow an increase in the level of liquid alloy.

Further research will be related to the implication of the proposed sequence of casting defect analysis, which is an effective way of solving quality problems within the production of other products offered by the company.

Materials Research Forum LLC
https://doi.org/10.21741/9781644902059-29

The analysis presented in this paper may be useful in many areas of research (non-destructive testing [24], special alloys [25], implants [26], tribological tests [27], welding [28], and biotechnology [29]) and industrial production (special coatings [30, 31], laser surface treatment [32, 33], machine regeneration [34], power hydraulics [35], steel industry [36]), where failures and improper actions may appear and should be repaired by corrective activities.

so the presented method may be inspiring for many recipients. Regardless, it is a development impulse both for the methods of experimental data analysis [37-39] and for the analysis of possible failure scenarios [40, 41].

Reference

[1] M. Łuszczak, R. Dańko. Stan zagadnienia w zakresie odlewania dużych odlewów strukturalnych ze stopów aluminium. Archives of Foundry Engineering 13 (2013) 113 -116.

[2] J. Pezda. Zastosowanie metody atnd do oceny właściwości mechanicznych okołoeutektycznego stopu AlSi12Cu2(Fe), Inżynieria Maszyn 22 (2017) 47-57.

[3] S. Kozakowski. Badanie odlewów – technologie odlewnicze, typowe dla nich wady i metody ich ujawniania. Biuro Gamma, Warszawa, 2001.

[4] K.E. Oczoś, A. Kawalec. Kształtowanie metali lekkich, PWN, Warszawa, 2012.

[5] Ł. Poloczek, A. Kiełbus. Wpływ czynników technologicznych na jakość odlewów ze stopów aluminium. Zarządzanie Przedsiębiorstwem 19 (2016) 14-19.

[6] Z. Falęcki. Analiza wad odlewów, AGH, Kraków, 1997.

[7] W. Łybacki, K. Zawadzka. Assistance of casting defects diagnosin by means of quality management tools, Archiwum Technologii Maszyn i Automatyzacji 28 (2008) 89-101.

[8] A. Pacana, L. Bednárová, J. Pacana. Wpływ wybranych czynników procesu produkcji folii orientowanej na jej odporność na przebicie, Przemysł Chemiczny 93 (2014) 2263-2264.

[9] A. Pacana, A. Gazda, D. Malindzak, R. Stefko. Study on improving the quality of stretch film by Shainin method, Przemysł Chemiczny 93 (2014) 243-245. https://doi.org/10.12916/przemchem.2014.243

[10] J.Tybulczuk, J. Seredyński, M. Szanda. Zarządzanie jakością w procesie produkcyjnym odlewów ze stopów Al-Si o specyficznych wymaganiach mechanicznych w „INNOWACJA" Sp. z o.o. w Nowej Dębie, Archiwum Odlewnictwa, Komisja Odlewnictwa Polskiej Akademii Nauk Oddział w Katowicach R. 6, nr 18/1, 2006.

[11] Thoni Alutec Ltd.. Unpublished papers, Stalowa Wola, 2021.

[12] PN-EN 1706:2011 (2011). Aluminum and aluminum alloys Castings. Chemical composition and mechanical properties, Warszawa: PKN

[13] Y. Briol. Effect of solution heat treatment on the age hardening capacity of dendritic and globular AlSi7Mg0.6 alloys, Int. J. Mater. Res. 101 (2010) 439-444.

[14] A. Pacana, K. Czerwinska, L. Bednarova. Comprehensive improvement of the surface quality of the diesel engine piston. Metalurgija 58 (2019) 329-332.

[15] A. Salomon, C. Voigt, O. Fabrichnaya, C. Aneziris, D. Rafaja. Formation of Corundum, Magnesium Titanate, and Titanium(III) Oxide at the Interface between Rutile and Molten Al or AlSi7Mg0.6 Alloy, Advanced Engineering Materials 19 (2017) art. 1700106. https://doi.org/10.1002/adem.201700106

[16] Q. Yang, C. Xia, Y. Deng, X. Li, H. Wang. Microstructure and mechanical properties of AlSi7Mg0.6 aluminum alloy fabricated by wire and arc additive manufacturing based on cold metal transfer (WAAM-CMT), Materials 12 (2019) art. 2525. https://doi.org/10.3390/ma12162525

[17] G. Ostasz, K. Czerwinska, A. Pacana. Quality management of aluminum pistons with the use of quality control points. Management Systems in Production Engineering 28 (2020) 29-33.

[18] P. Cavaliere, E. Cerri, P. Leo. Effect of heat treatment on mechanical properties and fracture behawior of a thixocas A356 aluminum alloy. Mater. Sci. 39 (2004) 1653- 1658. https://doi.org/10.1023/B:JMSC.0000016165.99666.dd.

[19] K. Czerwińska, P. Bełch, M. Hajduk-Stelmachowicz, D. Siwiec, A. Pacana. Doskonalenie procesu grafitowania wyrobów aluminiowych, Przemysł Chemiczny 100 (2021) 1191-1193. https://doi.org/10.15199/62.2021.12.8

[20] L. Hurtalová, J. Belan, E. Tillová, M. Chalupová. Changes in Structural Characteristics of Hypoeutectic Al-Si Cast Alloy after Age Hardening, Medziagotyra 18 (2012) 228-233. https://doi.org/10.5755/j01.ms.18.3.2430

[21] S. Pysz, M. Maj, E. Czekaj. High-Strength Aluminium Alloys and Their Use in Foundry Industry of Nickel Superalloys, Archives of Foundry Engineering 14 (2014) 71-76.

[22] L. Hurtalová, E. Tillová, M. Chalupová. The structure analysis of secondary (Recycled) AlSi9Cu3 cast alloy with and without heat treatment, Engineering Transactions 61 (2013) 197-218.

[23] M.G. Mueller, M. Fornabaio, G. Zagar, A. Mortensen. Microscopic strength of silicon particles in an aluminum-silicon alloy, Acta Materialia 105 (2016) 165-175. https://doi.org/10.1016/j.actamat.2015.12.006

[24] A. Pacana, D. Siwiec, L. Bednárová. Method of choice: A fluorescent penetrant taking into account sustainability criteria, Sustainability 12 (2020) art. 5854. https://doi.org/10.3390/su12145854

[25] A. Dudek, B. Lisiecka, R. Ulewicz. The effect of alloying method on the structure and properties of sintered stainless steel, Archives of Metallurgy and Materials 62 (2017) 281-287. https://doi.org/10.1515/amm-2017-0042

[26] A. Dudek, M. Klimas. Composites based on titanium alloy Ti-6Al-4V with an addition of inert ceramics and bioactive ceramics for medical applications fabricated by spark plasma sintering (SPS method), Materialwissenschaft und Werkstofftechnik 46 (2015) 237-247. https://doi.org/10.1002/mawe.201500334

[27] J. Brontček, P. Fabian, N. Radek. Tribological research of properties of heat-treated cast irons with globular graphite, Materials Science Forum 818 (2015) 209-212. https://doi.org/10.4028/www.scientific.net/MSF.818.209

[28] I. Miletić, A. Ilić, R.R. Nikolić, R. Ulewicz, L. Ivanović, N. Sczygiol. Analysis of selected properties of welded joints of the HSLA Steels, Materials 13 (2020) art.1301. https://doi.org/10.3390/ma13061301

[29] E. Skrzypczak-Pietraszek. Phytochemistry and biotechnology approaches of the genus *Exacum*. In: The Gentianaceae - Volume 2: Biotechnology and Applications, 2015, 383-401. https://doi.org/10.1007/978-3-642-54102-5_16

[30] A. Szczotok, N. Radek, R. Dwornicka. Effect of the induction hardening on microstructures of the selected steels. METAL 2018 – 27th Int. Conf. Metall. Mater. (2018), Ostrava, Tanger 1264-1269.

[31] N. Radek, J. Pietraszek, A. Gadek-Moszczak, Ł.J. Orman, A. Szczotok. The morphology and mechanical properties of ESD coatings before and after laser beam machining, Materials 13 (2020) art. 2331. https://doi.org/10.3390/ma13102331

[32] Ł.J. Orman Ł.J., N. Radek, J. Pietraszek, M. Szczepaniak. Analysis of enhanced pool boiling heat transfer on laser-textured surfaces. Energies 13 (2020) art. 2700. https://doi.org/10.3390/en13112700

[33] N. Radek, J. Konstanty, J. Pietraszek, Ł.J. Orman, M. Szczepaniak, D. Przestacki. The effect of laser beam processing on the properties of WC-Co coatings deposited on steel. Materials 14 (2021) art. 538. https://doi.org/10.3390/ma14030538

[34] S. Marković, D. Arsić, R.R. Nikolić, V. Lazić, B. Hadzima, V.P. Milovanović, R. Dwornicka, R. Ulewicz. Exploitation characteristics of teeth flanks of gears regenerated by three hard-facing procedures, Materials 14 (2021) art. 4203. https://doi.org/10.3390/ma14154203

[35] Barucca G. et al. PANDA Phase One: PANDA collaboration. European Physical Journal A 57 (2021) art. 184. https://doi.org/10.1140/epja/s10050-021-00475-y

[36] A. Maszke, R. Dwornicka, R. Ulewicz. Problems in the implementation of the lean concept at a steel works – Case study, MATEC Web of Conf. 183 (2018) art.01014. https://doi.org/10.1051/matecconf/201818301014

[37] T. Styrylska, J. Pietraszek. Numerical modeling of non-steady-state temperature-fields with supplementary data. Zeitschrift für Angewandte Mathematik und Mechanik 72 (1992) T537-T539.

[38] J. Pietraszek. Response surface methodology at irregular grids based on Voronoi scheme with neural network approximator. 6th Int. Conf. on Neural Networks and Soft Computing JUN 11-15, 2002, Springer, 250-255. https://doi.org/10.1007/978-3-7908-1902-1_35

[39] J. Pietraszek, R. Dwornicka, A. Szczotok. The bootstrap approach to the statistical significance of parameters in the fixed effects model. ECCOMAS 2016 – Proc. 7th European Congress on Computational Methods in Applied Sciences and Engineering 3, 6061-6068. https://doi.org/10.7712/100016.2240.9206

[40] G. Filo, J. Fabiś-Domagała, M. Domagała, E. Lisowski, H. Momeni. The idea of fuzzy logic usage in a sheet-based FMEA analysis of mechanical systems, MATEC Web of Conf. 183 (2018) art.3009. https://doi.org/10.1051/matecconf/201818303009

[41] A. Kubecki, C. Śliwiński, J. Śliwiński, I. Lubach, L. Bogdan, W. Maliszewski. Assessment of the technical condition of mines with mechanical fuses, Technical Transactions 118 (2021) art. e2021025. https://doi.org/10.37705/TechTrans/e2021025

Terotechnology XII
Materials Research Proceedings **24** (2022) 204-211

Materials Research Forum LLC
https://doi.org/10.21741/9781644902059-30

Improving the Quality of Products from Cast Alloy with the use of Grey Relational Analysis (GRA)

SIWIEC Dominika[1,a] , DWORNICKA Renata [2,b] and PACANA Andrzej [1,c]

[1]Rzeszow University of Technology, The Faculty of Mechanical Engineering and Aeronautics, al. Powstancow Warszawy 12, 35-959, Rzeszow, Poland

[2]Cracow University of Technology, Faculty of Mechanical Engineering, al. Jana Pawła II 37, 31-864 Krakow, Poland

[a]d.siwiec@prz.edu.pl, [b]renata.dwornicka@pk.edu.pl, [c]app@prz.edu.pl

Keywords: Mechanical Engineering, Quality Management, Decision Support, GRA Method, Grey Relational Analysis, Quality of Product, Product Improvement, Ishikawa Diagram

Abstract. As part of the quality of products, corrective actions are necessary. In this context, organizations should use different methods to skillfully analyze product incompatibilities and their causes. The aim of a study is to propose the method for determining the degree of impact of causes on the occurrence of incompatibilities. This method is a new combination of Grey Relational Analysis (GRA) with known methods: brainstorm, causes and effects diagram, and important technique. The method was tested for an example of a cast alloy product.

Introduction

In the context of effective improvement of the quality of products, determining the sequence of corrective actions is legitimate, e.g. by using the incompatibilities catalog or Rule 20/80 (Pareto-Lorenz) [1-4]. However, it still happens that organizations do not note the number of incompatibilities and causes of its occurrence. Therefore, it is problematic to indicate the main incompatibility of mentioned techniques. For this purpose, other techniques are used.

The literature review of the subject has shown that these techniques are, e.g. the Ishikawa diagram [2, 4, 5] and brainstorm [3, 6]. In these techniques, the mentioned main incompatibility mentioned was not mostly specified, but only the causes of incompatibility were specified [2, 6]. Despite that, it is necessary to use other techniques, mainly effective to solve the problems in the so-called fuzzy (uncertain) environment [7, 8]. One of these not complicated methods is, e.g. grey relational analysis (GRA) [9,10], which has not been used yet in combination with the mentioned techniques. For this reason, the proposal of a new combination of quality management techniques with the GRA method was adequate [11-15].

It was assumed that if it is impossible to clearly indicate the main incompatibility of the product (e.g., based on the number of incompatibilities), it is possible to indicate it based on the degree of impact of the causes of the incompatibility. It is possible by the integration of the techniques, i.e., GRA method, brainstorm, cause and effect diagram, and importance technique.

Problem of Research

In a production company located in south-eastern Poland, the problem of quality alloy products was identified. The problem was four types of incompatibilities that were often repeated, i.e.: porosity, nonmetallic inclusions, cracks, and brick. The incompatibilities were identified by NDT, i.e. the fluorescent method (FPI) and the magnetic-powder method (MT). The mentioned incompatibilities with different types of products were identified, e.g., mechanical seal or bearing

Materials Research Forum LLC

https://doi.org/10.21741/9781644902059-30

housing. However, in the company, the catalog of the number of incompatibilities was not made. Therefore, the problem was to unambiguous determine, which incompatibility is the most common. It made it difficult to accession effective corrective actions for incompatibility which generates the most source of waste. Despite it, methods that allow for a precise determine the causes of these incompatibilities were not used. It was claimed that the causes of these incompatibilities are, e.g. pollution and defective materials from the supplier. However, the main incompatibility was not yet clearly determined yet. To this aim, the integrated technique supported by grey relational analysis (GRA) was proposed.

Method

The integrated method is Grey Relational Analysis (GRA), which was combined with: a brainstorm, cause and effect diagram, and importance technique on the Likert scale. The choice of these methods was conditioned by their application to solving decision problems [5, 6, 9]. In turn, the choice of the GRA method was conditioned by its applications for a small number of criteria (that is, even 4 data) [9, 10, 12, 14]. The integrated method algorithm is shown in Figure 1.

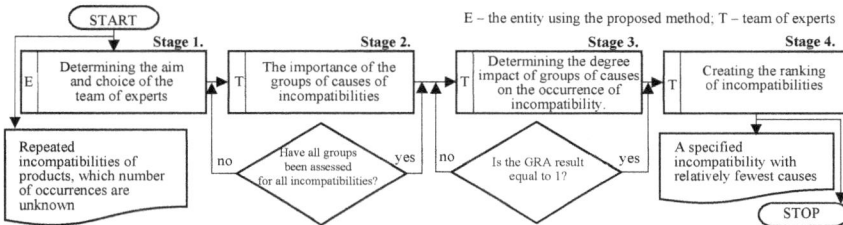

*Fig. 1. The algorithm of an integrated method of improving
the product quality supports by GRA. Own study.*

Stage 1. Determining the purpose and choice of the team of experts. The aim of the proposed method should apply to the need to identify the incompatibility on which the relatively fewest causes are influenced. The number of selected incompatibilities to analyze is at least 4 [9, 10]. The SMART method can be used to define the aim [15]. Members of the selected team of experts should have experience and knowledge in the area of product incompatibilities of products [16, 17].

Stage 2. The importance of groups of causes of incompatibility. This stage is realized by the team of experts using in a combined way: brainstorming (BM), cause and effects diagram, and importance technique. Initially, the team of experts reports the causes of incompatibility using the first of two stages of BM, as, e.g. [6]. It is recommended that the same team conducts the BM separately for each type of incompatibility. Then, the team of experts groups the causes of individual incompatibilities into groups of thematic (categories). In this aim, the members of a team determine any number and type of group, that is, 5M + E, man, machines, method, material, environment, and management [2]. Then, the team notes for each cause the name of its category. Next, the team of experts makes the cause and effect diagram according to the method shown, e.g. in [2, 4, 5]. This diagram should be made separately for each incompatibility and shown in a place visible to the team. Then, the team of experts validated the thematic (category) groups, in which the causes of incompatibility were noted. It consists of assessments in the Likert scale the causes of incompatibilities in the causes group [18]. By voting, the team assesses the importance (from 1 to 5) of the group of causes in a given category for the emergence of this type of incompatibility. Assessment 1 - It is the smallest influence (weight) of the group of causes on occurrence of the incompatibility, in turn, the assessment 5 - It is the greatest impact (weight) of the group of causes on occurrence of the incompatibility. Assessments should be noted, for example, in causes and effect diagrams [19-22]. At this stage, the grouped and visualized groups of causes and the weights of groups, which determine the impact the causes of occurrence the incompatibility, are achieved.

Stage 3. Determining the degree of impact of groups of causes on the occurrence of incompatibility. This stage is performed by the team of experts as part of using the Grey Relational Analysis (GRA) [9, 10, 12, 14]. Initially, the matrix M is created, i.e., $M = m \times n$, where m – alternative (that is, incompatibility), n is the criterion (i.e. group of causes). This matrix should be supplemented with the weights from stage 2. Subsequently, the ratings (weights) of the cause groups should be transformed (normalized) so that the rating values are between 0 and 1. It is accepted that $x_0^{(O)}(k)$ and $x_i^{(O)}(k)$ are, respectively sequence original and comparison; $i = 1, 2, \ldots, m$; $k = 1, 2, \ldots, n$; and m – alternative (i.e., incompatibility), n – criterion (i.e., group of causes) [9]. For causes of „lower assessment is better" the formula (1) is used [9, 12]:

$$x_i^*(k) = \frac{max\ x_i^{(O)}(k) - x_i^{(O)}(k)}{max\ x_i^{(O)}(k) - min\ x_i^{(O)}(k)} \tag{1}$$

In this context, it is accepted that the better is when fewer causes have influence on incompatibility. Then, based on normalized sequences, the grey relational coefficient is calculated (2) [9, 10, 14]:

$$\gamma[x_0^*(k), x_i^*(k)] = \frac{\Delta_{min} + \xi \Delta_{max}}{\Delta_{0i}(k) + \xi \Delta_{max}} \tag{2}$$

$$0 < \gamma[x_0^*(k), x_i^*(k)] \leq 1$$

where: $\Delta_{0i}(k)$ the sequence of deviations between the original sequence $x_0^*(k)$ and comparison sequence $x_i^*(k)$, which is calculated from formula (3) [9]:

$$\Delta_{0i}(k) = |x_0^*(k) - x_i^*(k)| \tag{3}$$

Similarly, the largest (4) and smallest (5) deviations are calculated [9]:

$$\Delta_{max} = \max_{\forall j \in i} \max_{\forall k} |x_0^*(k) - x_j^*(k)| \tag{4}$$

$$\Delta_{min} = \min_{\forall j \in i} \min_{\forall k} |x_0^*(k) - x_j^*(k)| \tag{5}$$

In turn, the coefficient ξ in formula (2) has value [0, 1]. Most often, it is assumed that $\xi = 0.5$ [9, 10]. The relational grey score is the weighted sum of the Grey coefficients as shown in formula (6) [9, 14]:

$$\gamma(x_0^*, x_i^*) = \sum_{k=i}^{n} \beta_k \gamma[x_0^*(k), x_i^*(k)] \tag{6}$$

where $\gamma(x_0^*, x_i^*)$, i.e. gray relational score, which is the level of correlation between the sequence of originally and the comparison, as if it were identical. The grey relational score should be equal to 1, that is, (7) [9]:

$$\sum_{k=1}^{n} \beta_k = 1 \qquad\qquad (7)$$

Consequently, based on the results from GRA, it is possible to determine the degree of impact of groups of causes of occurrence and the incompatibilities.

Step 4. Creating the ranking of incompatibilities. It consists of ordering the values obtained from the GRA from minimum to maximum. The minimum value (first position in the ranking) is incompatibility, which according to the team of experts is characterized by the lowest degree of impact of the group of causes on its occurrence. It is proposed that first corrective actions were taken for this incompatibility. Then, the corrective actions would be a concern for fewer causes. After the implementation of the corrective actions for this incompatibility, it is possible to make corrective actions for the incompatibility on the next position in the ranking.

Results
The aim was to identify the incompatibility of the product from which corrective actions are necessary. So, the incompatibility was affected by a relatively small number of causes. The four types of incompatibilities of alloy products were analyzed: porosity on the mechanical seal of the 410 alloy, non-metallic inclusions on the mechanical seal of the CPW 407 alloy, cracks in the outer casing from CPW-S 5616 alloy, and prick on the lever made of the CPW-S 5613 alloy.

An example after BM performed, grouped according to 5M + E, causes and their assessment on the Likert scale is shown in Fig. 2.

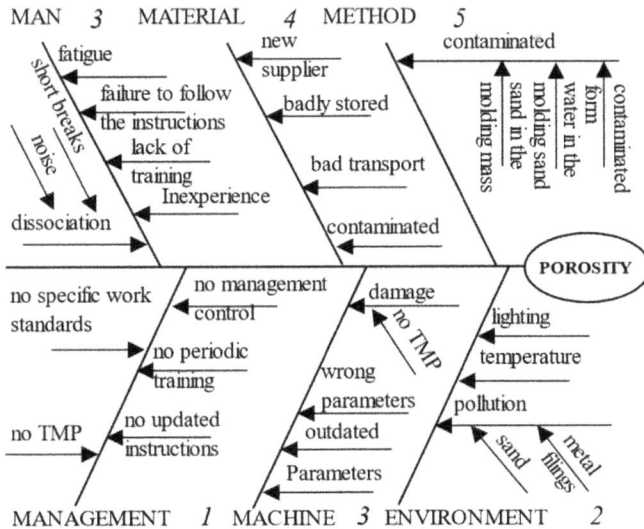

Fig. 2. *Example of a cause and effects diagram for the problem of porosity in the mechanical seal (own study).*

The matrix M with evaluations (weights) of groups of causes (M1-E) for selected incompatibilities (I1-I4) is shown in Table 1. The results obtained from the GRA method are shown in Table 2.

Table 1. *Assessment of groups of causes of the incompatibilities analyzed (own study).*

M	M1	M2	M3	M4	M5	E
I1	3	4	5	1	3	2
I2	2	5	5	2	3	1
I3	1	3	2	2	5	3
I4	5	2	4	1	2	1

Table 2. *The results of the GRA method (own study).*

G	M1	M2	M3	M4	M5	E	Rank	
I1	0,5	0,4	0,3	1,0	0,6	0,5	0,6	2
I2	0,7	0,3	0,3	0,3	0,6	1,0	0,5	3
I3	1,0	0,6	1,0	0,3	0,3	0,3	0,6	2
I4	0,3	1,0	0,4	1,0	1,0	1,0	0,8	1

It was shown that due to the I4 incompatibility (prick) it is necessary to make corrective actions are necessary.

Conclusion

The aim was to propose an integrated method to determine the degree of impact of causes of occurrence of incompatibilities. The incompatibilities analyzed in the company were: porosity on the mechanical seal of 410 alloy, non-metallic inclusions on the mechanical seal of CPW 407 alloy, cracks on the outer casing from CPW-S 5616 alloy, and prick on the lever made of CPW-S 5613 alloy. After using the proposed method, it was shown that the incompatibility of a prick on the lever of the CPW-S 5613 alloy was affected by the fewest causes. Therefore, corrective actions were proposed for this incompatibility were proposed. The analysis conducted confirmed that with the proposed method, it is possible to relatively precisely indicate the main incompatibility according to a certain degree of influence of the causes on their occurrence. Therefore, this method is an effective instrument to solve problems in an uncertain environment and to determine the main incompatibility of products.

The analysis presented in this paper may be useful in many areas of research (special alloys [23], implants [24], tribological tests [25], welding [26] and biotechnology [29]) and industrial production (special coatings [30, 31], laser surface treatment [32, 33], machine regeneration [34], power hydraulics [35], steel industry [36]), where failures and improper actions may appear, and should be repaired by corrective activities. Regardless, it is a development impulse both for the methods of experimental data analysis [37-39] and for the analysis of possible failure scenarios [40, 41].

References

[1] A. Hoła, M. Sawicki, M. Szóstak. Methodology of Classifying the Causes of Occupational Accidents Involving Construction Scaffolding Using Pareto-Lorenz Analysis, Applied Sciences 8 (2018) art. 48. https://doi.org/10.3390/app8010048

[2] A. Pacana, D. Siwiec. Universal Model to Support the Quality Improvement of Industrial Products, Materials 14 (2021) art. 7872. https://doi.org/10.3390/ma14247872

Terotechnology XII	Materials Research Forum LLC
Materials Research Proceedings **24** (2022) 204-211	https://doi.org/10.21741/9781644902059-30

[3] A. Pacana, D. Siwiec, L. Bednárová. Method of Choice: A Fluorescent Penetrant Taking into Account Sustainability Criteria, Sustainability 12 (2020) art. 5854. https://doi.org/10.3390/su12145854

[4] A. Pacana, D. Siwiec, L. Bednárová. Analysis of the incompatibility of the product with fluorescent method, Metalurgija 58 (2019) 337-340.

[5] L. Liliana. A new model of Ishikawa diagram for quality assessment, IOP Conf. Series: Materials Science and Engineering 161 (2016) art. 012099. https://doi.org/10.1088/1757-899X/161/1/012099

[6] J. Rossiter, G. Lilien. New "Brainstorming" Principles, Aust. J. Manag. 19 (1994) 61-72. https://doi.org/10.1177/031289629401900104

[7] R. Bellman, L. Zadeh. Decision-Making in a fuzzy environment. Management Science 17 (1970) b-141-64. https://doi.org/10.1287/mnsc.17.4.B141

[8] D. Siwiec, A. Pacana. Model Supporting Development Decisions by Considering Qualitative–Environmental Aspects, Sustainability 13 (2021) art. 9067. https://doi.org/10.3390/su13169067

[9] I. Ertugrul, T. Öztaş, A. Ozcil, G. Oztas. Grey Relational Analysis Approach In Academic Performance Comparison Of University: A Case Study Of Turkish Universities, European Scientific Journal 12 (2016) 128-139.

[10] S.A. Javed, K.A. Mashood, D. Wenjie, R. Adil, L. Sifeng. Systems evaluation through new grey relational analysis approach: An application on thermal conductivity-petrophysical parameters' relationships, Processes 7 (2019) art. 348. https://doi.org/10.3390/pr7060348

[11] A. Pacana, D. Siwiec. Model to Predict Quality of Photovoltaic Panels Considering Customers' Expectations, Energies 15 (2022) art. 1101. https://doi.org/10.3390/en15031101

[12] A.J. Liu, Q. Zhu, X. Ji, H. Lu, S-B. Tsai. Novel method for perceiving key requirements of customer collaboration low-carbon product design, International Journal of Environmental Research and Public Health 15 (2018) art. 1446. https://doi.org/10.3390/ijerph15071446

[13] D. Siwiec, A. Pacana. A Pro-Environmental Method of Sample Size Determination to Predict the Quality Level of Products Considering Current Customers' Expectations. Sustainability 13 (2021) art. 5542. https://doi.org/10.3390/su13105542

[14] F. Wang, H. Li, M. Dong. Capturing the key customer requirements for complex equipment design using Grey Relational Analysis, Journal of Grey System 27 (2015) 51-70.

[15] K. Lawlor, M. Hornyak. Smart Goals: How The Application of Smart Goals Can Contribute to Achievement of Student Learning Outcomes, Developments in Business Simulation and Experiential Learning 39 (2012) 259-267.

[16] D. Siwiec, A. Pacana. Model of Choice Photovoltaic Panels Considering Customers' Expectations, Energies 14 (2021) art. 5977. https://doi.org/10.3390/en14185977

[17] W. Kupraszewicz, B. Żółtowski. Dobór zespołu ekspertów do diagnozowania stanu maszyn, Diagnostyka 26 (2002) 94-100.

[18] A. Joshi et al., Likert Scale: Explored and Explained, Current Journal of Applied science and Technology 7 (2015) 4 396-403. https://doi.org/10.9734/BJAST/2015/14975

[19] G. Ostasz, K. Czerwinska, A. Pacana. Quality management of aluminum pistons with the use of quality control points, Manag. Systems Prod. Eng. 28 (2020) 29-33. https://doi.org/10.2478/mspe-2020-0005

[20] R. Ulewicz, D. Siwiec, A. Pacana, M. Tutak, J. Brodny. Multi-Criteria Method for the Selection of Renewable Energy Sources in the Polish Industrial Sector, Energies 14 (2021) art.2386. https://doi.org/10.3390/en14092386

[21] Wolniak R., Application methods for analysis car accident in industry on the example of power, Systemy Wspomagania w Inżynierii Produkcji 6 (2017) 34-40.

[22] A. Pacana, D. Siwiec, L. Bednarova, M. Sofranko, O. Vegsoova, M. Cvoliga. Influence of Natural Aggregate Crushing Process on Crushing Strength Index, Sustainability 13 (2021) art. 8353. https://doi.org/10.3390/su13158353

[23] A. Dudek, M. Klimas. Composites based on titanium alloy Ti-6Al-4V with an addition of inert ceramics and bioactive ceramics for medical applications fabricated by spark plasma sintering (SPS method), Materialwissenschaft und Werkstofftechnik 46 (2015) 237-247. https://doi.org/10.1002/mawe.201500334

[24] A. Dudek, R. Wlodarczyk. Structure and properties of bioceramics layers used for implant coatings, Solid State Phenom. 165 (2010) 31-36. https://doi.org/10.4028/www.scientific.net/SSP.165.31

[25] J. Brontek, P. Fabian, N. Radek. Tribological research of properties of heat-treated cast irons with globular graphite, Materials Science Forum 818 (2015) 209-212. https://doi.org/10.4028/www.scientific.net/MSF.818.209

[26] I. Miletić, A. Ilić, R.R. Nikolić, R. Ulewicz, L. Ivanović, N. Sczygiol. Analysis of selected properties of welded joints of the HSLA Steels, Materials 13 (2020) art.1301. https://doi.org/10.3390/ma13061301

[27] E. Skrzypczak-Pietraszek. Phytochemistry and biotechnology approaches of the genus *Exacum*. In: The Gentianaceae - Volume 2: Biotechnology and Applications, 2015, 383-401. https://doi.org/10.1007/978-3-642-54102-5_16

[28] A. Szczotok, N. Radek, R. Dwornicka. Effect of the induction hardening on microstructures of the selected steels. METAL 2018 – 27th Int. Conf. Metall. Mater. (2018), Ostrava, Tanger 1264-1269.

[29] N. Radek, J. Pietraszek, A. Gadek-Moszczak, Ł.J. Orman, A. Szczotok. The morphology and mechanical properties of ESD coatings before and after laser beam machining, Materials 13 (2020) art. 2331. https://doi.org/10.3390/ma13102331

[30] Ł.J. Orman Ł.J., N. Radek, J. Pietraszek, M. Szczepaniak. Analysis of enhanced pool boiling heat transfer on laser-textured surfaces. Energies 13 (2020) art. 2700. https://doi.org/10.3390/en13112700

[31] N. Radek, J. Konstanty, J. Pietraszek, Ł.J. Orman, M. Szczepaniak, D. Przestacki. The effect of laser beam processing on the properties of WC-Co coatings deposited on steel. Materials 14 (2021) art. 538. https://doi.org/10.3390/ma14030538

[32] S. Marković, D. Arsić, R.R. Nikolić, V. Lazić, B. Hadzima, V.P. Milovanović, R. Dwornicka, R. Ulewicz. Exploitation characteristics of teeth flanks of gears regenerated by three hard-facing procedures, Materials 14 (2021) art. 4203. https://doi.org/10.3390/ma14154203

[33] Barucca G. et al. PANDA Phase One: PANDA collaboration. European Physical Journal A 57 (2021) art. 184. https://doi.org/10.1140/epja/s10050-021-00475-y

[34] A. Maszke, R. Dwornicka, R. Ulewicz. Problems in the implementation of the lean concept at a steel works – Case study, MATEC Web of Conf. 183 (2018) art.01014. https://doi.org/10.1051/matecconf/201818301014

[35] T. Styrylska, J. Pietraszek. Numerical modeling of non-steady-state temperature-fields with supplementary data. Zeitschrift für Angewandte Mathematik und Mechanik 72 (1992) T537-T539.

[36] J. Pietraszek. Response surface methodology at irregular grids based on Voronoi scheme with neural network approximator. 6th Int. Conf. on Neural Networks and Soft Computing JUN 11-15, 2002, Springer, 250-255. https://doi.org/10.1007/978-3-7908-1902-1_35

[37] J. Pietraszek, R. Dwornicka, A. Szczotok. The bootstrap approach to the statistical significance of parameters in the fixed effects model. ECCOMAS 2016 – Proc. 7th European Congress on Computational Methods in Applied Sciences and Engineering 3, 6061-6068. https://doi.org/10.7712/100016.2240.9206

[38] G. Filo, J. Fabiś-Domagała, M. Domagała, E. Lisowski, H. Momeni. The idea of fuzzy logic usage in a sheet-based FMEA analysis of mechanical systems, MATEC Web of Conf. 183 (2018) art.3009. https://doi.org/10.1051/matecconf/201818303009

[39] A. Kubecki, C. Śliwiński, J. Śliwiński, I. Lubach, L. Bogdan, W. Maliszewski. Assessment of the technical condition of mines with mechanical fuses, Technical Transactions 118 (2021) art. e2021025. https://doi.org/10.37705/TechTrans/e2021025

Terotechnology XII

Materials Research Proceedings **24** (2022) 212-220

Materials Research Forum LLC

https://doi.org/10.21741/9781644902059-31

Alternative Fuels in Rail Transport and their Impact on Fire Hazard

RADZISZEWSKA-WOLIŃSKA Jolanta Maria

Instytut Kolejnictwa (the Railway Research Institute), 50, Chlopicki Street, 04-275 Warsaw, Poland

jradziszewska-wolinska@ikolej.pl

Keywords: Hydrogen Cells, Lithium Batteries, Natural Gas CNG and LNG, Fire Properties, Hybrid Rail Vehicles

Abstract. The interest in alternative fuels results from the depletion of crude oil resources and, thus, the search for new energy sources. The article discusses alternative power sources introduced into the drive of rail vehicles. Their advantages and disadvantages are presented, especially the uncontrolled ignitions cases are described. Attention is drawn to the need to develop European regulations on the fire safety of rail vehicles using hydrogen cells, lithium batteries, or natural gas. Such requirements are particularly relevant for rolling stock passing through tunnels. They should include alternative propulsion vehicles and the transport of such loads (e.g. discharged lithium batteries).

Introduction

Most of the discoveries of oil deposits had taken place until the 1960s. In the following decades, discoveries were smaller and smaller despite technological progress and ongoing intensive searches carried out using the most modern technologies, including satellite research. Moreover, they are found in more and more inaccessible regions that require much more complex and costly mining and processing techniques. This is due to the higher oil density and the exploitation of deposits supersaturated with toxic hydrogen sulfide (e.g. Kashgan deposits), which require special precautions. Currently, oil consumption is greater than new discoveries, and the availability of its resources is estimated at 40 to 100 years [1, 20].

Considering the above and striving to reduce environmental pollution by gases and solid particles emitted in the combustion processes of hydrocarbon fuels, the search for alternative fuels to be used in various areas, including means of transport, has been undertaken. Within the meaning of the Directive [28], these include fuels that serve, at least in part, as a substitute for crude oil-based energy in transport and which have the potential to reduce the dependence of EU Member States on oil imports. An important argument is also the desire to decarbonize transport and improve the environmental performance of this sector. Alternative fuels include, among others: electricity, hydrogen, biofuels, synthetic and paraffinic fuels, natural gas (including biomethane) in the form of compressed natural gas CNG and liquefied natural gas LNG and liquefied gas LPG.

The search activities were enforced by legislative acts, including the European Green Deal approved by the European Commission on December 13, 2019 [29]. It assumes, among other things, reducing greenhouse gases in the transport industry by 95% by 2050.

Alternative Fuels in Rail Transport

In rail vehicles, the following are primarily intended for energy supply: electric batteries, natural gas, hydrogen, and hybrid supply.

Hydrogen. The prospect of using hydrogen as a fuel has a good chance due to its unlimited resources (it accounts for 94% of the universe [4]) and because it is considered the cleanest fuel from an ecological point of view (the effect of burning hydrogen is water vapor). A feature of hydrogen is its high diffusion coefficient of H_2 in the air. As a result, it easily creates a homogeneous combustible mixture. The wide yields of the mixture (0.14 – 9.9) allow the use of qualitative regulation in the engine by changing the mixture composition. Hydrogen is the lightest element in any aggregation state; it has the highest fuel heating value per mass (120 MJ/kg). Moreover, it is a very reactive fuel with a high octane number and combustion speed. Therefore, it is considered the fuel of the future. It is used in fuel cells. Their operation scheme was developed as early as 1838 by the German-Swiss chemist Christian Friedrich Schönbein. The cell consists of two electrodes – cathode and anode – separated by an electrolyte or an electrolytic membrane. Typically the electrodes are in the form of carburized platinum-coated paper as a reaction catalyst. After supplying hydrogen to the cell, it undergoes oxidation (gives away electrons), producing hydrogen cations. At the cathode, oxygen reacts with the electrons, reducing to oxygen anions. The membrane inside allows protons to flow from the anode to the cathode while blocking other ions, including the formed oxygen anions. Upon reaching the cathode, the hydrogen cations react with these oxide anions to give water, and the electrons from the anode reach the cathode via an electrical circuit, producing energy. However, the breakthrough in the use of hydrogen cells did not come until the 1960s, when they became part of NASA spacecraft.

Hydrogen, unfortunately, also has its disadvantages. It is a colorless and odorless flammable gas that can form explosive mixtures with air. The following basic parameters describing the combustible and explosive properties of hydrogen include:
– explosion limits in the air: 4.1-74.2% vol.,
– limits of detonation in the air: 15-63.5% vol.,
– maximum pressure increase during an explosion in a mixture with air: 625 kPa,
– auto-ignition temperature: 580 ° C,
– minimum ignition energy: 0.011–0.02 mJ.

Hydrogen belongs to the T1 temperature class and the IIC explosion group. The above means that this gas has a relatively low lower explosive limit and can burn in the air over a very wide concentration range. Little energy is required to initiate combustion. The safety data sheet for hydrogen describes it as an extremely flammable gas and gives the following precautions: keep it away from heat sources (heating a pressurized container above 50 ° C may cause it to explode). It is also essential for safety that the flame of hydrogen burning in the air at atmospheric pressure is invisible. Moreover, hydrogen is characterized by a relatively low inversion temperature (about 205 K), which means that during a gas expansion (e.g. when flowing out of a leak), the temperature of the flowing gas stream increases (due to the negative Joule-Thompson effect) [2].

Safe use (including storage) of hydrogen requires knowledge of its specific properties and the effect of cryogenic temperatures on the material's behavior. Structural steels from which tanks and fasteners are made may undergo hydrogen corrosion in an environment containing hydrogen (low-temperature – below 100 ° C and high-temperature – above 200 ° C, steel decarburization, HTHA – High-temperature hydrogen attack), which results in a significant reduction in the mechanical properties of steel, in particular a decrease in strength, an increase in plasticity and creep rate, and leads to the formation of microcracks [2, 12]. Hydrogen also has a negative effect on non-metallic materials (elastomers for valve seats, seals), and their use should be verified by reliable test results confirming the ability to maintain the required physicochemical characteristics of the selected material under the expected conditions of contact with hydrogen. Welds are also prone to hydrogen embrittlement in any hydrogen environment. In the heat-affected zone, the so-called "Hard spots", residual stresses and microstructures contributing to brittleness, often arise. Post-weld annealing may be required to restore the structure. When building an installation for contact with hydrogen, the requirements for welding metals [12] should be met (e.g. ASME B31.12-2014 [36]. As demonstrated by the tests and numerical simulations carried out in SNL [11], hydrogen leakage from the tank may cause combustion deflagration and even detonation in the tunnel.

The significant disadvantages of hydrogen cells are still high production costs (cheaper methods force the combustion of fossil fuels, which is not indifferent to the environment).

Rail Vehicles with Hydrogen-Powered Drive. The term "hydrail" was first used on August 22, 2003, during a presentation at the Volpe Transportation Systems Center of the US Department of Transportation in Cambridge. Potential applications for hydrogen-powered railways include all types of rail transport: suburban, long-distance, high-speed, freight, mine, factory, and special vehicles in parks and museums. [14].

The first fuel cell train to be introduced into public transport was the Coradia iLint multiple unit manufactured by Alstom in Salzgitter. From September 17, 2018, operates a route of approximately 100 km from Cuxhaven to Buxtehude in Lower Saxony (Germany). It should be noted that the critical elements of the drive system, in addition to the set of fuel cells, also include the lithium-ion battery system, as well as an external converter, hydrogen tank, traction inverter, and traction motor [1, 3].

The Polish company PESA from Bydgoszcz also designed the SM42Dn hydrogen-powered locomotive. It includes four asynchronous electric motors with a total power of 720 kW, LTO (lithium titanate) battery with a capacity of 167.7 kWh, and ABB traction inverters with an auxiliary converter 3 x 400 V. The hydrogen fuel in the PESA SM42Dn locomotive will be stored in tanks under the pressure of 350 bar. The power sources will be 2 Ballard fuel cells produced in Canada with a total power of 170 kW [15].

In January 2021, the European FCH2RAIL Project was launched with partners from Belgium, Germany, Spain, and Portugal. Their task is to develop a new zero-emission train prototype. The vehicle is to be equipped with a hybrid and modular drive system that combines electric power from the overhead contact line with fuel cell power. The latter part will consist of hydrogen fuel cells and batteries. They must be connected and controlled so that the system meets all requirements and is cost-effective. [13].

Lithium Batteries. At the beginning of the 20th century, the great potential of lithium as a battery material was noticed. It is a metal with the lowest density, high electrochemical potential, and a high energy-to-mass ratio. The American chemical physicist George Newton Lewis began early experiments with lithium batteries in 1912. However, it was not until the 1970s when research (undertaken by John Goodenough, Stanley Whittingham, and Akira Yoshino) led to the development of lithium-ion batteries (for which they received the Nobel Prize in Chemistry). Exxon launched the first lithium-based battery in 1978. However, only after Sony produced a series of lithium-ion batteries in 1991 became it commonplace [21, 22].

A lithium-ion (Li-Ion) battery consists of a positively charged graphite cathode and a negatively charged lithium anode. The other two parts contain an electrolyte and a separator between the two charged electrodes. During charging, the lithium ions move from the carbon anode to the cathode made of lithium oxide and another metal and are stored there. This technology allows you to

Terotechnology XII   Materials Research Forum LLC
Materials Research Proceedings 24 (2022) 212-220   https://doi.org/10.21741/9781644902059-31

accumulate twice as much energy as in nickel-metal hydride (NiMH) batteries of the same weight and size. Because lithium-ion batteries are one of the lightest, their use began with all kinds of electronic equipment. However, there has also been growing interest in using large lithium-ion battery packs (typically 20 to 100 kWh) in electric automotive vehicles in recent years. At the same time, car batteries differ significantly from those used in electronic equipment. The differences result mainly from the more significant requirements related to working conditions and the more excellent required durability, reaching ten years. Moreover, the packages are equipped with special cooling and heating systems, ensuring optimal operating temperature [21-23].

The next step in developing lithium batteries was the development of lithium titanate Li4Ti5O12 batteries (LTO). By working on this technology, manufacturers obtained a somewhat developed nanocrystalline anode structure, which became the main advantage of the products. Unlike the porous carbon used to create other types of lithium batteries, the nanocrystalline structure makes the large anode surface "usable" ensuring surface stability. The LTO technology enables an effective anode area of approximately 100 m²/g, while for carbon anodes, only about 3 m²/g. Due to the large anode area, the charge is transferred much faster, and the characteristics of permissible currents are higher. All this ensures the device's operating time, stability, and safety of use [18].

The advantages of lithium batteries compared to other types of batteries (e.g. NiCd and NiMH) include 3 times greater capacity with the same battery size and lower weight, many times greater energy density, no memory effect, no heavy metals in the composition (harmful to the environment), a high operating temperature range (even from -20 ° C to 50 ° C) as well as a longer service life [23].

Li-ion batteries also have disadvantages. Dangerous incidents related to their use can still be heard about. In most cases, a fire and an explosion occurred due to a short circuit between the anode and the cathode due to overheating of the cells, e.g. a fire from spilled fuel or overheating due to improper charging or overcharging. The result is a rapidly progressing chain reaction, often out of control. The temperature rises quickly, and the separator between the electrodes melts, which fuels the further heating of the battery. The loss of control over the system temperature increases causes the so-called thermal runaway (Fig.1).

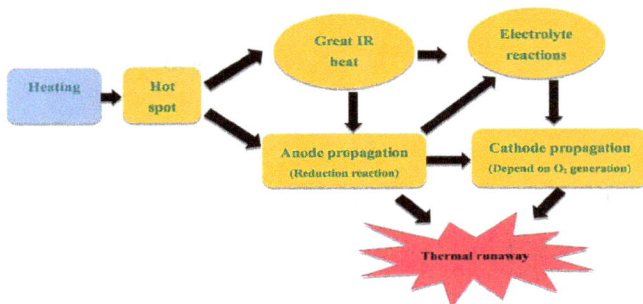

Fig.1 Mechanism of runaway in a lithium-ion battery under heating conditions [5]

The above causes a chain reaction, especially danger in batteries consisting of many cells because it initiates the reaction in subsequent, often undamaged, cells. As the temperature rises,

Terotechnology XII   Materials Research Forum LLC
Materials Research Proceedings **24** (2022) 212-220   https://doi.org/10.21741/9781644902059-31

the cathode emits oxygen which reacts with the organic electrolyte, eventually causing the battery to ignite or explode.

There are also battery design errors, among other non-mechanical causes of Li-ion battery fires. High-profile cases include explosions of batteries in telephones onboard airplanes, an explosion of an electric bicycle battery, or an outbreak of a battery in a Tesla car (which resulted in the death of the driver), where six days after the accident, there was a spontaneous explosion and another fire of the wreckage towed to the parking lot [16, 17]. While the largest fire in rail transport occurred on April 23, 2017, in the Union Pacific train, in which the container caught fire while carrying discharged lithium batteries to be recycled [19]. Although a long time has passed since these events, battery systems containing lithium-ion cells still pose a challenge in terms of fire safety. Extinguishing such fires is still a problem for emergency services. Research conducted by the Federal Aviation Administration (FAA) in the USA showed that water-based fire-fighting materials proved to be the most effective. Effective actions were based on the combination of the effect of extinguishing the burning electrolyte and the simultaneous cooling of the cell, so the role of as much water as possible as a cooling factor turned out to be very important: the more water, the better. Gas-based refrigerants showed good ability to extinguish the burning electrolyte but did not provide adequate ability to cool the cell. Therefore there is a risk (as in the case of the Tesla accident mentioned above) that another spontaneous fire may occur even many hours after the first fire is extinguished.

Natural Gas. It is often called the blue fuel and even the fuel of the 21st century. Its composition may change because it is influenced by the place where the deposits are exploited. However, the main component of natural gas is always methane. In addition, natural gas may contain various amounts of such gases as ethane, propane, butane, nitrogen, and organic and mineral compounds. In addition, noble gases (helium, argon) also appear there. Natural gas has no smell whatsoever. It is specially odorized to make it easier to smell when leaking.

Natural gas is used to power vehicles in compressed (CNG) or liquefied (LNG) form. This fuel has several advantages over other propulsion methods, including high calorific value and low exhaust emissions: reducing CO_2 emissions by 10% and particulate matter (soot) by 100%. Therefore, it has also been used in rail vehicles. For example, it is intended for use in the following vehicles:
- freight locomotive on the route Jacksonville - Miami (Florida, USA) [24],
- new railbuses in the Czech Giant Mountains on the Martinice v Krkonoších railway line from Rokytnice to Jizerou, in 2022 [25],
- a railbus used to transport tourists in the Rimini area (Italy) [26],
- Renfe trains (Spain) under the EU-funded RaiLNG program [27].

Fire Safety of Rail Vehicles with Alternative Drive
Passenger rail vehicles running on an interoperable TEN have to comply with the requirements of the LOC & PAS TSI [30]. On the other hand, the freight rolling stock is covered by the requirements of TSI WAG [31]. In addition, the SRT TSI [32] applies to rolling stock running in tunnels. In these specifications, the requirements of the EN 45545 series of standards (Part 1 to 7) [36] are referred to, depending on the area, in the scope of fire safety. The general purpose of these standards is to ensure the safety of passengers and train crew. The series of standards includes fire protection measures and requirements aimed at minimizing the likelihood of a fire occurring and controlling the speed and extent of fire spread, i.e., as a result: minimizing the impact of the products of a possible fire on passengers and crew. Part 7 specifies the requirements for flammable

Terotechnology XII
Materials Research Proceedings 24 (2022) 212-220

Materials Research Forum LLC
https://doi.org/10.21741/9781644902059-31

liquids and liquefied hydrocarbon gas installations, e.g. traction, auxiliary power units, heating or cooking, necessary to meet the purposes defined in Part 1. Its general requirements apply to tanks, piping, flexible connections, distribution system, ventilation devices, internal combustion engines, and heating devices. The requirements of EN 50153 [38] for containers and piping, EN 10204 [39] for each metal used in liquid or flammable gas installations, and EN 15227 [40] for LPG cylinders are also referenced. However, this standard does not cover the installation of alternative drives and does not contain any special requirements dedicated to these installations. This is because these systems started to be implemented in rail transport after the establishment of EN45545 [36]. Therefore, it seems necessary to analyze the current provisions concerning alternative drive installation elements and supplement them with requirements adequate to the above-described characteristic properties of individual fuels. The above applies, for example, to the requirements for hydrogen installations to prevent uncontrolled leaks. They should also be secured against a possible hazardous event, such as derailment, collision, or spreading of a fire in another train area. These events can cause damage to alternate propulsion components, which can lead to an explosion with potentially catastrophic consequences.

It is considered advisable to adapt the requirements for hydrogen-powered road vehicles, including Regulation of the European Parliament and Council (1243 of 2019) [33], and Regulation No. 134 (UNECE) [34]. However, due to the use of hybrid drives in rail vehicles, the approval procedures should also include tests and requirements for lithium batteries.

Incidents of spontaneous combustion and explosions of lithium batteries have prompted various laboratories to undertake tests to identify the magnitude of their fire hazard. These tests were carried out for batteries of multiple sizes with different charge levels and with the use of various methods and test stands (furnace calorimeter [5], Tewarson apparatus, CDG calorimeter [6], adapted decompression chamber [8], Single Burning Item (SBI) apparatus [7].

All tests confirmed that the more charged the battery, the faster ignition from the initiated source. The HRR value increases with the battery charge (for example, from 13 to 57 kW for a battery with an energy capacity of about 100 Wh [7]). In addition, these tests revealed the emission of toxic compounds (SO_2, HF, other - depending on the electrolyte composition of the batteries) that pose a risk to humans in the event of ignition in a closed/confined space. The above demonstrates the need to consider the specific requirements for passenger composite vehicles powered by lithium batteries and locomotives powered by lithium batteries (or carrying a battery load in a train) passing through tunnels. These studies allow for a more comprehensive understanding of phenomena occurring in uncontrolled situations using lithium batteries. However, it seems necessary to introduce a unified test and approval method for rail vehicles. The procedure according to R100.02 regarding the rechargeable energy storage system (REESS) [35] is considered the most adequate. At the same time, the tests carried out at the CTO showed the need to verify it to obtain the repeatability of the method (clarifying the size of the ignition source and environmental conditions). The above is necessary when using it in type approval tests [9].

In the area of natural gas supply, there are also no European regulations dedicated to this type of fuel on the fire safety of rail vehicles. In this case, it is considered advisable to adopt the following documents ISO 12991: 2012 [41], NFPA 52 [42], and NFPA 59A [43].

Summary
The increasing use of alternative fuels for propulsion has a great chance of expansion due to more and more innovative technologies. The introduction of new power supply solutions gives the undeniable benefits of environmental protection. However, it entails the need to develop European regulations regarding the fire safety of rail vehicles using hydrogen cells, lithium batteries, or

natural gas. Such requirements are essential for rolling stock passing through tunnels and should also apply to the carriage of this type of load.

References

[1] J. Siwiec. Zastosowanie wodorowych ogniw paliwowych w transporcie kolejowym, Problemy Kolejnictwa 190 (2021) 53-57. https://doi.org/10.36137/1906P

[2] M. Woliński. Zbiornik wodoru w samochodzie. Realne zagrożenie w pożarze? Zeszyty Naukowe SGSP 65 (2018) 47- 61.

[3] P. Daszkiewicz. Analiza wybranych napędów alternatywnych stosowanych w autobusach szynowych. Autobusy 6/2017 (2017) 143-146.

[4] W. Szada-Borzyszkowski. R. Bujaczek. Zagrożenia płynące ze stosowania paliw alternatywnych w samochodach. Autobusy 6/2014 (2014) 260-265.

[5] P. Huang, Q. Wang, Ke Li, P.Ping & J. Sun. The combustion behavior of large scale lithium titanate battery. Scientific Reports 5 (2015) art. 7788. https://doi.org/10.1038/srep07788

[6] A. Lecocq. G. Gebrekidan. G. Marlair. Scenario-based prediction of Li-ion batteries fire-induced toxicity. Journal of Power Sources 316 (2016) 197-206. https://doi.org/10.1016/j.jpowsour.2016.02.090

[7] F. Larsson. P. Andersson. Characteristics of lithium-ion batteries during fire tests. Journal of Power Sources 271 (2014) 414-420. https://doi.org/10.1016/j.jpowsour.2014.08.027

[8] N. S. Spinner. S. G. Tuttl. Physical and chemical analysis of lithium-ion battery cell-to-cell failure events inside custom fire chamber. Journal of Power Sources 279 (2015) 713-721. https://doi.org/10.1016/j.jpowsour.2015.01.068

[9] D. Darnikowski, M. Mieloszyk. Investigation into the Lithium-Ion Battery Fire Resistance Testing Procedure for Commercial Use. Batteries 7 (2021) art. 44. https://doi.org/10.3390/batteries7030044

[10] K. Leszczuk. Bezpieczne LNG. Przegląd Pożarniczy 1/2014 (2014) 29-33.

[11] A.M. Glover, A.R. Baird, C.B. LaFleur. Hydrogen Fuel Cell Vehicles in Tunnels. Sandia National Laboratories. SAND2020-4507 R. April 2020.

[12] Publikacja informacyjna 11/1 Bezpieczne wykorzystanie wodoru jako paliwa w komercyjnych zastosowaniach przemysłowych. Polski Rejestr Statków S.A. Gdańsk. czerwiec 2021 https://www.prs.pl/uploads/p11i_pl.pdf

[13] P. Farsewicz. Powstanie europejski pociąg na ogniwa wodorowe. 12.04.2021. https://www.rynek-kolejowy.pl/mobile/powstanie-europejski-pociag-na-ogniwa-wodorowe-101922.html

[14] M. Usidus. Wsiąść do pociągu wodorowego. https://mlodytechnik.pl/technika/30054-wsiasc-do-pociagu-wodorowego

[15] R. Przybylski. Wodororowy boom na horyzoncie. Logistyka (online). 30.09.2021 https://logistyka.rp.pl/szynowy/art18971851-wodorowy-boom-na-horyzoncie

[16] M. Kwiatkowski. Płonące ogniwa. 31.03.2022 https://icpt.pl/plonace-ogniwa/

[17] T. Trąd. Zmniejszenie zagrożenia wybuchu baterii litowo-jonowych. 10.05.2018
https://www.sgs.pl/pl-pl/news/2018/05/baterie-jonowo-litowe

[18] Akumulator LTO. 31.03.2022 https://technoluxpro.com/pl/akkumulyatory/batarei/lto.html

[19] M. Dempsey. Train explosion leads to chemical release in downtown Houston. Houston
Chronicle 24.04.2017. https://www.chron.com/news/houston-texas/houston/article/Train-
explosion-leads-to-chemical-release-in-11095738.php#photo-12779955

[20] Kiedy zabraknie ropy. Ziemia na rozdrożu. 31.03.2022
https://ziemianarozdrozu.pl/encyklopedia/74/kiedy-zabraknie-ropy

[21] K. Wolongiewicz. Bateria litowo-jonowa – wszystko co musisz o niej wiedzieć. Świat
Baterii. 31.03.2022. https://blog.swiatbaterii.pl/bateria-litowo-jonowa/

[22] A. Szulc. Jak działa akumulator litowo-jonowy? Teoria Elektryki. 31.03.2022.
https://teoriaelektryki.pl/jak-dziala-akumulator-litowo-jonowy/

[23] Baza Wiedzy: Wady i zalety akumulatorów Li-Ion. BatLit. 31.03.2022.
https://batlit.pl/zalety_i_wady_akumulatorow_liion

[24] Lokomotywa na LNG jeździ po Florydzie. gasHD.eu. 31.03.2022.
http://gashd.eu/2019/07/29/lokomotywa-na-lng-jezdzi-po-florydzie/

[25] Szynobusy na gaz ziemny pojadą w Czeskich Karkonoszach. gasHD.eu. 31.03.2022.
http://gashd.eu/2021/07/26/szynobusy-na-gaz-ziemny-pojada-w-czeskich-karkonoszach/

[26] Szynobus na LNG będzie jeździł we Włoszech. gadHD.eu. 31.03.2022.
http://gashd.eu/2021/04/26/szynobus-na-lng-bedzie-jezdzil-we-wloszech/

[27] Pociąg na LNG dla Renfe zaprojektuje Segula. gasHD.eu. 31.03.2022.
http://gashd.eu/2020/03/21/pociag-na-lng-dla-renfe-zaprojektuje-segula/

[28] Regulation (EC) No 79/2009 of the European Parliament and of the Council of 14 January
2009 on type-approval of hydrogen-powered motor vehicles, and amending Directive
2007/46/EC

[29] Communication from the Commission to the European Parliament, the European Council,
the Council, the European Economic and Social Committee and the Committee of the Regions,
the European Green Deal, Brussels, 11.12.2019 COM(2019) 640 final.

[30] TSI LOC&PAS – Commission Regulation (EU) No 1302/2014, of 18 November 2014
concerning a technical specification for interoperability relating to the 'rolling stock —
locomotives and passenger rolling stock' subsystem of the rail system in the European Union,
02014R1302 — EN — 11.03.2020 — 004.001 — 1

[31] TSI WAG - Commission Regulation (EU) No 321/2013, of 13 March 2013, concerning the
technical specification for interoperability relating to the subsystem 'rolling stock — freight
wagons' of the rail system in the European Union and repealing Decision 2006/861/EC,
02013R0321 — EN — 11.03.2020 — 004.001 — 1

[32] Commission Regulation (EU) No 1303/2014 of 18 November 2014 concerning the technical
specification for interoperability relating to 'safety in railway tunnels' of the rail system of the
European Union. 02014R1303 — EN — 16.06.2019 — 002.001 — 1

Materials Research Forum LLC

https://doi.org/10.21741/9781644902059-31

[33] Regulation 2019/1243 - Adaptation of a number of legal acts providing for the use of the regulatory procedure with scrutiny to Articles 290 and 291 of the Treaty on the Functioning of the EU

[34] Regulation No 134 of the Economic Commission for Europe of the United Nations (UN/ECE) — Uniform provisions concerning the approval of motor vehicles and their components with regard to the safety-related performance of hydrogen-fuelled vehicles (HFCV) [2019/795]

[35] Regulation No 100 of the Economic Commission for Europe of the United Nations (UNECE) — Uniform provisions concerning the approval of vehicles with regard to specific requirements for the electric power train [2015/505]

[36] ASME B31.12-2014 Hydrogen Piping and Pipeline Code

[37] EN 45545-1÷7- Railway applications – Fire protection on railway vehicles

[38] EN 50153 - Railway applications - Rolling stock - Protective provisions relating to electrical hazard

[39] EN 10204 – Metallic products – Types of inspection documents

[40] EN 15227 - Railway applications - Crashworthiness requieement for railway vehicle bodies

[41] ISO 12991:2012 Liquefied natural gas (LNG) - Tanks for on-board storage as a fuel for automotive vehicles

[42] NFPA 52 Vehicular Gaseous Fuel Systems Code. 2013

[43] NFPA 59A Standard for the Production, Storage and Handling of LNG. 2013

Terotechnology XII
Materials Research Proceedings 24 (2022) 221-226

Materials Research Forum LLC
https://doi.org/10.21741/9781644902059-32

Comparative Analysis of the Mobility Assessment Methods for Tracked Vehicles

PARTYKA Jacek[1,a] *

[1]Military Institute of Engineer Technology, ul. Obornicka 136, 50-961 Wroclaw, Poland

[a]partyka@witi.wroc.pl

Keywords: Mobility, Trafficability, Tracked Vehicle, Mobility Index

Abstract. The article presents the methods of determining the mobility of tracked vehicles, as well as a comparative analysis of the mobility of these vehicles on dirt roads and soil characterised by low bearing capacity. As part of the comparative analysis, a general description of the individual mobility assessment methods is presented, along with the indicator characteristics, as well as the advantages and disadvantages of the methods, when compared to others presented in the article. In the comparative analysis of the military vehicle mobility assessment, the means, by which mobility parameters were determined, the possibility of the practical application of a given approach, as well as its accuracy were considered.

Introduction

The ability of tracked vehicles to negotiate the terrain depends primarily on the dimensions, structure and shape of a single element of the vehicle's propulsion system – continuous track. The dimensions of the tracks and their plates have a significant impact on the value of the unit pressure exerted. Pressures too high, that is exceeding the bearing capacity, cause the soil to deform, increase the rolling resistance, and thus lower the vehicle's traction capabilities [1].

There are several known methods of assessing the impact of a tracked vehicle on the ground. They serve the purpose of determining the mobility of a tracked vehicle during its movement on dirt roads and in rough, slow-go terrain.

This article includes a review of the comparative analysis of the following metrics: Vehicle Cone Index (VCI), Mean Maximum Pressure (MMP), and Vehicle Limiting Cone Index (VLCI).

VCI Method

General Description of the Method. In the Vehicle Cone Index (VCI) methods, the measure used to assess the mobility of tracked vehicles travelling on dirt roads and slow-go terrain (soft-soil) is the parameter characterizing the soil. The soil withstands the load of the tracked vehicle so that it can successfully complete the specified number of passes on the same track, usually, this means one pass or fifty passes [2].

In this method, the Mobility Index (MI) is the basis for empirically determining mobility. It takes into account the weight of the tracked vehicle, parameters related to contact pressure and grouser factor, as well as loads concentrated under the drive wheels and ground clearance. The MI is calculated from the following formula [3,4]:

$$MI = \left(\frac{p_n K_m}{0,01bK_0} + K_k - K_p \right) K_{ss} K_{sb},$$

(1)

where:
p_n — nominal (average) ground contact pressure, lbs/in^2;
b — track width, ins;

K_m – vehicle weight factor;
K_0 – grouser factor;
K_k – factor describing loads concentrated under road wheels;
K_p – ground clearance factor;
K_{SS} – engine factor;
K_{SB} – transmission factor.

Depending on the MI value, a single-pass VCI is also calculated according to the following formula:

$$VCI_1 = 7.0 + 0.2MI - \left(\frac{39.2}{MI+5.6}\right), \tag{2}$$

where: MI – tracked vehicle mobility index.

In [5], the authors state that VCI values for a single pass and 50 passes can be determined from MI estimation based on empirical equations. In research experiments, it is the Rating Cone Index (RCI) which is the parameter used to estimate the strength of the soil in a given area. It is defined as the product of the Cone Index (CI), representing the resistance to penetration into the terrain per unit cone base area, and the measure of the sensitivity of soil to strength losses under vehicular traffic called Remold Index (RI). These parameters are used to calculate the terrain trafficability and mobility of individual vehicles and vehicle columns.

Advantages of the Method. The VCI method makes it possible to easily determine the mobility of the vehicle by calculating the mobility index on the basis of technical data, as well as comparing the indicators calculated for different vehicles.

According to the US Army Field Manual [6], the knowledge of the VCI_1 and VCI_{50} parameters and the critical layer's depth adapted depending on the type of tracked vehicle, the soil and the number of passes by a given vehicle can be applied practically. Based on these data, it is possible to estimate the number of vehicles that can traverse the terrain, as shown in the example. The following were assumed:
– vehicle type: Abrams M1A1 main battle tank;
– $CI = 65$;
– $RI = 0.8$;
– $VCI_1 = 25$ [10];
– $VCI_{50} = 58$ [10].

$$RCI_{50} = 65 \cdot 0.8 = 52$$

The increase in the VCI per tracked vehicle is the following:

$$VCI_{50} - VCI_1 = 58 - 25 = 33$$

$$\frac{VCI}{50} = \frac{33}{50} = 0.66$$

In order to determine the number of tanks that could pass under the specified terrain conditions, the calculation should be carried out in such a way that the VCI_1 was equal to the RCI or was greater than the RCI.

$$\frac{52 - 25}{0,66} = 40,09 = 41 \; tracked \; vehicles$$

$$41 \cdot 0{,}66 = 27{,}06$$

$$27{,}06 + 25 = 52{,}06 > RCI_{50} = 52$$

Considering the above, 41 tanks could move in the given terrain conditions.

Thus, the comparison of the RCI with VCI indicates the ability of a given vehicle to negotiate the given soil condition for a given number of passes. The VCI method is very useful for determining the mobility of moving vehicles off-road, as it is a function of the potential vehicle ground contact pressure and the soil strength.

Disadvantages of the Method. The mobility index does not take into account soil parameters (its type and properties), on which the tracked vehicle travels.

The cone index has a limited range of applicability for the soil critical layer and only pertains to fine-grained and coarse-grained soils. This method is not used in the case of frozen ground, as well as for the road surface covered with a layer of snow [2].

Comparison to Other Methods. The VCI method, similarly to the Mean Maximum Pressure (MMP) method presented in the article, later on, takes into account the value of the maximum pressures and the geometrical dimensions of the vehicle undercarriage. Currently, these two basic analytical methods, developed based on experimental studies, are used alternatively for military vehicle cross-country mobility. Both methods are equivalent to each other and can be used interchangeably, taking into account that given acceptable mobility in the VCI method corresponds to a specific limit value of the MMP parameter.

The MMP Method

General Description of the Method. The Mean Maximum Pressure method concerns the assessment of tracked vehicles' ability to overcome terrain and roads with low load capacity and is based on the analysis of the maximum pressures occurring under the wheels of the vehicle track system. For a low bearing capacity soil terrain to be traversed, its bearing capacity must not be exceeded by the average value of the peak stresses under the drive wheels of a track system (MPP). This value was determined from the following empirical relationship [7,8]:

$$MMP = \frac{1.26W}{2nb(td)^{0.5}} \ [kPa] \tag{3}$$

where:
W – vehicle weight, kN;
n – the number of roadwheel per one track of the vehicle;
b – track width, m;
d – road wheel outer diameter, m;
t – track pitch, m.

Advantages of the Method. The MMP values are well-correlated with the $VCI_{(RCI)}$ values obtained from multiple passes for high plasticity clayey soils according to the following formula:

$$VCI_{(RCI)} = 0.83 \ MMP \tag{4}$$

According to the author of [4], the MMP calculation method is less labour-intensive in comparison with the VCI method. Because the MMP method was developed to determine the peak pressures

Terotechnology XII   Materials Research Forum LLC
Materials Research Proceedings 24 (2022) 221-226   https://doi.org/10.21741/9781644902059-32

and is based on the results of the interaction of the drive train and power transmission systems on the ground, it should be considered more reliable.

Disadvantages of the Method. In [5], Wong et al. are of the opinion that the MMP parameter is insufficient to fully assess the possibility of movement of tracked vehicles on dirt roads.
In this situation, it is necessary to use a correlation approximation between the VCI and MMP, to facilitate the assessment of cross-country mobility, according to the following relationships:

$$VCI_1 = 0.096 \, MMP \tag{5}$$

$$VCI_{50} = 0.27 \, MMP \tag{6}$$

The usefulness of the method is limited to the design and upgrade (modernization) of track systems, as well as the evaluation of the existing designs in terms of their capability to negotiate low bearing capacity soil.

Comparison to Other Methods. After taking into account the RI, the MMP index is comparable to the $VCI_{1(CI)}$ value. The MMP method enables a comparative assessment of the MI with the parameters of the VCI method, as well as VCI_1 and VCI_{50}, for selected tracked vehicles of different weights, as shown in Table 1.

Table 1. Comparison of the parameters determining the relative mobility of tracked vehicles [5]

TV 1, 2, 3 = Tracked Vehicle	Weight of the System [kg]	MMP [kPa]	MI	VCI₁	VCI₅₀
TV 1	26 000,00	148	66	24	55
TV 2	28 000,00	165	77	26	60
TV 3	29 000,00	211	89	29	66

As a result of the analysis of the data in Table 1, it was possible to conclude that the value of all the analysed indicators used in the presented methods is increasingly dependent on the weight of a given vehicle. The largest percentage increase (around 17%) was noted for the MMP index and the vehicles marked as TV2 and TV3.
The weight-dependent increase in the remaining parameters, namely MI, VCI_1 and VCI_{50}, relative to TV1, TV2 and TV3 vehicles is insignificant, and for these parameters amounts to 1%, 4%, and 1%, respectively.

VLCI Method
General Description of the Method. The Vehicle Limiting Cone Index (VLCI) is an analytical method in which the tractive force (traction) is determined depending on the measured soil bearing capacity. In this method, the Mobility Number (MN) for tracked vehicles is defined in the following manner [9]:

$$MN_{DERA} = \frac{CI \cdot n \cdot b \cdot t^{0,5} \cdot d^{0,5}}{W} = \frac{1.26 \cdot CI}{MMP}, \tag{7}$$

where, after transformations, the minimum bearing capacity of the soil which would ensure that the terrain could be traversed (VLCI) by tracked vehicles, assumes the form:

$$VLCI = \frac{1.56 \cdot W}{2 \cdot n \cdot b \cdot t^{0.5} \cdot d^{0.5}},$$ (8)

where:
W – vehicle weight, kN;
n – the number of roadwheel per one track of the vehicle;
b – track width, m;
d – road wheel outer diameter, m;
t – track pitch, m.

Advantages of the Method. High correlation of estimated tractive forces and the values recorded during tests using a mobile tester are possible thanks to the mobility number (MN) formula adopted in the VLCI method.

Disadvantages of the Method. The VLCI method can be used to estimate the available tractive forces on clayey soils. However, when using the method for other types of soil, for example, the sandy and the clayey ones, the indicators produced may be significantly flawed.

Comparison to other methods. The VLCI parameter as represented by the formula (8), points to values higher than those calculated using the MMP method.

In the case of tracked vehicles, the mobility parameters determined via the VLCI method was proven to be approximately 30% higher than the one calculated using the VCI method, and 26% higher than those resulting from the use of the MMP method.

The VLCI method, based on the measurement of the tractive force, clearly overstates the value of the necessary bearing capacity of the soil in relation to the value determined using the MMP method, based on the trafficability tests using real tracked vehicles.

Summary
Mobility analysis is an important element of military logistics planning and execution activities. The integrated data on the terrain and operational parameters of the vehicle is used by the commanding officer to make decisions regarding the movement of tracked vehicles or their columns.

The methods of determining the mobility of tracked vehicles on dirt roads or in cross-country terrain presented in the article are based on empirical dependencies.

A serious limitation in the development of an empirically-reliable model of a tracked vehicle is the lack of reliable experimental data from the tests performed on low bearing capacity soils [10].

The Polish Army utilizes field methods, based on (probably experimental) mapping of vehicle mobility as a function of directly or indirectly determined soil bearing capacity (known as natural soil strength, shear force, soil resistance and California Bearing Ratio (CBR)) and vehicle ground pressure [11] to determine the permissible number of vehicles crossings over a given area.

Currently, the tracked vehicle mobility is determined with the use of the available methods for determining soil trafficability under field conditions. These include methods utilizing instruments which enable the determination of soil parameters, mainly the bearing capacity of the soil measured directly or indirectly, as well as the densitometry, for determining soil trafficability with the use of a self-recording penetrometer. A professional assessment of trafficability is also supported by interesting technological solutions that are a part of military vehicle equipment. One such example is an on-board device for determining the terrain trafficability which is part of the on-board equipment of the Wheeled Engineer Reconnaissance Vehicle.

The use of novel measuring instruments and IT solutions as part of the mobility assessment methods will significantly affect the quality and speed of land trafficability measurements, as well as facilitate optimal mobility assessment for military vehicles.

References

[1] A. Surowiecki, Z. Golemo, P. Saska. Wpływ zbrojenia tymczasowej nawierzchni drogowej na rozkład naprężeń w podłożu, WSOWL, Wrocław, 2004.

[2] J.Y. Wong, P. Jayakumarb, E. Tomac, J. Preston-Thomas. A review of mobility metrics for next generation vehicle mobility models, Journal of Terramechanics 87 (2020) 11-20. https://doi.org/10.1016/j.jterra.2019.10.003

[3] Y.J. Wong. Theory of ground vehicles 3rd Ed., McGrow-Wiley, Hoboken, 2001.

[4] M. Łopatka. Metody określania zdolności pokonywania terenu o niskiej nośności, Biuletyn WAT 10 (2004) 49-66.

[5] Y.Ch.D. Wong, H.H.S. Lim, W.Q.W. Chan. An Assessment of Land Vehicles' Trafficability, DSTA Horizons (2016), pp. 54-63.

[6] FM 5-430-00-1 AFJPAM 32-8013, Vol I. Planning and design of roads, airfields, and heliports in the theater of operations – road design, Headquarters, Department of the Army Department of the Air Force, August 1994.

[7] D. Rowland. Tracked vehicle ground pressure and its effect on soft ground performance, Proc. 4th Int. Conf. of the International Society for Terrain-Vehicle Systems, Stockholm, 1972.

[8] D. Rowland, J.W. Peel. Soft ground performance prediction and assessment for wheeled and tracked vehicles, Institute of Mechanical Engineering 205 (1975) 81.

[9] E.B. Maclurin. The use of mobility numbers to describe the in-field tractive performance of pneumatic tyres, Proc. 10th Int. of the International Society for Terrain-Vehicle Systems Conf., Harper Adams University, 2003.

[10] M. Łopatka, W. Płocharz, A. Rubiec, K. Zembrowski. Badania nośności terenów podmokłych w aspekcie ich przejezdności, Logistyka i Nauka 3 (2015) 2929-2936.

[11] T. Ratajczak, Z. Kamyk. Metody wyznaczania przejezdności gruntu w warunkach polowych, Inżynieria wojskowa, problemy i perspektywy, V Konferencja Naukowo-Techniczna, 22-24.09.2008 (2008), Wrocław, 151 – 162.

Terotechnology XII   Materials Research Forum LLC
Materials Research Proceedings **24** (2022) 227-232   https://doi.org/10.21741/9781644902059-33

Concept of Laser Welding of Concentric Peripheral Lap Joint

DANIELEWSKI Hubert[1,a*] and ZRAK Andrej[2,b]

[1] Faculty of Mechatronics and Mechanical Engineering, Kielce University of Technology, Tysiąclecia Państwa Polskiego 7, 25-314 Kielce, Poland

[2] Faculty of Mechanical Engineering, Žilinská Univerzita v Žiline Univerzitná 1, 010 26 Žilina, Slovakia

[a] hdanielewski@tu.kielce.pl, [b] andrej.zrak@fstroj.uniza.sk

Keywords: Laser Welding, Numerical Simulation, Concentric Lap Joint Concept, Sealed Circumferential Joint

Abstract. This paper presents the concept of concentric peripheral joint. Designed concentric joint was projected to laser welding process, where using keyhole effect deep penetration trough three materials was presented. Concept of concentric joint includes using of elastic stainless steel material as a connector between two pipes. Stainless steel in form of a ring was used as an additional distance element, joined with two pipes. Presented concept was investigated using numerical simulation based on finite element method with Simufact Welding software. Performed investigation of laser welding presents possibility of using single and double beam welding. Performed simulation included a sealed joint, where only partial penetration of bottom material was obtained and full penetration joint [1]. The authors presented a comparative study of the joints using single and double laser beam welding. The welding parameters for the assumed joints were estimated via numerical simulations [2]. The study of the concentric lap joint shows the possibility of using laser beam welding in single pass welding for obtained assumed joint geometry [3].

Introduction

Transporting different kind of gas or liquid mediums require using of different type of materials. Some type of medium required stainless steels with good corrosion and chemical resistance, other material with creep resistance. Moreover, often transporting installation are operating under severe weather conditions which may required additional coating surface or some kind of covers [4,5]. Alternative approach is using concentric installation where inside pipe are made of material dedicated for transporting medium, and outside material are dedicated for weather conditions. This type of joint however required using advanced joining technology. Traditional arc welding methods indicated high temperature into elements which can lead to overheat welded materials and chance its properties [6]. Therefore, some advanced welding methods such as laser beam welding (LBW) can be used. Laser welding using keyhole effect allow to performed deep penetration in welded material in single pass and full automation of welding process [7, 8]. Moreover capillary with ionized plasma can penetrates more than one material, therefore, more advanced joint types such as lap joint can be performed [9].

When dissimilar joint are considered, differences in properties and chemical composition of welded materials must be take into account. Differences in crystallization dynamics of welded materials can affect in welding defects or damage during installation exploitation [10, 11]. In concentric joint important is to maintain constant distance between both pipes. Placing elastic connectors freely without joining with pipes effect in friction and wear of those elements, especially when material chance it's volume during heating and cooling. Joining those elements permanently can prevent it, however reduces moves of joined pipes. Therefore, important is to use

Terotechnology XII   Materials Research Forum LLC
Materials Research Proceedings 24 (2022) 227-232   https://doi.org/10.21741/9781644902059-33

connector made of elastic steels with good plastic deformability, such as stainless steels [12]. In transporting installations there is important to achieve sealed joints to avoid abuse flow of transported medium. In lap joints dominant force affected joint strength is shearing, therefore, increase of weld width is essential to improve joint strength. Multi beam welding system which increase spot size of beam interaction without decrease surface power density can be used for achieve this effect [13]. Concentric installation have additional advantages, over single pipe system, leakage of inside pipe not affect in leak of medium outside transporting installation.

Multiple articles consider welding of dissimilar materials, including in lap configurations, however using laser beam for welding of concentric lap joint with welding over more than two materials is new approach for projects transfer installation [14,15]. Some results of newly published studies show the application potential of special technological coatings. These coatings can be used to improve the tribological properties of materials from which conveying systems are made [5,16]. However, the coating technology has some limitations and on an industrial scale its application is difficult.

Numerical Simulation
Projecting of concentric peripheral joint was performed using numerical simulation within Simufact Welding software environment. Model consist two elements in form of pipes with wall thickness equal to 2 mm and ring connector with same thickness, which were meshed using ring mesh procedure with hexahedral type of elements [17]. Diameter of outside piper is equal to 50 mm. Finite element method (FEM) with established model was used to investigate possibility of use laser beam welding for performed concentric joint (Fig. 1).

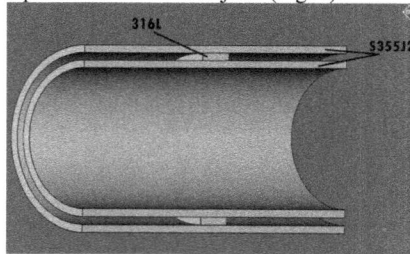

Fig. 1. Numerical model of concentric peripheral lap joint in half view.

Concentric joint simulation was performed in two configurations, first with joining edges of two materials and second with joining connector to both pipes. Moreover, possibility of use double beam welding optics was shown.

Heat flow in numerical simulation are based on Fourier's law. Solving the governing heat equation for three-dimensional heat conduction (1), with a partial differential equation in a nonlinear form [18], is done with the following equation:

$$\rho c(T)\frac{\partial T}{\partial t} = \frac{\partial}{\partial x}\left(k(T)\frac{\partial T}{\partial x}\right) + \frac{\partial}{\partial y}\left(k(T)\frac{\partial T}{\partial y}\right) + \frac{\partial}{\partial z}\left(k(T)\frac{\partial T}{\partial z}\right) + q_v \qquad (1) \,(1)$$

where $c(T)$ - temperature-dependent specific heat capacity; $k(T)$ - temperature-dependent thermal conductivity; q_v -volumetric internal energy; x, y, z - space coordinates; T - temperature; ρ - density; and t - time.

The conical heat source (HS) can be described as follows:

$$q_l(x,y,z) = \frac{9\eta_i P_i e^3}{\pi(e^3-1)(z_t-z_i)(r_t^2+r_t r_i+r_i^2)} \exp\left(-\frac{3[(x-vt)^2+y^2]}{\left(r_t-(r_t-r_i)\frac{z_t-z}{z_t-z_i}\right)}\right) \qquad (2)$$

where q_l - heat flux, $z_t - z_i$ - z coordinates (heat source depth), $r_t - r_i$ - upper and lower conical radius, e - natural logarithm, $r_t - (r_t - r_i) ((z_t-z)/(z_t-z_i))$ - linear decrease in distribution along the conical heat source, P_l - laser power, η_l - laser heat source efficiency.

Thermal conductivity, specific heat, and emissivity in the heat transfer analysis depend on the temperature, however, the used model is based on the assumption that mass density is constant [19]. Heat source efficiency was established as a 0.75. Calibration of HS was adopted for literature [20], and procedure description was presented in separate study. In case of welding using twin spot system distance between focal points was 1.4 mm.

Results

The performed thermal simulation give a results in form of temperature distribution, where shape of molten area was investigated. Results for single and double beam welding optics was presented (Fig. 2-4).

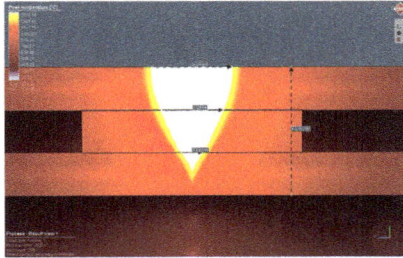

Fig. 2. Results for single beam laser welding optics.

Simulation shows assumed partial penetration of bottom pipe, which were achieved for laser power equal to 6 kW with welding speed 1 m/min, reducing welding velocity lead to full penetration of lower material. Width of obtained weld for surfaces of every welded materials is equal to 3.6, 2.8 and 0.95mm.

Second joint was calculated for double beam welding optics.

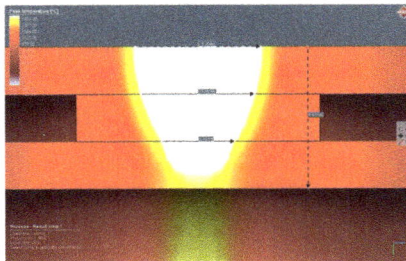

Fig. 3. Results for double beam laser welding optics,
with beams moving side by side.

Results also shows achieve partial penetration of bottom material, however for this welding system reduction of welding velocity was required, therefore, showed on figures 3 weld was achieved for laser power equal to 6 kW with welding speed 0.8 m/min. Calculated width is equal to 5.15, 4.66 and 3 mm respectively.

Concentric peripheral joint assume not only lap type of joint, but more complex, in form of butt-lap joint as well. In butt-lap joint two type of pipes are welded with ring connector between. While full penetration welding is not required for lap welding with connectors, quality level requirements for butt welding require complete penetration of both welded pipes [21]. For improving joint strength into projected joint double beam welding optics was considered. Comparing to previous research, full penetration was achiever reducing welding speed to 0.6 m/min (figure 4).

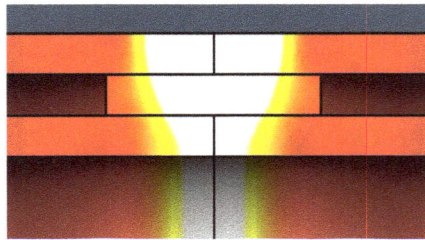

Fig. 4. *Result of full penetration concentric butt-lap joint welded using laser beam and double beam welding optics.*

Due to welding process high ratio of welded materials mixing occur, therefore, used connector have crucial role because, it's chemical composition will affect joint strength.

Summary
Authors presents concept of concentric pipe joint with inter pipes connector welded with joined pipes. Conceptualization was projected joint was investigated using numerical simulation modeling using Simufact Welding software. Established model was used for simulated single and double beam welding optics. Comparing shape and dimensions of calculated weld zone parameters for laser welding process were estimated. Partial penetration with assumed sealed joint was achieved with output power equal to 6 kW and welding speed 1 m/min for single and 0.8 m/min for double beam welding optics. For double beam welding optics full penetration was achieved with speed reduced to 0.6 m/min. Concept of projected joint was confirmed, and possibility of use double beam welding optics for improve width of weld zone was shown. Future planned research assume performing trial joints, and perform thermo-mechanic simulation method, where stress-strain analysis was planned, with additional mechanical and metallographic tests.

The presented results may be useful for researchers concerning a laser welding or a laser treatment of a surface layer [22-24] and related industry activities [25-29].

References
[1] M. Khan, L. Romoli, M. Fiaschi, F. Sarri, G. Dini. Experimental investigation on laser beam welding of martensitic stainless steels in a constrained overlap joint configuration. J. Mater. Process. Technol. 210 (2010) 1340-1353. https://doi.org/10.1016/j.jmatprotec.2010.03.024W.

[2] L. Mayboudi, A. Birk, G. Zak, P. Bates. A Three-Dimensional Thermal Finite Element Model of Laser Transmission Welding for Lap-Joint. Int. J. Model. Simul. 29 (2009) 149-155. https://doi.org/10.1080/02286203.2009.11442520.

[3] A. Lisiecki, P. Wójciga, A. Kurc-Lisiecka, M. Barczyk, S. Krawczyk. Laser welding of panel joints of stainless steel heat exchangers. Weld. Technol. Rev. 91 (2019) 7-19. https://doi.org/10.26628/wtr.v91i7.1037.

[4] N. Radek, A. Sladek, J. Bronček, I. Bilska, A. Szczotok. Electrospark alloying of carbon steel with WC-Co-Al_2O_3: deposition technique and coating properties. Adv. Mater. Res. 874 (2014) 101-106. https://doi.org/10.4028/www.scientific.net/AMR.874.101

[5] N. Radek, K. Bartkowiak. Laser treatment of Cu-Mo electro-spark deposited coatings. Phys. Proc. 12 (2011) 499-505. https://doi.org/10.1016/j.phpro.2011.03.061

[6] G. Phanikumar, K. Chattopadhyay, P. Dutta. Joining of dissimilar metals: Issues and modelling techniques. Sci. Technol. Weld. Join. 16 (2011) 313–317. https://doi.org/10.1179/1362171811y.0000000014

[7] W. Steen, J. Mazumder. Laser Material Processing, 4th Ed., Springer, 2010.

[8] J. Pietraszek, N. Radek, A.V. Goroshko. Challenges for the DOE methodology related to the introduction of Industry 4.0. Prod. Eng. Arch. 26 (2020) 190-194. https://doi.org/10.30657/pea.2020.26.33

[9] P.S. Mohanty, J. Mazumder. Workbench for keyhole laser welding. Sci. Technol. Weld. Join. 2 (1997) 133–138. https://doi.org/10.1179/stw.1997.2.3.133

[10] M. Dal, R. Fabbro. An overview of the state of art in laser welding simulation. Opt. Laser Technol. 78 (2016) 2–14. https://doi.org/10.1016/j.optlastec.2015.09.015

[11] Y. Miyashita, Y. Mutoh, M. Akahori, H. Okumura, I. Nakagawa, X. Jin-Quan. Laser welding of   dissimilar metals aided by unsteady thermal conduction boundary element method analysis. Weld. Int. 19 (2005) 687–696. https://doi.org/10.1533/wint.2005.3487

[12] N.S. Shanmugam, G. Buvanashekaran, K. Sankaranarayanasamy, K. Manonmani. Some studies on   temperature profiles in AISI 304 stainless steel sheet during laser beam welding using FE simulation. Int. J. Adv. Manuf. Technol. 43 (2008) 78–94. https://doi.org/10.1007/s00170-008-1685-0

[13] A. Matsunawa. Problems and solutions in deep penetration laser welding. Sci. Technol. Weld. Join. 6 (2001) 351–354. https://doi.org/10.1179/stw.2001.6.6.351

[14] J. Mazumder. Laser Welding: State of the Art Review. JOM: Journal of The Minerals, Metals & Materials Society 34 (1982) 16–24. https://doi.org/10.1007/BF03338045

[15] T. Kik, J. Górka. Numerical Simulations of Laser and Hybrid S700MC T-Joint Welding. Materials 12 (2019) art. 516. https://doi.org/10.3390/ma12030516

[16] N. Radek, A. Szczotok, A. Gądek-Moszczak, R. Dwornicka, J. Bronček, J. Pietraszek. The impact of laser processing parameters on the properties of electro-spark deposited coatings. Arch. Met. Mater. 63 (2018) 809-816. https://doi.org/10.24425/122407

Terotechnology XII   Materials Research Forum LLC
Materials Research Proceedings **24** (2022) 227-232   https://doi.org/10.21741/9781644902059-33

[17] W. Sudnik, D. Radaj, S. Breitschwerdt, W. Erofeew. Numerical simulation of weld pool geometry in   laser beam welding. J. Phys. D Appl. Phys. 33 (2000) 662–671. https://doi.org/10.1088/0022-3727/33/6/312

[18] W.Guo, A. Kar. Determination of weld pool shape and temperature distribution by solving three- dimensional phase change heat conduction problem. Sci. Technol. Weld. Join. 5 (2000) 317–323. https://doi.org/10.1179/136217100101538371

[19] A. Evdokimov, K. Springer, N. Doynov, R. Ossenbrink, V. Michailov. Heat source model for laser beam welding of steel-aluminum lap joints. Int. J. Adv. Manuf. Technol. 93 (2017) 709-716. https://doi.org/10.1007/s00170-017-0569-6

[20] C. Chen, Y.J. Lin, H. Ou, Y. Wang. Study of Heat Source Calibration and Modelling for Laser Welding Process. Int. J. Precis. Eng. Manuf. 19 (2018) 1239–1244. https://doi.org/10.1007/s12541-018-0146-4

[21] F. Farrokhi, B. Endelt, M. Kristiansen. A numerical model for full and partial penetration hybrid laser welding of thick-section steels. Opt. Laser Technol. 111 (2019) 671-686. https://doi.org/10.1016/j.optlastec.2018.08.059

[22] Ł.J. Orman Ł.J., N. Radek, J. Pietraszek, M. Szczepaniak. Analysis of enhanced pool boiling heat transfer on laser-textured surfaces. Energies 13 (2020) art. 2700. https://doi.org/10.3390/en13112700

[23] N. Radek, J. Pietraszek, A. Gadek-Moszczak, Ł.J. Orman, A. Szczotok. The morphology and mechanical properties of ESD coatings before and after laser beam machining, Materials 13 (2020) art. 2331. https://doi.org/10.3390/ma13102331

[24] N. Radek, J. Konstanty, J. Pietraszek, Ł.J. Orman, M. Szczepaniak, D. Przestacki. The effect of laser beam processing on the properties of WC-Co coatings deposited on steel. Materials 14 (2021) art. 538. https://doi.org/10.3390/ma14030538

[25] J. Pietraszek, A. Szczotok, N. Radek. The fixed-effects analysis of the relation between SDAS and carbides for the airfoil blade traces. Archives of Metallurgy and Materials 62 (2017) 235-239. https://doi.org/10.1515/amm-2017-0035

[26] Barucca G. et al. PANDA Phase One: PANDA collaboration. European Physical Journal A 57 (2021) art. 184. https://doi.org/10.1140/epja/s10050-021-00475-y

[27] A. Kubecki, C. Śliwiński, J. Śliwiński, I. Lubach, L. Bogdan, W. Maliszewski. Assessment of the technical condition of mines with mechanical fuses, Technical Transactions 118 (2021) art. e2021025. https://doi.org/10.37705/TechTrans/e2021025

[28] H. Danielewski, A. Skrzypczyk, W. Zowczak, D. Gontarski, L. Płonecki, H. Wiśniewski, D. Soboń, A. Kalinowski, G. Bracha, K. Borkowski. Numerical analysis of laser-welded flange pipe joints in lap and fillet configurations, Technical Transactions 118 (2021) art. e2021030. https://doi.org/10.37705/TechTrans/e2021030

[29] I. Nová, K. Fraňa, T. Lipiński. Monitoring of the Interaction of Aluminum Alloy and Sodium Chloride as the Basis for Ecological Production of Expanded Aluminum. Physics of Metals and Metallography 122 (2021) 1288-1300. https://doi.org/10.1134/S0031918X20140124

Terotechnology XII
Materials Research Proceedings **24** (2022) 233-239

Materials Research Forum LLC
https://doi.org/10.21741/9781644902059-34

Ideas and Assumptions of a New Kind Helical Metal Expansion Joints

KURP Piotr[1, a *]

[1] Faculty of Mechatronics and Mechanical Engineering, Kielce University of Technology, Al.Tysiąclecia Państwa Polskiego 7, 25-314 Kielce, Poland

[a]pkurp@tu.kielce.pl

Keywords: Laser Forming, Metal Expansion Joints, Laser Manufacturing

Abstract. The author of this paper presented the ideas and assumptions with regard to manufacturing and application of a new metal expansion joints type. The manufacturing technology based on a hybrid mechanically assisted laser forming method was briefly discussed. The exculpation for the development of this expansion joints type and their effect on the torques and distortions compensation occurring in industrial pipelines installations was discussed as well. Furthermore, the preliminary experimenter's research results about manufacturing of this expansion joints type were presented as well.

Introduction

In every industrial pipelines installations we deal with variable parameters of the transmitted medium. These are mainly changes in pressure and temperature. As a result of changing working conditions, such an installation is exposed to damage related to a change in its geometry due to e.g. thermal expansion of the material. Therefore, there is a need to compensate for such an installation. The simplest solution is to compensate when changing the direction of the pipeline, e.g. at elbows. However, it is not always possible, e.g. due to collisions with other installations, lack of space, etc. In such a case, expansion joints integrated with that installations are used. Depending on installation operating parameters these may be cloth, rubber or metal expansion joints.

In the case of high pressures, metal expansion joints are used. They are made from a pipe, on which there are upsets in the form of a bellows or a lens. These upsets act as a kind of "spring" which compensate for the above-mentioned deformations. Their design ensures compensation of axial, lateral and angular deformations [1]. On the other hand, they do not compensate for the deformations resulting from the torque moments, which are the operation result of the equipment installed in the installation, e.g. valves, pumps, etc. Therefore, the idea of helical metal expansion joints was born, which was designed to compensate mainly for this type of deformation.

Currently, standard expansion joints are made mainly by plastic cold working, using roller systems, hydroforming methods, etc. [1]. The author of this paper proposed the use of a hybrid mechanically assisted laser forming method to produce helical expansion joints. Currently, laser techniques are widely used, e.g. for cutting [2], welding [3], industrial coatings applications [4], creating textures on the material's surface [5], additive manufacturing [6], etc. Instead the laser forming method uses the phenomenon of distortion in the material as a result of the element's local temperature change. The element is heated by the energy from the laser beam acting on it. The appropriate geometry and trajectory of the laser beam leads to the desired shapes of the element. In this case, the local change in the shape of the element is achieved due to the difference in thermal expansion of the "cool" and "warm" parts of the material. Plastic deformation is obtained as a result of causing internal thermal stresses in the material without the participation of external forces, by means of three main mechanisms: temperature gradient mechanism (TGM), upsetting mechanism (UM) and buckling mechanism (BM) [7].

The free-form laser forming method (as we can call it) is, however, a very time-consuming process. That is why the idea of a hybrid method of mechanically assisted laser forming was born. In this method, in order to accelerate the process, apart from laser heating, external forces are also used to create plastic deformations. This approach significantly speeds up the forming process, the process becomes effective, and the method allows for obtaining complex shapes. An example of such an approach is, for example, shaping thin-walled elements made from pipes [8]. Of course, the ideas and assumptions should predict the experiment [9] and make the necessary macroscopic measurements of the obtained elements [10].

Bearing in mind the above, the ideas and assumptions for making helical expansion joints using this method are presented. A preliminary experiment was performed to confirm the validity of the assumptions.

Ideas and assumptions
The necessity to compensate for deformations of industrial piping installations resulting from torque moments led to the idea of making helical expansion joints. The helical expansion joint has bellows on its circumference, but not in the form of classic rings, but in the form of a helix. This helix is similar in appearance to the thread. It can be both dextrorotatory and laevorotatory - Fig. 1.

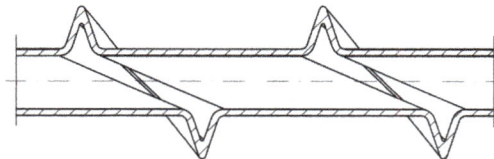

Fig. 1. Helical expansion joint cross section (idea).

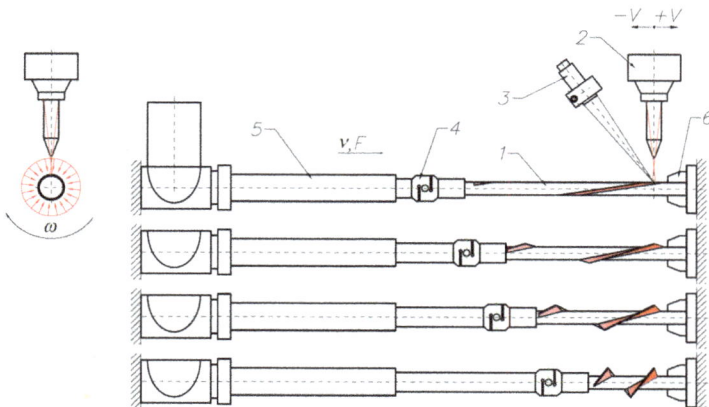

Fig. 2. Individual steps of helical expansion joints forming (idea): 1 - quickly rotating around its axis pipe, 2 - laser head (pipe heating), 3 - pyrometer, 4 - force sensor, 5 - axial thrust actuator, 6 - swivel handle. The red line shows laser incidence path with pipe surface.

Terotechnology XII
Materials Research Proceedings **24** (2022) 233-239

Materials Research Forum LLC
https://doi.org/10.21741/9781644902059-34

Helical metal expansion joints manufacturing process with a hybrid mechanically assisted laser forming method is based on the following assumptions (Fig. 2.):
- the laser beam heats a given area of the element to a certain, preset temperature, which improves the plastic properties in this area,
- evenly and uniformly heating of the element around its entire circumference is achieved by its quick rotation around its axis,
- the laser head moves in a direction parallel to the axis of the rotating pipe at a given speed, creating a helix on its surface,
- an axial force acts on the element simultaneously, which causes its upsetting in the plasticized zone (heated by the laser beam),
- only the part of the pipe that is exposed to the laser beam at a given moment is deformed,
- the remaining part of the formed element, which has a lower temperature, does not deform.

So formulated assumptions were verified as to their validity during the initial experimental study. The description of the experiment and its results are presented in the following chapters.

Experiment
In order to perform the experiment, an execution and measurement stand was built based on the TRUMPF TruFlow 6000 CO_2 laser, generating laser radiation with a wavelength of $\lambda=10.6$ μm with a maximum power of 6kW and operating in CW laser mode. Stainless steel grade X5CrNi18-10 pipes with dimensions of $\phi20x1$ mm (diameter x wall thickness) and length of 250 mm were used for the experiment. The element was installed between axial actuator with a maximum pressure force of 5kN (4) and swivel handle (6). The surface of the sample was covered with a special absorber (matt black enamel) in order to increase and uniform the absorption coefficient of the laser radiation. The treatment parameters were as follows:
- laser power: P=1200 W,
- process temperature: approx. T=1100°C (X5CrNi18-10 steel hyperquenching temperature which guarantees the preservation of the austenitic microstructure after process),
- pipe compressive force: max. F=1.2 kN,
- compressive length: s=30 mm,
- pipe rotation speed: ω=10000 °/min,
- pipe compressive speed: v=10 mm/s.
- initial beam pitch: p=80 mm.

The circularly polarized laser beam was incident perpendicularly on the rotating pipe surface. Furthermore the laser beam traveled along the pipe axis, creating a heated, convoluted thread-like strip on its surface. It was a zone of heating and its plasticization. After obtaining the appropriate temperature T, recorded by the sensor (3), the actuator (5) was started. The actuator pressed the pipe axially with the force F (4) and by velocity v.

Results, Discussion and Conclusions
The final result of the preliminary experiment was the helical expansion joint manufacturing. Fig. 3 and Fig. 4 show an exemplary element with a macrograph gained shape analysis.

Terotechnology XII   Materials Research Forum LLC
Materials Research Proceedings **24** (2022) 233-239   https://doi.org/10.21741/9781644902059-34

Fig. 3. *Helical expansion joint view after processing – a graphic*
filter was used for better image analysis.

a)

b)

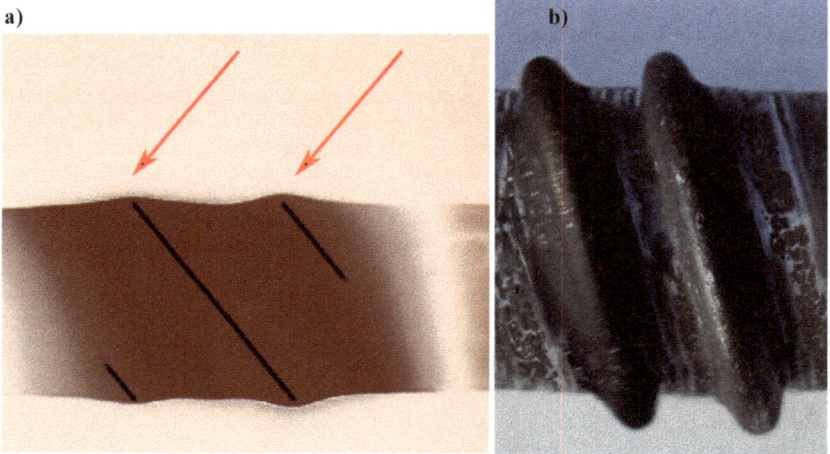

Fig. 4. *Magnification of the image with marked peaks a) and helix coils b).*

Analyzing the photographic documentation presented in Fig. 4a), we can observe the creation
of bulges formed on the circumference of the pipe (marked by red arrows). These are the upsets of
the emerging helical expansion joint. Moreover the upset path is marked with a black lines.
Appropriate selection of the process parameters leads to the formation of a helix. After many
experimental tests with the selection of parameters, it was possible to make a helical expansion
joint with two coils on the perimeter. The final product is shown on Fig. 4b). However, a number
of trials were required to obtain such a shape. In particular, it is problematic to control the
temperature of the process and to start upsetting at the appropriate time. In many cases, the tubular
element was burned or the element was squeezed out. In both cases, it resulted in the expansion
joint tearing, loss of material cohesion or unforeseen deformations. It is also difficult to select an
appropriate initial pitch and initial pitch speed. Too small pitch leads to a slight upsetting of the
pipe on the perimeter. On the other hand, too large pitch causes the necessity to use a lot of force
in order to upset.

Summary

The results of the experiment seem to confirm the idea validity of manufacturing metal helical expansion joints with the use of a hybrid mechanically assisted laser forming method. However, mastering this process caused some problems, mainly technological. Many of the attempts made have been unsuccessful. Appropriate selection of process parameters turned out to be very important. The process is started and ended manually, not automatically. The final effect therefore largely depends on the experience of the operator. However, this approach leads to the lack of repeatability of the final elements. Moreover the final shape of element depends on many variable factors.

The improvement of the process will be the aid of further works planned in the future. Furthermore, it is planned to carry out a FEM analysis in order to recognize thermoelastic and thermoplastic phenomena occurring during the process. It is planned to study the microstructure of the obtained expansion joints. In order for the expansion joint to have the expected strength properties, the microstructure should not change and remain an austenitic. The selection of parameters (temperature) should ensure this, however, detailed research in this matter will give the answer what the real microstructure is. It is also planned to perform strength tests of the created expansion joints in order to determine their utilitarian suitability in industrial applications.

The obtained results may be interesting for analogous cases of laser processing [11-13], preparing parts for highly loaded devices in waste treatment [14, 15] or producing responsible parts of military equipment [16, 17]. It would be also very interesting to consider the thermomechanical analysis of the problem using the adjustment calculus with supplementary data [18, 19] and non-parametric methods [20, 21].

Acknowledgement

The research reported herein was supported by a grant from the National Centre for Research and Development (NCBiR).

Program title: Lider XI, grant title "Development of new type metal expansion joints and their manufacturing technology", contract number: : LIDER/44/0164/L-11/19/NCBR/2020.

References

[1] Standards of the Expansion Joint Manufacturers Association, Tenth Edition, Expansion Joints Manufacturer Association INC., 2020

[2] H. Danielewski, J. Meško, R. Nigrovič, R. R. Nikolić, B. Hadzima, N. Gubeljak. Laser cutting of ductile cast iron, Materialpruefung/Materials Testing 62 (2020) 820-826. https://doi.org/10.3139/120.111548

[3] S. Tofil, H. Danielewski, G. Witkowski, K. Mulczyk, B. Antoszewski. Technology and Properties of Peripheral Laser-Welded Micro-Joints, Materials 14 (2021) art. 3213. https://doi.org/10.3390/ma14123213

[4] N. Radek, A. Szczotok, A. Gądek-Moszczak, R. Dwornicka, J. Bronček, J. Pietraszek. The impact of laser processing parameters on the properties of electro-spark deposited coatings. Archives of Metallurgy and Materials 63 (2018) 809-816. https://doi.org/10.24425/122407

[5] S. Tofil, R. Barbucha, M. Kocik, R. Kozera, M. Tański, N. Arivazhagan, J. Yao, A. Zrak. Adhesive Joints with Laser Shaped Surface Microstructures, Materials 14 (2021) art. 7548. https://doi.org/10.3390/ma14247548

[6] B. Antoszewski, H. Danielewski, J. Dutkiewicz, Ł. Rogal, M.S. Węglowski, K. Kwieciński, P. Śliwiński. Semi-Hybrid CO2 Laser Metal Deposition Method with Inter Substrate Buffer Zone, Materials 14 (2021) art. 720. https://doi.org/10.3390/ma14040720

[7] F. Vollertsen F. Mechanisms and Models for Laser Forming, Laser Assisted Net Shape Engineering, Proc. LANE'94 1. Meisenbach-Verlag, Bamberg (1994) 345-360.

[8] J. Widłaszewski, M. Nowak, Z. Nowak, P. Kurp. Laser-assisted thermomechanical bending of tube profiles, Archives of Metallurgy and Materials 64 (2019) 421-430. https://doi.org/10.24425/amm.2019.126268

[9] J. Pietraszek, N. Radek, A. V. Goroshko. Challenges for the DOE methodology related to the introduction of Industry 4.0, Production Engineering Archives 26 (2020) 190-194. https://doi.org/10.30657/pea.2020.26.33

[10]A. Gądek-Moszczak, N. Radek, S. Wroński, J. Tarasiuk, Application the 3D image analysis techniques for assessment the quality of material surface layer before and after laser treatment. Advanced Materials Research 874 (2014) 133-138. https://doi.org/10.4028/www.scientific.net/AMR.874.133

[11]N. Radek, J. Pietraszek, A. Gadek-Moszczak, Ł.J. Orman, A. Szczotok. The morphology and mechanical properties of ESD coatings before and after laser beam machining, Materials 13 (2020) art. 2331. https://doi.org/10.3390/ma13102331

[12]N. Radek, J. Konstanty, J. Pietraszek, Ł.J. Orman, M. Szczepaniak, D. Przestacki. The effect of laser beam processing on the properties of WC-Co coatings deposited on steel. Materials 14 (2021) art. 538. https://doi.org/10.3390/ma14030538

[13]H. Danielewski, A. Skrzypczyk, W. Zowczak, D. Gontarski, L. Płonecki, H. Wiśniewski, D. Soboń, A. Kalinowski, G. Bracha, K. Borkowski. Numerical analysis of laser-welded flange pipe joints in lap and fillet configurations, Technical Transactions 118 (2021) art. e2021030. https://doi.org/10.37705/TechTrans/e2021030

[14]M. Zenkiewicz, T. Zuk, J. Pietraszek, P. Rytlewski, K. Moraczewski, M. Stepczyńska. Electrostatic separation of binary mixtures of some biodegradable polymers and poly(vinyl chloride) or poly(ethylene terephthalate), Polimery/Polymers 61 (2016) 835-843. https://doi.org/10.14314/polimery.2016.835

[15]M. Dobrzański. The influence of water price and the number of residents on the economic efficiency of water recovery from grey water, Technical Transactions 118 (2021) art. e2021001. https://doi.org/10.37705/TechTrans/e2021001

[16]B. Szczodrowska, R. Mazurczuk. A review of modern materials used in military camouflage within the radar frequency range, Technical Transactions 118 (2021) art.e2021003. https://doi.org/10.37705/TechTrans/e2021003

[17]A. Kubecki, C. Śliwiński, J. Śliwiński, I. Lubach, L. Bogdan, W. Maliszewski. Assessment of the technical condition of mines with mechanical fuses, Technical Transactions 118 (2021) art. e2021025. https://doi.org/10.37705/TechTrans/e2021025

Terotechnology XII   Materials Research Forum LLC
Materials Research Proceedings **24** (2022) 233-239   https://doi.org/10.21741/9781644902059-34

[18] T. Styrylska, J. Pietraszek. Numerical modeling of non-steady-state temperature-fields with supplementary data. Zeitschrift für Angewandte Mathematik und Mechanik 72 (1992) T537-T539.

[19] G. Majewski, Ł.J. Orman, M. Telejko, N. Radek, J. Pietraszek, A. Dudek. Assessment of thermal comfort in the intelligent buildings in view of providing high quality indoor environment, Energies 13 (2020) art. 1973. https://doi.org/10.3390/en13081973

[20] J. Pietraszek. Response surface methodology at irregular grids based on Voronoi scheme with neural network approximator. 6th Int. Conf. on Neural Networks and Soft Computing JUN 11-15, 2002, Springer, 250-255. https://doi.org/10.1007/978-3-7908-1902-1_35

[21] J. Pietraszek, R. Dwornicka, A. Szczotok. The bootstrap approach to the statistical significance of parameters in the fixed effects model. ECCOMAS 2016 – Proc. 7th European Congress on Computational Methods in Applied Sciences and Engineering 3, 6061-6068. https://doi.org/10.7712/100016.2240.9206

Terotechnology XII

Materials Research Proceedings **24** (2022) 240-246

Materials Research Forum LLC

https://doi.org/10.21741/9781644902059-35

Microstructure and Mechanical Properties of Dissimilar Friction Welded B500B-1.4021 Steel Joint

OSTROMĘCKA Małgorzata

Instytut Kolejnictwa (The Railway Research Institute), Materials & Structure Laboratory, 50 Chłopicki Street, 04-275 Warsaw, Poland

mostromecka@ikolej.pl

Keywords: Rotary Friction Welding, Dissimilar Joint, Hardness, Tensile Test, Microstructure, Martensitic Stainless Steel, Reinforcing Steel

Abstract. In this study, unalloyed reinforcing steel (B500B) is merged with martensitic stainless steel (1.4021). The paper presents microstructural observation, hardness measurements, and tensile test results of this dissimilar joint as-welded. In this specific case, no heat treatment is demanded. The joint is a part of the pile foundation for railway traction lines.

Introduction

Joining dissimilar materials offers the potential to utilize the advantages of different materials and provide unique solutions to engineering requirements. The main reason for dissimilar joining is to combine good mechanical properties of one material and, for example, good corrosion resistance of the second material. But very often, the most important is an economic reason [1,2].

Friction welding is well established as highly productive and one of the most economical methods in joining similar and dissimilar metals. The process is readily used by producers, especially since it enables joint creation in circumstances that exclude other welding methods.

One complex combination of the dissimilar joint is between unalloyed and high alloyed martensitic steel [2]. Such a combination is not mentioned in PN-EN ISO 15620: 2019 [3] as weldable with a friction welding method. Not many publications concerning such a joint, but this is an application in railway infrastructure. The reinforcing steel, which offers good weldability, is merged with martensitic stainless steel that is not easily welded, often needs to be preheated, and uses special welding consumables. This joint is used in concrete piles that act as foundations and anchors for traction lines. The pile is manufactured of reinforced concrete where reinforcing is made of B500B ribbed steel and anchor of 1.4021 martensitic stainless steel. The joints are covered with the concrete and are not subject to shear stresses until the concrete cover is still. The alternate solution for this application is to combine reinforcing steel and stainless austenitic steel, but this kind of joint is well described in the literature. Therefore, this is not a subject of the paper. The only quality control of these joints is conducted according to Factory Production Control. The final pile's quality guidelines do not provide for testing these joints, even at the stage of authorization for placing in service.

Fig. 1. The joint after welding

Research Methodology

Materials and Welding Parameters. The manufacturer company prepared the joints (Fig.1) according to approved welding technology using the ZT3-22 machine. Each joint was made between a ribbed reinforcing steel bar and martensitic steel 1.4021/X20Cr13 (AISI 420) anchor. The chemical composition of base materials and technological conditions are presented in Tables 1, 2, and 3. The microstructure of the base materials is shown in Fig.2.

Table 1. *Chemical composition (% by mass) for B500B steel according to PN-EN 10080:2007 [4], the base material 1*

B500B steel	C [%]	N [%]	S [%]	P [%]	Cu [%]	Carbon equivalent C_{eq}
Cast analysis	$\leq 0,22$	$\leq 0,012$	$\leq 0,050$	$\leq 0,050$	$\leq 0,80$	$\leq 0,50$
Product analysis	$\leq 0,24$	$\leq 0,013$	$\leq 0,055$	$\leq 0,055$	$\leq 0,85$	$\leq 0,52$

Table 2. *Chemical composition (% by mass) for martensitic stainless steel 1.4021/X20Cr13 according to PN-EN 10088-3: 2015 [5], the base material 2.*

C [%]	Mn [%]	Si [%]	P [%]	S [%]	Cr [%]	Ni [%]	Cu [%]
0.16 - 0.25	<1.5	<1.0	<0.04	<0.03	12.0 - 14.0	-	-

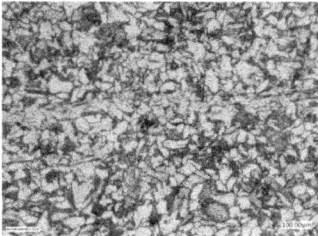

Fig. 2. *Microstructure of base materials: reinforcing steel on the left and martensitic stainless steel on the right (description in the text).*

Table 3. *Welding parameters.*

Rotational speed V [rpm]	Friction pressure F_t [kG/cm²]	Forging pressure F_s [kG/cm²]	Friction time t_t [s]	Forging time t_s [s]	Welded area [mm²]
800	115	140	5	6	490

Microstructural observation, hardness, and tensile test. The metallographic investigations were carried out on polished and etched surfaces. Etching was executed using 4% Nital solution on the B500B steel side and then with Adler's reagent to reveal the martensitic microstructure and plastic deformation lines. Keyence VHX-900F microscope was used to characterize microstructure.

HV0.1 Vicker's microhardness measurements were executed on polished surfaces across the weld. KB Prüftechnik GmbH automatic hardness tester type KB50BYZ-FA was used according to PN-EN ISO 6507-1: 2018-05 [6] standard. Three lines were made on each of the samples, measuring 101 points at a

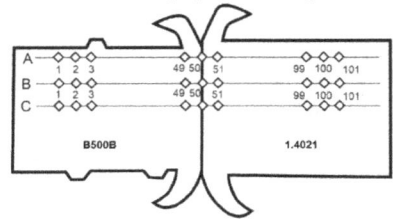

Fig. 3. Microhardness measurement scheme

distance of 0.2 mm, while the measuring lines were spaced 5 mm apart. The measuring scheme is presented in Fig.3.

Instron Schenk Testing System LFV 4000-2500 was used on two types of samples: on round tensile specimens where the weld center was marked with red lacquer and on specimens without any special preparation just as they work.

Results and Discussion.

Microstructure and hardness. The microstructure of base material 1 - B500B steel comprises outer tempered martensite, an inner core of ferrite-pearlite (Fig. 2 on the left), and between a narrow bainitic transition zone. The maximum hardness is achieved near the ribs side (Fig. 9 left side of the A line), decreasing towards the center. The macrostructure presented in Fig.4 reveals a darker area of peripheries by the rebar side. In the core of the bar close to the weld line, grain refinement can be observed, and the grain size increases toward the distance of the weld line (Fig.5). Closer to the ribs side, grains growth appears instead of refinement, and near the weld, several secondary surfaces are part of a partially mixed region (Fig.6). The thermo-mechanically affected zone of the reinforcing steel side was presented in Fig.8.

The microstructure of base material 2 – stainless steel consists of tempered martensite with carbide precipitations (Fig. 2 – right side). The weld line is very narrow, but the thermo-mechanically affected zone (TMAZ) is about 400 μm near the center and broader on the peripheries (Fig. 4). Next to the flash, many partially mixed regions can be observed. In the core, there are many deformation lines and bands (the darker areas in Fig.7).

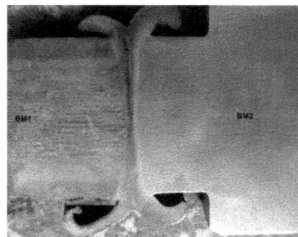

Fig. 4. Macrostructure of the joint with asymmetrical flashes

Materials Research Forum LLC

https://doi.org/10.21741/9781644902059-35

Fig. 5. *Microstructure of reinforcing steel in the axis of the specimen.*
The white region is an unetched martensite steel side.

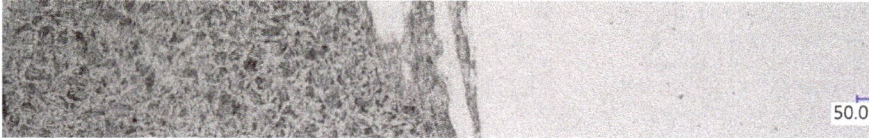

Fig. 6. *Microstructure of reinforcing steel in the area closer to the edge of the specimen.*

Fig. 7. *Thermo-mechanically affected zone (TMAZ) in the 1.4021 steel (on the right from weld)*

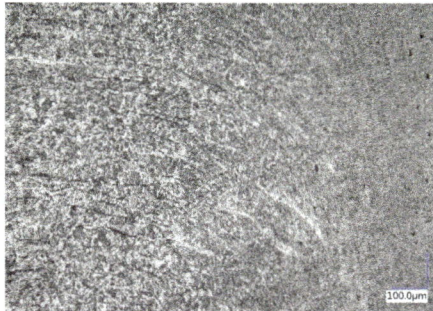

Fig. 8. *Thermo-mechanically affected zone (TMAZ) on B500B reinforcing steel side.*

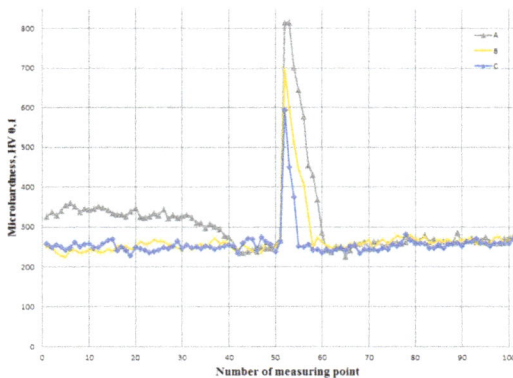

Fig. 9. *Microhardness profiles*

The hardness of the base materials cores exhibits similar values of about 260 HV_{01}. The hardness increases rapidly near the weld line by the martensitic steel side. Line A presents measurement near the specimen's outer side, and the highest value is 814 HV_{01}. In line C, the highest measurement is 594 HV_{01}, and in line B – 696 HV_{01}. In the TMAZ of reinforcing steel, the hardness slightly increases, but close to the outer area, it decreases to the base material value. The difference in microhardness value between the center and outer region can be attributed to the supplied heat distribution, and therefore differences in microstructure arise [7]. The hardness increase in the area close to the weld is a common situation, and it can be explained by quenched martensite microstructure formation [1, 8]. In special quality applications such as the automotive industry, such a joint would need a post-weld treatment to lower the hardness in the weld. Another possibility is to use a transition layer [2, 9]. The hardness is the end value in the investigated joint, and technology does not demand a post-weld treatment or flash removal.

Tensile test. The chosen tensile test specimens after fracture are presented in Fig.10. The welded joints represent a brittle fracture with no necking and plastic deformation. The microstructure observation and hardness measurements let us expect that the presented specimens are broken close to the weld in the thermo-mechanically affected zone of the stainless steel 1.4021.

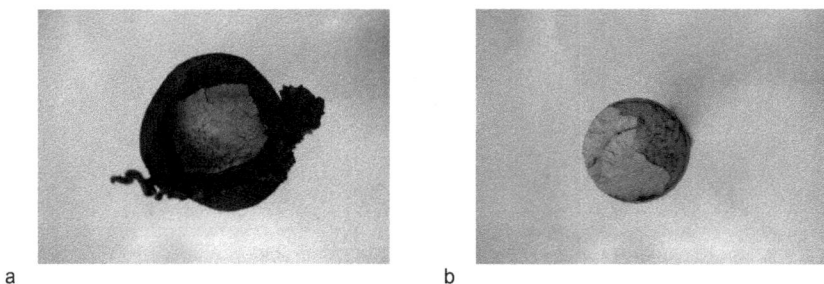

a   b

Fig. 10. The view of fractured sample numbers 2 (a) and 8 (b) (Table 4,5)

Table 4. *Tensile test samples as-welded with flashes*

Sample number	Tested area of weld cross-section [mm²]	Tensile force F_m [kN]	Tensile strength R_m,[MPa]	Fractured in
1		332.50	678	base material BM1
2		327.28	668	weld
3	490	333.19	680	base material BM1
4		328.15	670	base material BM1
5		328.26	670	base material BM1

Terotechnology XII
Materials Research Proceedings **24** (2022) 240-246

Materials Research Forum LLC
https://doi.org/10.21741/9781644902059-35

Table **5** *Round polished tensile test samples without flashes*

Sample number	Tested area of weld cross-section [mm^2]	Tensile force F$_m$ [kN]	Tensile strength R$_m$,[MPa]	Fractured in
6		240.82	630	weld
7	380	253.10	666	weld
8		245.70	647	weld

Conclusions

The rotary friction welding process successfully carried out dissimilar material joining B500B reinforcing to 1.4021 martensitic steel, and the microstructure and mechanical properties were studied. The summary of the observations is as follows:

- the affected zone near the weld joint is not homogenous and presents several areas with different microstructure morphologies and mechanical properties.
- the hardness measurements performed on the joint present the highest hardness at the thermo-mechanically deformed zone of the stainless steel. It causes a reduction in ductility of the weld joint.
- the observations confirm that, in this case, flash removal is not recommended.

References

[1] M.R. Abbasai, Kh. Gheisari. Microstructure and Mechanical Properties of the Friction Welded Joint between X53CrMnNiN219 and X45CrSi93 Stainless Steel, Journal of Advanced Materials and Processing 5 (2017) 81-92.

[2] A. Ambroziak. Zgrzewanie tarciowe materiałów o różnych właściwościach. Oficyna Wydawnicza Politechniki Wrocławskiej, Wrocław, 2011.

[3] PN-EN ISO 15620:2019-07 Friction welding of metallic materials.

[4] PN-EN 10080:2007 Stal do zbrojenia betonu – Spajalna stal zbrojeniowa – Postanowienia ogólne.

[5] PN-EN 10088-3: 2015-01 Stale odporne na korozję – Część 3: Warunki techniczne dostawy półwyrobów, prętów, walcówki, drutu, kształtowników i wyrobów o powierzchni jasnej ze stali nierdzewnych ogólnego przeznaczenia.

[6] PN-EN ISO 6507-1: 2018-05 Metale – Pomiar twardości sposobem Vickersa – Część 1: Metoda badania.

[7] B.S Taysom, C.D. Sorensen. Controlling Martensite and Perlite Formation with Cooling Rate and Temperature Control in Rotary Friction Welding. International Journal of Machine Tools & Manufacture, 150 (2020) art. 103512. https://doi.org/10.1016/j.ijmachtools.2019.103512

[8] M. Demouche, E.H. Ouakdi, R. Louahdi. Effect of Welding Parameters on the Microstructure and Mechanical Properties of Friction-Welded Joints of 100Cr6 Steel. Iranian Journal of Materials Science and Engineering 16 (2019) 24-31. https://doi.org/10.22068/ijmse.16.2.24

[9] N. Sridharan, E. Cakmak, B. Jordan. Design, Fabrication, and Characterization of Graded Transition Joints. Welding Journal 96 (2017) 295-306.

[10] C.S. Paglia, R.G. Buchheit. Microstructure, microchemistry and environmental cracking susceptibility of friction stir welded 2219-T87. Materials Science and Engineering A 429 (2009) 107-114. https://doi.org/10.1016/j.msea.2006.05.036

Terotechnology XII
Materials Research Proceedings **24** (2022) 247-253

Materials Research Forum LLC
https://doi.org/10.21741/9781644902059-36

Testing of Flash Butt Welded Rail Joints

MIKŁASZEWICZ Ireneusz

Instytut Kolejnictwa (Railway Research Institute), Chłopicki 50, Street, 04-275 Warsaw, Poland

imiklaszewicz@ikolej.pl

Keywords: Bending, Fracture of The Joint, Fracture Discontinuity

Abstract. This article presents selected aspects of testing flash butt welded rail joints that have a decisive impact on the rails' operational strength, including cracking in the track. Bending tests of rail joints, fatigue tests, hardness HV30, residual stress results, and microstructure of welded joints are presented. The introduction shows sporadically occurring joint defects revealed on broken fractures related to the technological process of flash butt welding. The above tests were based on the requirements of PN-EN 14587-1:2019-03 [1] and PN-EN 13674-1+A1:2017-07 [2].

Introduction

The safety of running trains and the increasing speed of travel forces the use of suitable quality materials, including rails and flash butt welded rail joints. The existing railway superstructure, especially on the routes with the volume of train traffic, requires monitoring mainly in the form of visual and ultrasonic tests in the tracks, whose purpose is, among others, to assess the condition of rails and rail joints as one of the most important factors of safe transport [3].

In addition to rail defects revealed during operation, including hazardous defects of the head check type [4], which arise on the surface of the railhead, then penetrate deep into the material and constitute a dangerous source of cracks, we may also include occasionally occurring defects of rail joints arising in the process of flash butt welding with stationary and mobile welding machines (Fig.1), also posing a potential danger of distribution cracks.

Rail welding is performed using the flash butt welding method with the process of flashing the surfaces to be joined and subsequent upsetting with force, enabling the front surfaces to be welded [5]. The rail joining process is fully automatic, operating according to computer-controlled programs, depending on the grade and profile of the flash butt welded rails, with a complete registration of technological parameters. An example of the rail flash butt welding diagram for the R260 grade is shown in Fig. 2.

Fig. 1. Stationary rail flash butt welding machine during flashing (Photo: Author)

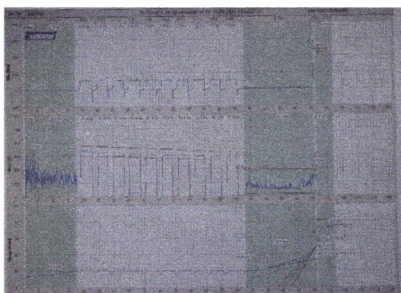

Fig. 2. Flash butt welding parameters of R260 grade rail (Photo: Author)

Materials Research Forum LLC

https://doi.org/10.21741/9781644902059-36

The course of force, current and pre-heating period paths, and flashing and upsetting periods are represented by the corresponding lines in the diagram. Abrupt differences in the above values are visible. This is due to the developed flash butt welding technology. Several times, the flashing period is characterized by the ejection of molten metal due to the formation of current bridges of the emerging electric arc and metal vapor pressure. This results in cleaning and leveling the surfaces of the rails to be flash butt welded and further heating of the ends and upsetting in the final part of the joint flash butt welding process. As a result of joining, the rails are shortened by about 35-45 mm, and the flash butt welding time closes in the range of 80-110 seconds.

The rail flash butt welding diagrams of the R350HT and R400 grades are similar, with a slight difference in the amount of the applied current strength and path.

Rail joints flash butt welded to the shape of the rails to be joined by straightening and grinding treatment to ensure smooth running. In addition, penetrant, magnetic powder, and ultrasonic tests are used to evaluate the flash butt welded rail joints. However, there are occasional cases of irregularities caused mainly by unforeseeable jumps in the technological parameters of flash butt welding, related, among others, to inadequately prepared rail ends for flash butt welding. This is manifested through the appearance of so-called dull spots in the flash butt welding line. These spots are revealed at the fractures of broken rail joints during bending tests. Fig. 3 and Fig. 4 show instances of dull spots at different locations of the fractures.

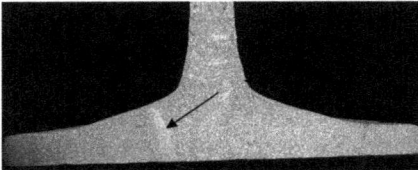

Fig. 3. Dull spots on the fracture of a broken joint (source: author)

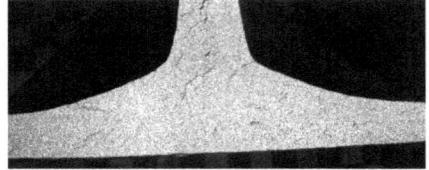

Fig. 4. Dull spots on the fracture of a broken joint (source: author)

Fatigue testing of flash butt welded rail joints is a perfect way of verifying the correctness of the rail flash butt welding process. Any defects occurring during that process affect the strength and ductility of the joints, causing them to crack. These cracks were observed at different times during the dynamic fatigue tests. In the presence of larger dull patches, joint cracking occurs in the range up to 1.0 million fatigue cycles, while with small patches above 3.0 million cycles. These discontinuities are the source of fatigue cracking of rail joints. Figures 5 and 6 show the fractures of flash butt welded joints that underwent separation cracks during fatigue tests at cycles up to 3.0 million.

Internal stresses in rails and rail joints are also crucial for the behavior of these components during further operation in tracks [6] [7]. High stresses in rails and rail joints, i.e. above 250 MPa [1], maybe the source of fatigue cracking during operation in tracks.

Terotechnology XII

Materials Research Proceedings **24** (2022) 247-253

Materials Research Forum LLC

https://doi.org/10.21741/9781644902059-36

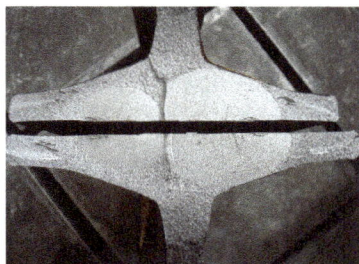

Fig. 5. *Fatigue crack of flash butt welded joint of R260 grade rails (source: author)*

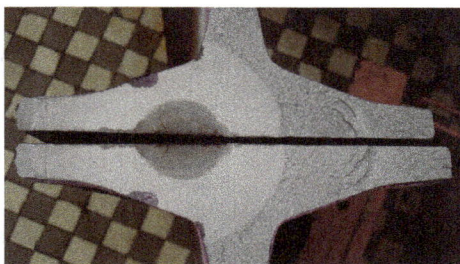

Fig. 6. *Fatigue crack of a flash butt welded joint of R350HT grade rail (source: author)*

Strain gauge testing of R260 and R350HT grade rails and rail joints before installation on the track is shown in Fig. 7 and Fig. 8. The graphs show the presence of tensile stresses in rails and compressive stresses in rail joints. The highest stresses in the rails were recorded in the railhead above ±200 MPa and the rail foot at +195 MPa and +230 MPa, respectively, while in the joints at -175 MPa and -90 MPa, respectively.

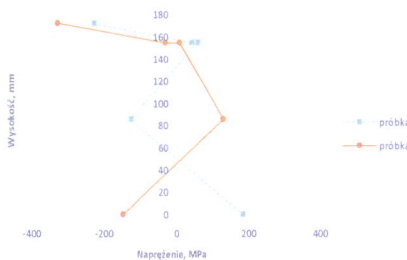

Fig. 7. *Stress distribution in a rail and flash butt welded joint of R260 grade (source: author)*

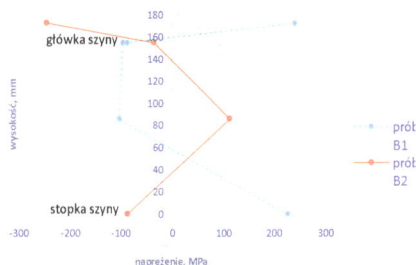

Fig. 8. *Stress distribution in a rail and a flash butt welded joint of R350HT (source: author)*

Plot description: A1, B1 – rail; A2, B2 – rail joint. Plots labels: Wysokość – height; Próbka – speciment; Stopka szyny – rail foot; Naprężenie – stress; Główka – head rail.

Requirements and Test Results for Flash Butt Welded Rail Joints
Testing of 60E1 profile rail joints of R260, R350HT, and R400HT steel grades was carried out by the PN-EN 13674-1:2019 standard. Chemical analysis and strength properties of rails of these grades showed compliance with the above standard. The quality requirements of the joints included in PN-EN 14587-1;2019 concern joint geometry, penetration, magnetic particle, ultrasonic tests, HV30 hardness, strength tests i.e. bending test and fatigue strength tests, also metallographic tests. The geometry profile significantly impacts the comfort of the train journey, so the straightness of flash butt welded joints should be within a maximum of 0.30 mm. The external quality of the joints was tested using the penetration method, while the internal quality of

the flash butt welded rails was examined using ultrasonic testing. The result of the tests was the absence of internal discontinuity of the joints and the ascertained quality of the rail joints of the investigated steel grades as required by the standards.

The bending strength tests of flash butt welded rail joints of the above grades are presented in Table 1. Three-point support with swing support was used during the bending test. The highest mean forces needed to fracture the joint were recorded for the R400HT grade, i.e. 2081.0 kN, while the displacement for the R350HT grade joints was 48.8 mm. The required bending force for these grades should be at least 1600kN, and the displacement should be at least 15.0 mm and 20.0 mm, respectively. A diagram of the bending test for specimen No. 13 is shown in Fig. 9.

Fatigue tests on rail joints of three steel grades were carried out using the "past-the-post" method, i.e. three specimens of each grade were tested. The forces for dynamic tests were determined using 1500 mm spacing between supports and stress in the foot of solid rail 190 MPa. Table 2 shows the values of applied dynamic forces at 5.0 Hz and the fatigue testing results of 9 specimens.

The stresses in the joints of R260 and R350HT grade rails after fatigue tests simulating the operation of the rails during train running were examined using the strain gauge method. The magnitudes of the internal stresses were continuously recorded while cutting the joints with glued strain gauges. Fig. 10 shows the residual stresses in the joints' foot, head, and web. They do not exceed the 250MPa value required by the standard.

As a result of the tests, it was found that the stresses in the feet and heads were tensile, while in the web of the joints, they were compressive. This is consistent with the stress direction in the new rails coinciding with the stresses shown in Fig. 7 and Fig. 8.

Table 1. Results of bending tests on rail joints.

Steel grade/requirements	Specimen number	Bending force, kN	Total bending, mm	Mean value
R260	1	1653.3	36.4	1746.5kN 46.9 mm
	2	1861.7	62.5	
	3	1795.2	53.1	
	4	1638.3	30.4	
	5	1784.1	51.9	
R350HT	6	1686.0	21.0	2026.8 kN 48.8 mm
	7	2291.0	77.0	
	8	1788.0	25.0	
	9	2309.0	80.0	
	10	2060.0	41.0	
R400HT	11	2064.0	30.7	2081.0 kN 31.6 mm
	12	2181.0	37.7	
	13	2160.0	34.2	
	14	2029.0	28.5	
	15	1975.0	27.1	
PN-EN 14587-1:2019		Min. 1600 kN for 60E1/E2	R260, R350HT-min 20.0 mm R400HT- min 15.0 mm	Positive result

Terotechnology XII   Materials Research Forum LLC
Materials Research Proceedings **24** (2022) 247-253   https://doi.org/10.21741/9781644902059-36

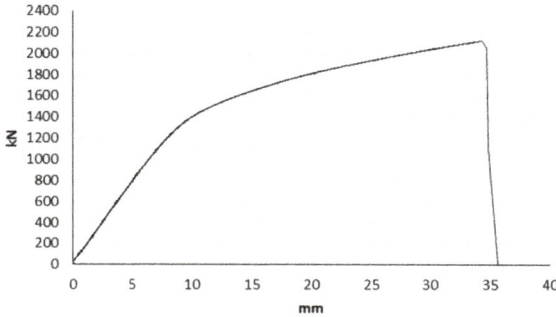

Fig. 9. *Bending diagram of rail joint of specimen No. 13 (source: author).*

Table 2. *Fatigue test results of rail joints*

Steel grade	Specimen number	Bending force, kN	Number of cycles, mln	Result
R260	3	215.0	3 x 5.0	positive
R350HT	3	225.0	3 x 5.0	positive
R400HT	3	224.5	3 x 5.0	positive

The HV30 hardness tests of the rail joints were carried out on the longitudinal section at a distance of 5.0 mm from the rail rolling surface with a measurement every 2.0 mm. The results compliant with the standard requirements [1] were obtained on all samples. The course of HV30 hardness in specimen No. 13 is shown in Fig. 11.

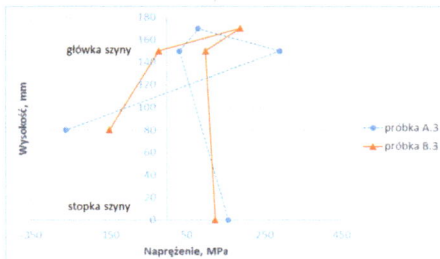

Fig. 10. *Internal stresses in R260 and R350HT joints after fatigue tests, A3 - R260 joint, B3 - R350HT joint (source: author).*

Plots labels: Wysokość – height; Próbka – speciment; Stopka szyny – rail foot; Naprężenie – stress; Główka – head rail.

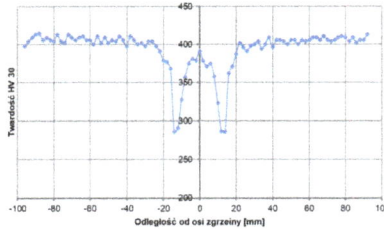

Fig. 11. *HV30 hardness on the longitudinal section of the flash butt welded joint (source: author).*

Plots labels: Twardość – hardness; Odległość od osi zgrzeiny – distance from weld axis.

Examination of the macrostructure for the above grades revealed no internal defects such as nonmetallic inclusions, internal cracks, or unwelded places. The macrostructure of the joint is shown in Fig. 12, with the flash butt weld line and the heat-affected zone visible.

Fig. 12. *Macrostructure of the R260 grade rail joint after flash butt welding (source: author).*

The examination of the microstructure of the joints included the study of the flash butt weld line and the heat-affected zone. In the joints of R260 and R350HT steels, with the carbon content at the lower limit provided by the standard, a differentiated pearlitic structure was found in the flash butt weld line, with separations of ferrite on the borders of former austenite grains (Fig. 13 and 14). In contrast, separations of sorbitol and fine perlite characterized the heat-affected zone.

In rail joints with a higher carbon content of about 0.80 % and above, a pearlitic structure with cementite separations at the boundaries of former austenite grains was found in the flash butt weld line. At the same time, the heat-affected zone had a sorbitic structure. Bainitic-martensitic structures were not found in the joints.

Materials Research Forum LLC

https://doi.org/10.21741/9781644902059-36

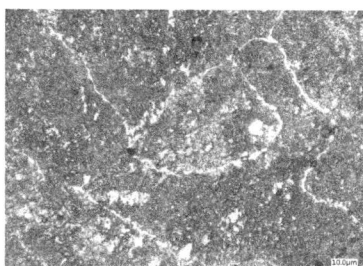

Fig. 13. Microstructure of the flash butt welded joint head of R260 grade rail after fatigue tests (source: author)

Fig. 14. Microstructure of the flash butt welded joint head of R350HT grade rail after fatigue tests (source: author)

Summary

The conducted tests of flash butt welded rail joints of the R260, R350HT, and R400HT grades after stationary flash butt welding in the form of profile geometry tests, penetration and ultrasonic tests, bending strength, fatigue strength tests, and hardness tests showed that the tested flash butt welded rail joints met the requirements included in the standards in question and could be used in railway tracks. At the same time, it was found that the stresses in the tested rail joints after fatigue testing are consistent with the stresses of the rails and rail joints before installation on the tracks. Metallographic tests of the rail joints did not reveal the presence of a martensitic-bainitic structure.

References

[1] PN-EN 14587-1:2019-03. Railway application – Infrastructure – Flash butt welding of new rails – part 1: R220, R260, R320Cr, R350HT, R350LHT, R370CrHt and R400HT grade rails in a fixed plant, 2019.

[2] PN-EN 13674-1+A1:2017-07. Railway application – Track – Rail – Part 1: Vignole railway rails 46kg/m and above, 2019.

[3] H. Bałuch. Trwałość i niezawodność eksploatacyjna nawierzchni kolejowej, Wyd. Komunikacji i Łączności, Warszawa, 1980.

[4] Rolling Contact Fatigue in Rails: a Guide to Current Understanding and Practice, Railtrack PLC Guide-lines: RT/PWG/001, Issue 1, 2001, (13).

[5] I. Mikłaszewicz. Wykonywanie i badanie kolejowych złączy szynowych, Problemy Kolejnictwa 158 (2013) 35-50.

[6] S.J. Wineman, F.A. McClintock. Residual Stresses Near Rail End, Theoretical and Applied Fracture Mechanics 13 (1990) 29-37.

[7] I. Mikłaszewicz, J. Siwiec Badanie naprężeń wewnętrznych w połączeniach szyn kolejowych, Problemy Kolejnictwa 177 (2017) 45-50.

Terotechnology XII
Materials Research Proceedings **24** (2022) 254-261

Materials Research Forum LLC
https://doi.org/10.21741/9781644902059-37

Analysis of Pressure Distribution in Non-Contacting Impulse Gas Face Seals

BŁASIAK Slawomir[1,a*] and TAKOSOGLU Jakub[1,b]

[1]Kielce University of Technology, Faculty of Mechatronics and Mechanical Engineering, Al. 1000-lecia Państwa Polskiego 7, 25-314 Kielce, Poland

[a]sblasiak@tu.kielce.pl, [b]qba@tu.kielce.pl

Keywords: Impulse Gas Face Seals, Dynamic, Reynolds Equation, Numerical Method

Abstract. Strict norms regarding the emission of harmful substances into the natural environment impose stringent requirements on engineers designing sealing units, especially when designing new mechanical seals. This especially concerns mechanical non-contacting seals with various surface layer modifications. Complex mathematical models that enable analyzing complex physical phenomena are developed to support the designing works. The presented paper includes specification of the impulse gas face seals mathematical model which takes into consideration the non-linear Reynolds equation. The mathematical model was solved based on the author's computer program developed in the C++ language, thereby enabling a series of numerical tests and analyses on the phenomena taking place in the radial clearance during the seal's operation. The paper also includes the final conclusions and a series of features specific for the subject impulse gas face seals.

Introduction

Mechanical seals were firmly established as structural elements of sealing units. In recent years, the literature on sealing technology featured a series of papers on non-contacting face seals. It is important to note that such seals are used in virtually all industry branches and operate in extremely different operating and environmental conditions. They are used in rotor machines, including compressors, high-speed pumps, mixers, etc. They play a crucial role in separating the sealed working mediums from the external environment. Many variants of non-contacting face seals that enabled achieving the assumed goal to a lesser or higher degree were developed over the years. The recurring problem in virtually all designs is to maintain a stable separation clearance between work rings. In the case of non-contacting gas face seals, one of the methods that enable maintaining a stable medium layer is to introduce various geometric modifications (micro-structures) to the work rings' tracks. These can include radial or spiral grooves [1,2] or face surface texturing [3]. Introducing changes to the sealing rings' face surfaces causes changes in the fluid film's dynamic properties. Ensuring balance between the forces acting on the ring system allows for maintaining a stable medium layer and prevents contact between the rings during operation. The results of testing on the impact of geometric changes in the applied face surface modifications on the fluid film's dynamic properties are included in [4–10], among others.

In papers [11–15] the authors present the results of numerical analyses of a complex mathematical model encompassing the ring vibration dynamics' equations and the non-linear Reynolds equation for a compressible medium. The calculation of forces and hydrodynamic moments requires solving the Reynolds equation and designating the pressure distribution in the fluid film. In the discussed papers, the two-dimensional Reynolds equation is solved with the use of numerical methods.

Terotechnology XII   Materials Research Forum LLC
Materials Research Proceedings **24** (2022) 254-261   https://doi.org/10.21741/9781644902059-37

The most common methods used include: Finite Volume Method (FVM) [16,17], Finite Element Method (FEM) [18], while the Finite Difference Method is used less often. The correct designation of the pressure distribution, especially in a clearance with complex topography, determines the accuracy of the obtained forces and (hydro-)gaso-dynamic moments obtained.

This paper presents model testing of little-known non-contacting impulse gas face seals. It includes a mathematical model solved using numerical methods and procedures collected in the author's computer program. The paper also includes results of tests mainly concerning pressure changes in the radial clearance at various stages of operation of non-contacting impulse gas face seals.

Subject of Testing

The specificity of non-contacting gas face seals assumes their operation with a certain minimum leakage. A feature specific to non-contacting seals is that they maintain a stable clearance restricting the leak between the interoperating face surfaces of the rings.

Based on [19–23] Fig. 1 presents a drawing of a non-contacting impulse gas face seal.

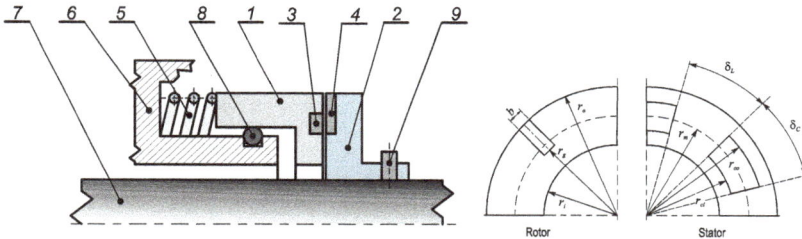

Fig. 1. *Cross section scheme of a non-contacting impulse gas face seal; 1 – stator, 2 – rotor, 3 – chamber, 4 – parallel groove, 5 – spring, 6 – casing, 7 – shaft, 8 – O-ring, 9 – steady pin.*

This seal, similarly to traditional non-contacting gas face seals, consists of two rings. The stator (1) is embedded flexibly in the casing and its face surface features peripherally closed chambers (3). On the other hand, the rotor (2) turns along with the rotor machine's shaft (6) and its face surface includes open radial grooves (4). The grooves are facing the process side (gland) space filled with the sealed medium with a set pressure P_0. The operating principle for the impulse gas face seal was provided in the papers of the creators of such seals [24].

During operation, the sealed medium quickly flows (is injected) from the gland to the aforementioned chamber when the groove (4) overlaps with the chamber (3). The chamber features a rapid pressure increase at that time. When the groove passes a peripheral section equal to the chamber's length, the pressure in the chamber decreases due to the medium's leakage via the outer peripheral area. The pressure decrease in the chamber lasts until the next medium injection (pressure impulse).

The presented mode of operation of non-contacting impulse gas face seals ensures the maintenance of a continuity of the gas film of a low height (several micrometers) and prevents contact between the work rings during the device's operation.

Mathematical Model

It can be stated that the mathematical model for the non-contacting impulse gas face seal is widely known. The subject literature (e.g. [25]) describes it as a discrete and continuous system of differential equations, e.g. equations describing the vibration of a ring mounted flexibly in a casing, equations describing the film fluid's motion and equations describing the stream continuity (mass conservation equation). By applying widely known simplifying assumptions, the motion equation and continuity equation were brought down to a single equation that described the pressure distribution in the radial clearance, i.e. the so-called Reynolds equation [26].

$$\vec{\nabla}\left[\, p\,h^3\,\vec{\nabla}p - 6\mu\omega r\,p\,h\vec{e}_\theta\right] = 12\,\mu\frac{\partial\left(p\,h\right)}{\partial t} \tag{1}$$

It is possible to designate the pressure distribution when adopting the following boundary conditions and periodicity condition:

$$p(r,\theta)\big|_{r=r_i} = p_i\,;\ \ p(r,\theta)\big|_{r=r_o} = p_o\,;\ \ p(r,\theta)\big|_{\theta=0} = p(r,\theta)\big|_{\theta=2\pi} \tag{2}$$

The Reynolds equation (1) is a non-linear equation. It specifies pressure changes in the radial clearance for compressible media. When conducting a dependency analysis (1), it can be stated that the pressure distribution in the clearance depends mainly on the function describing the radial clearance's height. With the assumption of a parallel position of the interoperating rings, the distance of any point located on the ring surfaces in regards to the beginning of the inertial coordinates system is presented as a dependency for the stator and rotor, respectively:

$$h^s = -h_c\left(r,\theta\right) \tag{3}$$

$$h^r = h_o + h_g\left(t,r,\theta\right) \tag{4}$$

The formula describing the simplified radial clearance height function was obtained by subtracting equation (4) from (3).

$$h(t,r,\theta) = h^r - h^s = h_o + h_g\left(t,r,\theta\right) + h_c\left(r,\theta\right) \tag{5}$$

The nominal radial clearance height h_o derives from the balance of forces acting on the work ring system and is determined during the designing of the given non-contacting impulse gas face seal. In the discussed example, the calculations featured the assumption that $h_o = 6\,\mu m$. Special attention must be paid to the surface functions h_c and h_g that describe the geometry of the chambers and grooves on the face surfaces of the stator and rotor, respectively. The function $h_g\left(t,r,\theta\right)$ also takes into account the rotational motion of the rotor along with the radial grooves (provided on the rotor's face surface) with the angular velocity ω.

Numerical Solution of the Reynolds Equation
In the presented paper, the equation (1) was solved by using the numerical method specified in the literature in more detail, i.e. the Finite Volume Method. The method is often used in numerical solutions of differential equations with partial derivatives, especially in the case of notions concerning convection and diffusion. The figure below presents a scheme for the FVM calculation grid.

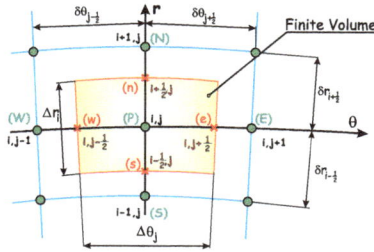

Fig. 2. Finite volume grid scheme.

Virtually two Reynolds equations were solved in the presented FVM calculation algorithm. The first case concerned the designation of the steady state pressure distribution and the calculations did not feature the derivative $\frac{\partial p}{\partial t}$.

$$\int_{s}^{n}\int_{w}^{e}\frac{\partial}{\partial r}\left(ph^{3}r\frac{\partial p}{\partial r}\right)dr\,d\theta + \int_{s}^{n}\int_{w}^{e}\frac{\partial}{\partial \theta}\left(\frac{ph^{3}}{r}\frac{\partial p}{\partial \theta}-6\mu\omega r\,ph\right)dr\,d\theta = 12\int_{s}^{n}\int_{w}^{e}\mu r\,p\frac{\partial h}{\partial t}dr\,d\theta \tag{6}$$

In the second case, the complete Reynolds equation (1) was solved:

$$12\int_{s}^{n}\int_{w}^{e}\mu r\,h\frac{\partial p}{\partial t}dr\,d\theta = \int_{s}^{n}\int_{w}^{e}\frac{\partial}{\partial r}\left(ph^{3}r\frac{\partial p}{\partial r}\right)dr\,d\theta +$$

$$+\int_{s}^{n}\int_{w}^{e}\frac{\partial}{\partial \theta}\left(\frac{ph^{3}}{r}\frac{\partial p}{\partial \theta}-6\mu\omega r\,ph\right)dr\,d\theta - 12\int_{s}^{n}\int_{w}^{e}\mu r\,p\frac{\partial h}{\partial t}dr\,d\theta . \tag{7}$$

The following dependency was obtained by integrating particular components of the equation (6) on the elementary finite volume:

$$\left(A_N + A_S + A_E + A_W + F_e - F_w + \bar{S}\right)p_P = A_N p_N + A_S p_S + A_E p_E + A_W p_W \tag{8}$$

where:

$$A_N = D_n \quad A_S = D_s \tag{9}$$

$$A_E = D_e + \max(0, -F_e) \quad A_W = D_w + \max(F_w, 0) \tag{10}$$

The elements present in the aforementioned components depend on the pressure in the N,S,W,E and P grid nodes and refer to convection and diffusion on each finite volume surface.

$$D_n = \left(\frac{ph^{3}r\Delta\theta}{\Delta r}\right)_n \; ; \; D_s = \left(\frac{ph^{3}r\Delta\theta}{\Delta r}\right)_s \; ; \; D_e = \left(\frac{ph^{3}\Delta r}{r\Delta\theta}\right)_e \; ; \; D_w = \left(\frac{ph^{3}\Delta r}{r\Delta\theta}\right)_w .$$

$$F_e = \left(6\mu\omega r h \Delta r\right)_e,, F_w = \left(6\mu\omega r h \Delta r\right)_w,, \overline{S} = \left(12\mu r \frac{\partial h}{\partial t}\Delta r \Delta \theta\right)$$

The main deficiency of the presented central differential scheme (equation (8)) is its dependency on the direction of flow. The upwind differential scheme, the general form of which was written as dependency (10), was introduced to improve the solution's stability.

$$\left(12\mu r h \Delta r \Delta \theta\right)\frac{\partial p}{\partial t} = \left(A_N + A_S + A_E + A_W + F_e - F_w + \overline{S}\right)p_P - A_N p_N - A_S p_S - A_E p_E - A_W p_W \qquad (11)$$

In the case of the complete non-stationary Reynolds equation (7) (encompassing the pressure derivative over time), it will assume a discretized form as in (11). In the case of the Finite Volume Method, the boundary conditions written with dependency (2) can be easily implemented into the calculations.

Results and Discussion
The solution of the developed numerical model required the introduction of a series of parameters that describe the geometry and working conditions of the impulse gas face seals.

In the case of the impulse face seals, the numerical designation of pressure distribution in the fluid film can cause some problems. As opposed to traditional gas face seals which feature modifications to the sealing rings' tracks in the form of geometric structures (micro-channels, texturing, etc.) applicable only to the (e.g.) stator's surface, the impulse gas face seals also require modeling of the grooves located on the rotor's face surface and move along with the rotor. This means that the aforementioned grooves must move continuously and periodically on the generated calculation grid throughout the entire computer simulation's duration. The results obtained based on the developed numerical algorithm and the author's computer program are presented below. The pressure value was designated at the checkpoint placed on the ring's circumference, in the central part of the chamber located on the stator's surface. The parameters presented in Table 1 were used in the numerical calculations.

It was assumed that the nominal radial clearance height amounted to $h_o = 6\cdot10^{-6}\,(m)$, whereas the depth of the chambers and grooves amounted to $h_g = h_c = 6\cdot10^{-6}\,(m)$.

Table 1. Gas face seal geometry- and performance-related parameters.

Parameter	Value	Parameter	Value
Shaft angular velocity	$\omega = 2094\,[rad\,/\,s]$	Inner radius	$r_i = 0.048\,[m]$
Gas viscosity	$\mu = 1.8\left(10^{-5}\right)[Pas]$	Outer radius	$r_o = 0.060\,[m]$
Design clearance	$h_o = 6\left(10^{-6}\right)[m]$	Chamber inner radius	$r_{ci} = r_m$
Groove depth	$h_g = 6\left(10^{-6}\right)[m]$	Chamber outer radius	$r_{co} = 0.8\cdot r_o$
Chamber depth	$h_c = 6\left(10^{-6}\right)[m]$	Groove radius	$r_g = r_m = 0.5\left(r_i + r_o\right)$
Pressure on the inner radius	$p_i = 1\left(10^{5}\right)[Pa]$	Number of grooves	$n_g = 3$
Pressure on the outer radius	$p_o = 2\left(10^{5}\right)[Pa]$	Number of chambers	$n_c = 6$

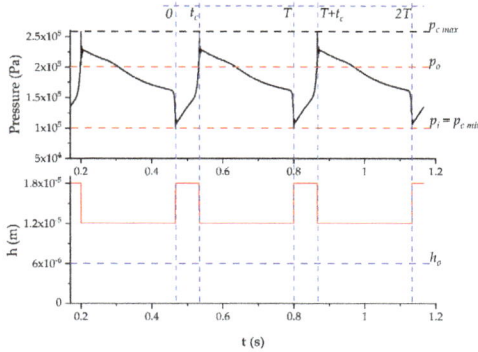

Fig. 3. *Pressure changes in a single chamber*

When analyzing the pressure changes in a single chamber (Fig. 3), it is possible to note that an impulse pressure increase of up to $2.25 \cdot 10^5 (Pa)$ takes place at the instance of time t_c. It is also necessary to note the phenomenon related to the pressure instability of $2.5 \cdot 10^5 (Pa)$. When the groove passes the chamber area, the chamber features a pressure drop of up to $p_{c\min}$. For the adopted sealing ring geometry, it was noted that for the nominal radial clearance height of $h_o = 6 \cdot 10^{-6} (m)$, the pressure value is $p_{c\min} \approx \sim 1 \cdot 10^5 (Pa)$.

In the subject literature, non-contacting impulse gas face seals are often described as semi-active or self-adjusting seals [22, 24]. In the discussed structural solution, the self-adjustment of the radial clearance height is generated through the selection (during the designing phase) of the geometric parameters of the modifications applied to the rings' face surfaces. The expected radial clearance height is obtained by "adjusting" the hydrodynamic force (opening force) generated in the fluid film. In terms of structural assumptions, the perfect operation of a pair of sealing rings should feature the maintenance of the radial clearance height of ($h/h_o \approx 1$), whereas the vibration amplitude should be as small as possible. Maintaining this regime also ensures minimum leakage. In addition, it reduced the probability of contact between the rings, thereby preventing their excessive wear.

The pressure distribution in the radial clearance of non-contacting impulse gas face seals is presented graphically for the parameters given in Table 1.

a)   *b)*

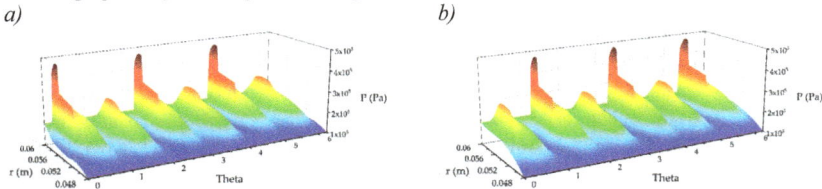

Fig. 4. *Pressure changes in a single chamber*

Fig. 4 presents pressure distributions in the radial clearance in subsequent phases of pressure operation. When analyzing the presented results, it is possible to note the groove motion along the counter ring's track which causes an increase in the medium's pressure along the entire groove length.

Summary

This paper presents the structure, working principle and model testing of little-known non-contacting impulse gas face seals. It also includes the development of a mathematical model as well as numerical calculations featuring the solution of the non-linear Reynolds equation. The obtained results are presented in the form of plots featuring pressure distributions in the radial clearance during the discussed seal's various operation cycles. It can be stated that the reliability and durability of any non-contacting seal, not only the impulse gas face seals, depends on the radial clearance height occurring between the surfaces of a pair of work rings. The above deliberations feature clear advantages of the discussed impulse gas face seals, including: effective heat dissipation, reversibility in regards to the rotation directions, substantial reduction of dimensions when compared to conventional mechanical face seals. The main disadvantage is the presence of moving feed grooves made on the rotor, which pose the risk of clogging. The computer program developed by the author can be a good tool for supporting the designing of new impulse gas face seals. It can also be useful in the selection of adequate geometric parameters of chambers and grooves enabling the achievement of the set operating conditions.

References

[1] R.A. Shellef, R.P. Johnson. A Bi-Directional Gas Face Seal, Tribology Trans. 35 (1992) 53–58. https://doi.org/10.1080/10402009208982088.

[2] N. Zirkelback. Parametric study of spiral groove gas face seals, Tribol. Trans. 43 (2000) 337-343. https://doi.org/10.1080/10402000008982349.

[3] Y. Feldman, Y. Kligerman, I. Etsion. A Hydrostatic Laser Surface Textured Gas Seal, Tribology Lett. 22 (2006) 21-28. https://doi.org/10.1007/s11249-006-9066-z.

[4] S. Blasiak. Influence of Thermoelastic Phenomena on the Energy Conservation in Non-Contacting Face Seals, Energies 13 (2020) art. 5283. https://doi.org/10.3390/en13205283

[5] S. Blasiak. Heat Transfer Analysis for Non-Contacting Mechanical Face Seals Using the Variable-Order Derivative Approach, Energies 14 (2021) art. 5512. https://doi.org/10.3390/en14175512

[6] I. Green. A Transient Dynamic Analysis of Mechanical Seals Including Asperity Contact and Face Deformation, Tribol. Trans. 45 (2002) 284–293. https://doi.org/10.1080/10402000208982551

[7] B. Wang. Numerical Analysis of a Spiral-groove Dry-gas Seal Considering Micro-scale Effects, Chinese Journal of Mechanical Engineering 24 (2011) 146 https://doi.org/10.3901/CJME.2011.01.146

[8] B.A. Miller, I. Green. Semi-Analytical Dynamic Analysis of Spiral-Grooved Mechanical Gas Face Seals, Journal of Tribology 125 (2003) 403–413. https://doi.org/10.1115/1.1510876

[9] B.A. Miller, I. Green. Numerical Formulation for the Dynamic Analysis of Spiral-Grooved Gas Face Seals, Journal of Tribology 123 (2001) 395–403. https://doi.org/10.1115/1.1308015

[10] B.A. Miller, I. Green. Numerical Techniques for Computing Rotordynamic Properties of Mechanical Gas Face Seals, Journal of Tribology 124 (2002) 755–761. https://doi.org/10.1115/1.1467635

[11] I. Green, R.M. Barnsby. A Simultaneous Numerical Solution for the Lubrication and Dynamic Stability of Noncontacting Gas Face Seals, Journal of Tribology 123 (2001) 388–394.

https://doi.org/10.1115/1.1308020

[12] S. Hu, W. Huang, X. Liu, Y. Wang. Stability and tracking analysis of gas face seals under low-parameter conditions considering slip flow, Journal of Vibroengineering 19 (2017) 2126–2141. https://doi.org/10.21595/jve.2016.17111

[13] S. Blasiak. An analytical approach to heat transfer and thermal distortions in non-contacting face seals, International Journal of Heat and Mass Transfer 81 (2015) 90–102. https://doi.org/10.1016/j.ijheatmasstransfer.2014.10.011

[14] W. Xu, J. Yang. Spiral-grooved gas face seal for steam turbine shroud tip leakage reduction: Performance and feasibility analysis, Tribology International 98 (2016) 242–252. https://doi.org/10.1016/j.triboint.2016.02.035

[15] D. Bonneau, J. Huitric, B. Tournerie. Finite Element Analysis of Grooved Gas Thrust Bearings and Grooved Gas Face Seals, Journal of Tribology 115 (1993) 348-354. https://doi.org/10.1115/1.2921642

[16] S. Blasiak. Numerical modeling and analysis of the noncontacting impulse gas face seals, Proc. Institution of Mechanical Engineers Part J – Journal of Engineering Tribology 233 (2019) 1139-1153. https://doi.org/10.1177/1350650118817188

[17] S. Blasiak, A.V. Zahorulko. A parametric and dynamic analysis of non-contacting gas face seals with modified surfaces, Tribology International 94 (2016) 126–137. https://doi.org/10.1016/j.triboint.2015.08.014

[18] B. Ruan. Finite Element Analysis of the Spiral Groove Gas Face Seal at the Slow Speed and the Low Pressure Conditions – Slip Flow Consideration, Tribology Trans. 43 (2000) 411-418. https://doi.org/10.1080/10402000008982357

[19] A.V. Zahorulko. Theoretical and experimental investigations of face buffer impulse seals with discrete supplying, Eastern-European Journal of Enterprise Technologies 4 (2015) 45-52. https://doi.org/10.15587/1729-4061.2015.48298

[20] A. Zahorulko. Experimental investigation of mechanical properties of stuffing box packings, Sealing Technology 8 (2015) 7-13. https://doi.org/10.1016/S1350-4789(15)30244-0

[21] A. Zahorulko, D.V. Lisovenko, V. Martsynkovskyy. Development and investigation of face buffer impulse seal of centrifugal compressor, East.-Eur. J. Enterp. Technol. 1 (2016) 30-39. https://doi.org/10.15587/1729-4061.2016.59884

[22] V. Martsynkovskyy, A. Zahorulko, S. Gudkov, S. Mischenko. Analysis of Buffer Impulse Seal, Procedia Engineering 39 (2012) 43-50. https://doi.org/10.1016/j.proeng.2012.07.006

[23] V.A. Martsynkovskyy, S. Hudkov, C. Kundera. Influence of feeders on operating characteristics of the impulse seals, IOP Conf. Ser.: Mater. Sci. Eng. 233 (2017) art. 012033. https://doi.org/10.1088/1757-899X/233/1/012033

[24] V. Martsynkovskyy, C. Kundera, A. Khalizeva. The Calculation of Static Characteristics of Impulse Gas Seal, Applied Mechanics and Materials 630 (2014) 267-276. https://doi.org/10.4028/www.scientific.net/AMM.630.267

[25] B. Ruan. Numerical Modeling of Dynamic Sealing Behaviors of Spiral Groove Gas Face Seals, Journal of Tribology 124 (2002) 186-195. https://doi.org/10.1115/1.1398291

[26] I. Green, R.M. Barnsby. A Parametric Analysis of the Transient Forced Response of Noncontacting Coned-Face Gas Seals, Journal of Tribology 124 (2002) 151-157. https://doi.org/10.1115/1.1401015

Terotechnology XII
Materials Research Proceedings **24** (2022) 262-265

Materials Research Forum LLC
https://doi.org/10.21741/9781644902059-38

Influence of Operating Parameters and Selected Design Parameters of Ball Bearing on the Friction Torque

ADAMCZAK Stanisław[1,a], MAKIEŁA Włodzimierz[1,b] and GORYCKI Łukasz[1,c*]

[1] Kielce University of Technology, Poland

[a]adamczak@tu.kielce.pl, [b]wmakiela@tu.kielce.pl, [c]lgorycki@tu.kielce.pl

Keywords: Ball Bearing, Friction Torque, Inner and Outer raceway, Curvature Ratio, Roundness, Waviness, Speed, Load

Abstract. Tests were carried out to determine the influence of selected parameters under designated working conditions on ball bearings friction torque. Such parameters include roundness outline and waviness of inner and outer raceways/rings, as well as curvature ratio. In the case of operating parameters, speed and axial load were taken into account. The research presented in the article shows that these parameters have a significant impact, mainly in the case of low loads and medium speeds, and that the parameters of the inner raceway have a much greater impact on the final result of the friction torque than the parameters of the outer raceways. The studies presented below are part of a larger study aimed at creating a mathematical model to determine the theoretical values of the friction torque in rolling bearings.

Introduction

Roller ball bearings are simple structural elements that consist of four elements: two rings (inner and outer), rolling elements, and a cage. Such a simple structure and yet so many different factors that can affect the length of work (durability) and the quality of work (vibrations and friction torque).

The friction torque relates to the performance of the bearings. The greater the value of the friction torque, the worse the quality of work - movement resistance between the different parts of the bearing. The greater the resistance, the greater the loss of energy in the form of heat. As a consequence, this may lead to faster wear of the bearing [6,7].

The factors that theoretically affect the value of the friction torque are for example: rotation speed; axial and radial load; clearances; curvature ratio (wo_{pz}, wo_{pw}, wo_c); runout (Kia, Kea, Sia, Sea, Sd); raceway roundness outline (ΔZ, Zp, Zv, O, $Z3$, Za, Zq); raceway waviness (Wt, Wp, Wv, $W5$, Wa) [8-14].

Research and Results

In order to analyze the influence of selected design parameters on the friction torque, tests were carried out on 100 pieces of rolling bearings 6203. On each of the bearings, 12 measurements were carried out in which the rotational speed (8 000 rpm, 10 000 rpm, 12 000 rpm, 16 000 rpm) and axial load (100 N, 150 N, 200 N) were changed.

The main goal is to determine which parameters (operating and design) and how they influence the final value of the friction torque in rolling bearings. Based on a series of studies, some of which are presented in the following articles [2-4], the following parameters were taken into account for the analysis: wo, ΔZ, $W5$, Wa for both the inner raceway (*ir*) and outer raceway (*or*). The raceways are located on the inner and outer rings, correspondingly. Table 1 contains the results of the performed analysis. It contains the results of coefficient of determination R^2, adjusted R^2 and p for

Terotechnology XII
Materials Research Proceedings **24** (2022) 262-265

Materials Research Forum LLC
https://doi.org/10.21741/9781644902059-38

particular combinations of parameters, speed and axial load. The most important are the combinations of parameters which R^2 and adjusted R^2 results are the highest [1].

Table 1. Sample results of the preliminary analysis of selected parameters.

		8 000 [rpm]	10 000 [rpm]	12 000 [rpm]	16 000 [rpm]
or 100N	R^2	0.1906	0.2042	0.1593	0.0355
	Adjusted R^2	0.1430	0.1581	0.1106	-0.0153
	p	0.0056	0.0030	0.0163	0.5951
ir 100N	R^2	0.0885	0.1132	0.0727	0.0449
	Adjusted R^2	0.0348	0.0618	0.0190	-0.0054
	p	0.1719	0.0777	0.2592	0.4722
or 150N	R^2	0.1833	0.1884	0.1919	0.0732
	Adjusted R^2	0.1359	0.1414	0.1451	0.0245
	p	0.0068	0.0056	0.0049	0.2102
ir 150N	R^2	0.5222	0.5731	0.5541	0.0901
	Adjusted R^2	0.4945	0.5483	0.5282	0.0422
	p	0.0000	0.0000	0.0000	0.1224
or 200N	R^2	0.1282	0.1347	0.1358	0.0782
	Adjusted R^2	0.0791	0.0859	0.0871	0.0296
	p	0.0425	0.0341	0.0327	0.1802
ir 200N	R^2	0.3312	0.4000	0.3340	0.1811
	Adjusted R^2	0.2935	0.3662	0.2964	0.1380
	p	0.0000	0.0000	0.0000	0.0040

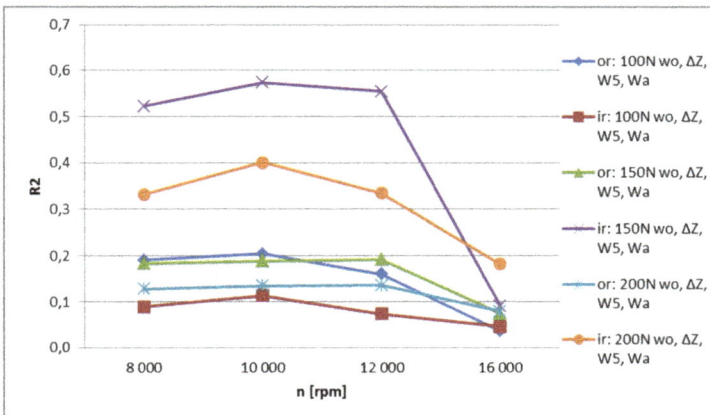

Fig. 1. *Values of R^2 linear regression analysis.*

Materials Research Forum LLC

https://doi.org/10.21741/9781644902059-38

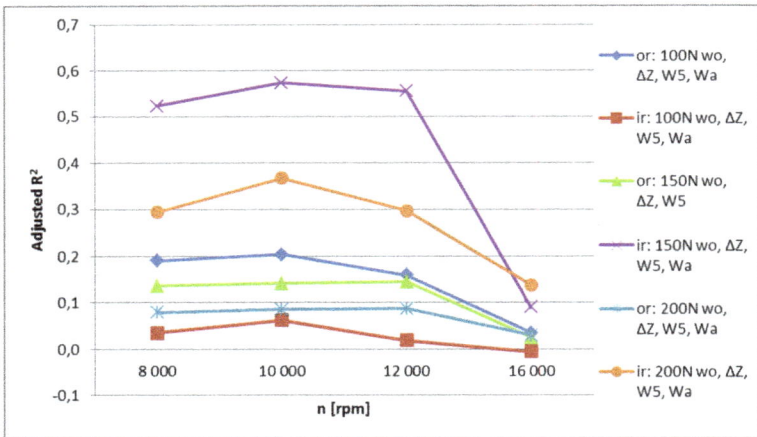

Fig. 2. *Values of adjusted R^2 linear regression analysis.*

Fig. 1 and Fig. 2 show the impact (R^2 and adjusted R^2) of different combinations of operating parameters and design parameters: *wo, ΔZ, W5, Wa – ir* and *or* on the final results of the friction moment.

Conclusions

Based on the results presented above, the following conclusions were drawn:

1. The parameters obtained in the measurement of the inner raceway are more important than the parameters of the outer raceway. The values of R^2 for the analysis of the outer raceway parameters did not exceed 0.2. This means that the selected combinations of parameters did not explain even 20% of the friction torque results.

2. The selected design parameters do not have a significant influence on the resistance torque when the rotational speed is very high. The selected design parameters for the rotational speed of 16 000 rpm reach values of R^2 about: ~ 0.037; ~ 0.074 and ~ 0.166 respectively for 100 N, 150 N, 200 N. This does not completely exclude these parameters from further analysis for operating conditions, including 16 000 rpm. The results of R^2 for the inner raceway at operating conditions of 200 N and 16 000 rpm are significantly higher than the others.

3. As in the case of the rotation speed of 16 000 rpm, the parameters for the axial load of 100 N have little influence on the final result. The values of R^2 for 100 N are below 0.2. On the one hand, these results show that, at low loads, the selected parameters have a minimal effect on the final value of the friction torque. But on the other hand, these results can still be used in further analysis to determine the causes of such significant differences between the design parameters of the inner and outer raceways, such as curvature ratio, raceway roundness outline, raceway waviness.

4. As previously observed, the selected design parameters do not have a significant impact on the final results of the torque with an axial load of 100 N. But comparing the results between the inner and outer rings we can observe a different trend than in the case of the other two loads. Namely, the values of R^2 and adjusted R^2 are twice as large in favor of the outer ring.

5. Depending on the operating conditions, the selected design parameters have a different influence on the friction moment. This can be seen from the values of R^2 of the inner raceway. 3 groups of results were obtained: ~ 0.07; ~ 0.55; ~ 0.35, which correspond to the loads of 100 N, 150 N and 200 N respectively (without taking into account the speed of 16 000 rpm). This shows that for low and medium operating conditions, these parameters explain about 7% of the final value of the friction torque. R^2 reaches its highest value for the operating parameters of 150 N and 10 000 rpm, then with the increase of both operating parameters, the values of R^2 decrease.

References

[1] A.D. Aczel. Statystyka w zarządzaniu. PWN, Warszawa, 2009.

[2] S. Adamczak, Ł. Gorycki, W. Makieła. The analysis of the impact of the design parameters on the friction torque in ball bearings, Tribologia 5 (2016) 11-19. https://doi.org/10.5604/01.3001.0010.6577

[3] Ł. Gorycki, S. Adamczak, W. Makieła, M. Wrzochal. Investigation the Influence of the Curvature Ratio on the Frictional Moment in Rolling Bearings, Procedia Engineering 192 (2017) 255-258. https://doi.org/10.1016/j.proeng.2017.06.044

[4] Ł. Gorycki, R. Domagalski, P. Zmarzły. Measurement of frictional moment in roller bearings, in terms of accuracy geometrically – dimensional elements cooperating, Mechanik 3 (2015) 187-193.

[5] C. Jermak, J. Dereżyński, M. Rucki. Measurement system for assessment of motor cylinder tolerances and roundness, Metrology and Measurement Systems 25 (2018) 103-114. https://doi.org/10.24425/118164

[6] J. Kaczor. The influence of the shape of races for the ball bearings on the life of bearings, Tribologia 5 (2013) 33-44.

[7] J. Kaczor. The influence of clearances in ball bearings on work failure of three-bearing shafts. Tribologia 5 (2014) 89-100.

[8] A. Kowal, Influence of raceway shape on ball rolling movement resistance, Proc. 5th Int. Conf. Zastosowanie Mechaniki w Górnictwie. Dzierżno k. Gliwic 14-15.04.2005.

[9] R. Xiaoli, Z. Jia, R. Ge. Calculation of radial load distribution on ball and roller bearings with positive, negative and zero clearance, Int. J. Mech. Sci. 131-132 (2017) 1-7. https://doi.org/10.1016/j.ijmecsci.2017.06.042

[10] Low-Torque Bearing for Fan Motors and Vacuum Cleaner Motors, Motion & Control 10 (2001) NSK Technical Journal.

[11] Production catalog SKF, 2011.

[12] Bearings catalog. PBF Fabryka Łożysk Tocznych – Kraśnik S.A.

[13] GMN Ball Bearings website. [viewed 31.03.2022] https://www.gmn.de/en/ball-bearings/

[14] Hong Kong Nodes Technology Co., Ltd. – Bearing Distributor. [viewed 31.03.2022] https://www.nodeshk.com/

Terotechnology XII

Materials Research Proceedings **24** (2022) 266-272

Materials Research Forum LLC

https://doi.org/10.21741/9781644902059-39

Locking Mechanism of a Slider with Self-Adjusting Backlash: Design and Dynamic Analysis

JASIŃSKI Wiesław[1,a] *, KRYSIAK Piotr[1,b] and PICHLAK Cezary[1,c]

[1]Military Institute of Engineer Technology, Obornicka 136 Str., 50-961 Wroclaw, Poland

[a]jasinski@witi.wroc.pl, [b]krysiak@witi.wroc.pl, [c]pichlak@witi.wroc.pl

Keywords: Locking Mechanism, Kinematic Analysis, Dynamic Analyses, Design Optimisation

Abstract. In mechanical designs which are to perform linear or rotary motion, there are forces resulting from the system inertia, speed, velocity of moving parts, friction resistance, etc. Therefore, the essence of modern design lies in predicting, in as much as possible, all the phenomena impacting the design and simulating them appropriately with the use of the available numerical tools. This paper presents both the modelling and kinematic and dynamic analyses for a mechanism blocking the linear movement of a slider connected to a movable element weighing approx. 1000 kg. The structure of the locking mechanism, essentially composed of two moving parts joined with the sliding element, was analysed. The developed mechanism consists in the use of moving parts to eliminate backlash (play), which during long-term operation, would be a disadvantage and lead to excessive dynamic forces and structural failure. The studies were performed in the "dynamic analysis" environment of the Inventor Professional 2021 3D CAD software.

Introduction

In the construction of mechanisms that implement movement based on linear guide rails, there is a need to lock the moving objects in specific, desired positions. For this purpose, engineers employ complex braking systems, as well as stoppers and locking mechanisms, simple in their construction. Stoppers restrict movement in one direction while locking mechanisms allow the movement to be limited in both directions thanks to the opening and closing mechanism. These mechanisms may be powered manually, electrically or using hydraulic or pneumatic systems. The drawbacks of such locking include difficulties in establishing the locking in both directions, which ensures a backlash-free connection, especially during operation [1-10]. This paper describes an innovative type of locking mechanism that performs its function with no backlash and a simulation was carried out to verify its operation. Simulation can not only replace costly practical tests and tests after the mechanism has been manufactured, but also the performance of multiple tests in a dedicated software environment.

Materials and Methods

Autodesk Inventor Professional 2021, version 2021.3.3 was used for the design works. This program contains the necessary standards and norms relative to mechanics and a number of useful calculation modules and wizards. The process of designing a given mechanism originally required the preparation of pre-designed components, and joining them together via nods at the later stage. This procedure allowed the authors to verify the matching of components and interconnections. For the purposes of the study, a dynamic simulation module was utilised, which enabled the simulation of the mechanism's operation, considering concurrent phenomena in the form of driving forces, forces in springs and friction in kinematic pairs. The results of the analyses conducted yielded the desired results in the form of the driving force runs as the function of the locking mechanism displacement and the spring forces as the function of time. Graphs in Figs. 1 and 2 serve as an illustration of these results.

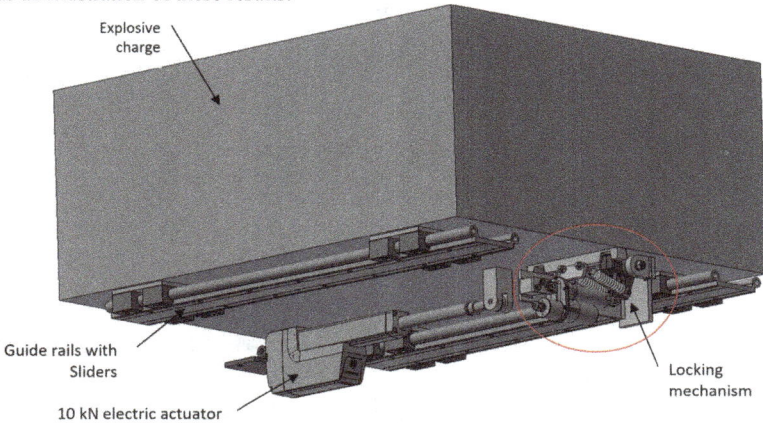

Fig.1. *General view of the studied mechanism.*

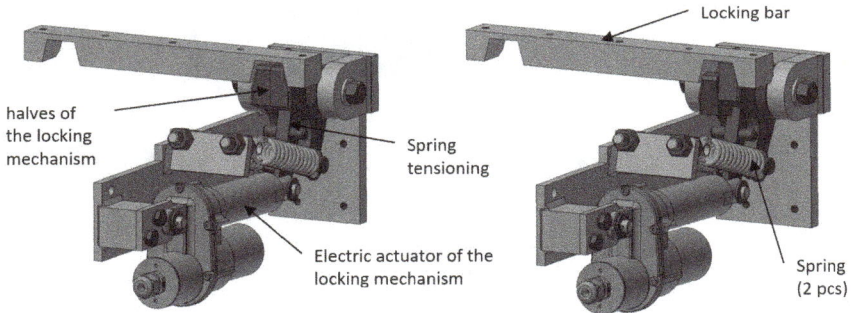

Fig.2. *Diagram of the designed locking mechanism.*

Results

The authors strived to verify the correct operation of the backlash-free locking mechanism. This generally takes place after the product has been manufactured and its operation – initiated. Autodesk Inventor's Dynamic Simulation module enables product testing at the digital prototype stage.

Design of the Backlash-Free Locking Mechanism. Consisting of two halves, the locking mechanism automatically compensates for any play resulting from inaccuracies of the slider braking distance (displacement) within the range of $\pm$ 1.5 mm. The wear and tear on the mating surfaces can also be compensated. The operation of the mechanism is presented in Fig. 3.

Terotechnology XII
Materials Research Proceedings **24** (2022) 266-272

Materials Research Forum LLC
https://doi.org/10.21741/9781644902059-39

State: LOCKED with no backlash on the locking mechanism
Displacement: equals the nominal value (0 mm)

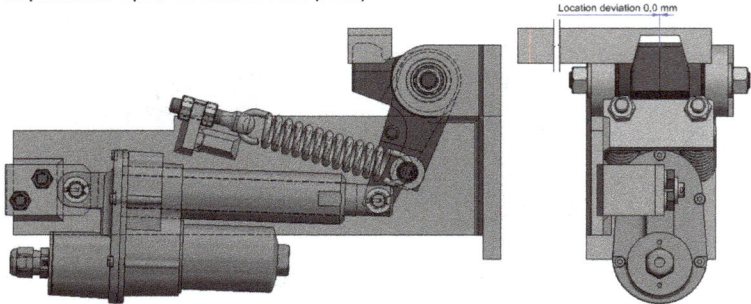

Location deviation 0,0 mm

State: LOCKED with 0.3 mm backlash on the locking mechanism
Displacement: within the nominal value with 1.5 mm location deviation

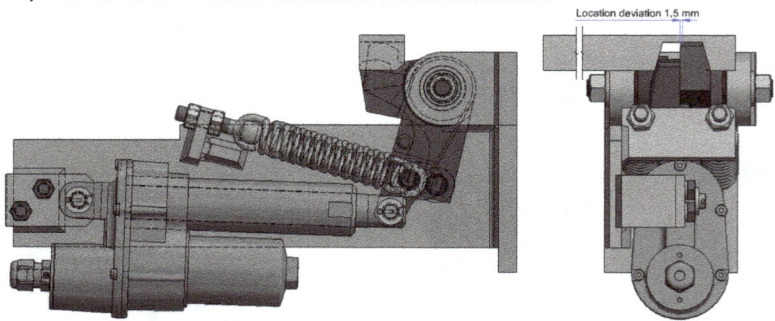

Location deviation 1,5 mm

Fig.3. *Self-adjustment of a backlash-free locking mechanism.*

Such a mechanism should lock the position effectively and with no backlash, regardless of the location deviation within the range of ± 1.5 mm. In addition, the locking strength within this distance should be approximately the same. Such a solution has the advantages of applications based on brake mechanisms (brake blocks or brake shoes) while being simpler and more economic.

Verification of the Mechanism's Operation Using a Digital Model. The necessary elements of the locking mechanism were introduced into the digital model and the proper loads were applied. It was assumed that the mechanism operates correctly correct if the momentum of the slider's movement does cause the locking mechanism to disengage. This is an extreme case (e.g. an automation error), but certainly, such a force would not occur under normal operation and it is a safe assumption. In the structure under consideration, the drive used was an electric actuator with a nominal pulling force of 10 kN. The springs pressing the locking mechanism's halves were responsible for the effective operation of the mechanism, and hence the presence of these springs with the designed characteristics, presented in Fig. 4, in the dynamic simulation.

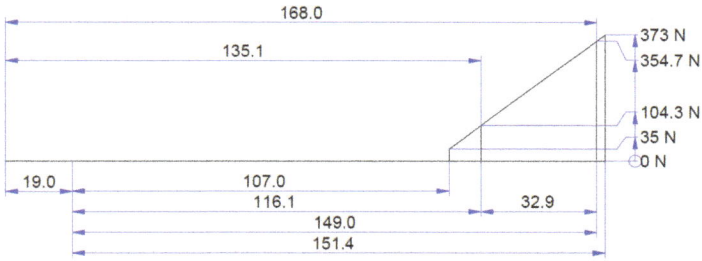

Fig.4. Spring characteristics.

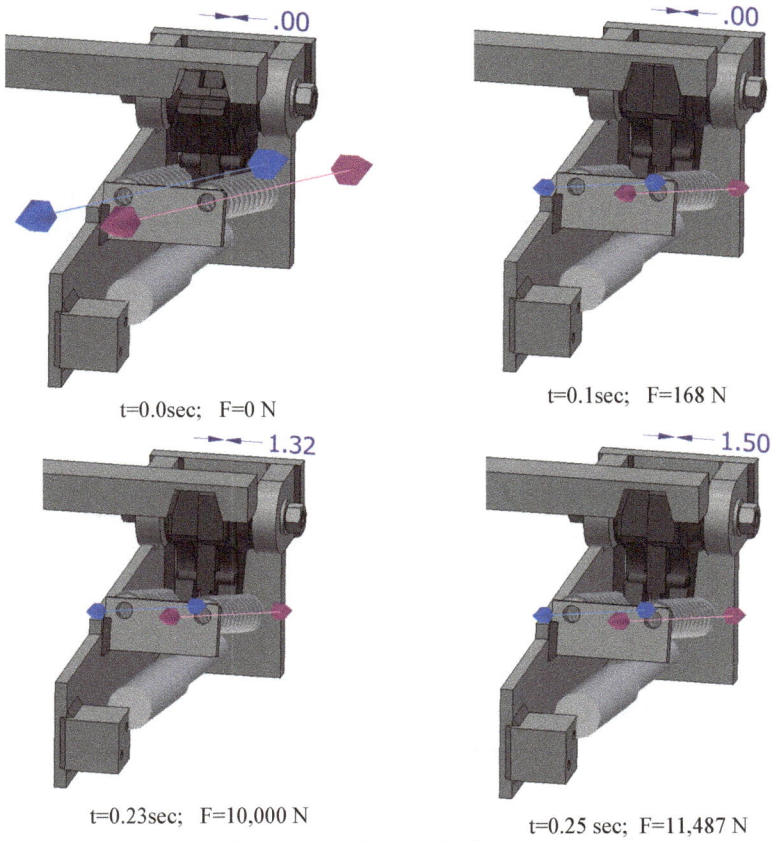

t=0.0sec; F=0 N

t=0.1sec; F=168 N

t=0.23sec; F=10,000 N

t=0.25 sec; F=11,487 N

Fig.5. Driving force as the function of time.

The dynamic simulation was carried out with the following assumptions in mind:

1. Simulation of the loading over time:
   - the time interval of 0.0 ÷ 0.6 sec,
   - the coefficient of friction equal $\mu = 0.2$ was assumed for the mating surfaces (steel/steel);

2. The locking mechanism halves (the P-block and the L-block) are pressed to the recess (cut out) by spring forces in accordance with their characteristics throughout the operating range;

3. The displacement of the locking bar was being applied across the slider in the range from 0 mm to 5 mm during 0.6 s. This movement, which generates the driving force necessary for the displacement, and is shown in Fig. 5 and on the graph in Fig. 6.

4. The displacement of the slider is monitored. A displacement of up to ± 1.5 mm is allowed, as it is within the range of position adjustment;

5. The slider displacement value for a driving force of 10 000 N is to be read out from Fig. 7;

6. The mechanism will work properly; the slider displacement needs to be under 1.5 mm if the driving force equals 10 kN.

Fig.5 illustrates the simulation at t = 0; 0.1; 0.23; 0.25 sec.

A graph of the driving force as the function of time was drafted based on the simulation (Fig. 6). What could be read out from the graph, is that the permissible slider driving force of 10 kN will occur at 0.23 sec.

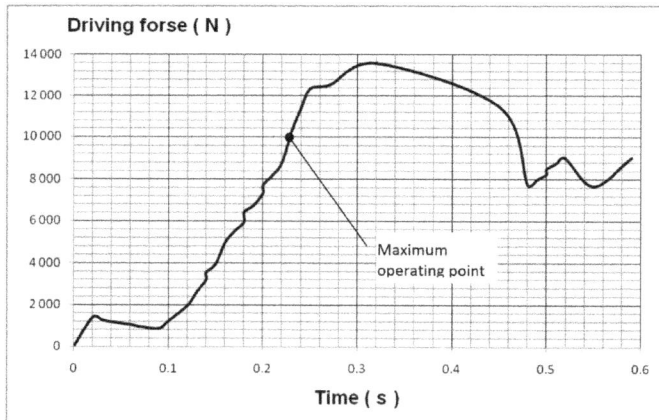

Fig.6. *Diagram of driving force as the function of time.*

A graph of the driving force as the function of time (Fig. 7) was drafted based on the simulation. What could be read out from the graph, is that the permissible slider driving force of 10 kN will occur at 1.23 mm of the locking bar movement.

A graph of the driving force as the function locking mechanism displacement (Fig. 7) was also based on the simulation. The graph reveals that the permissible slider driving force of 10 kN will occur for the slider displacement equal 1.23 mm.

The spring force as the function of time is presented in Fig. 8. The graph was drafted based on the simulation conducted. It can be seen that for displacement of 1.5 mm the force of the P-spring equals 302 N, and that of the L-spring is 274 N, which is well within their performance characteristics.

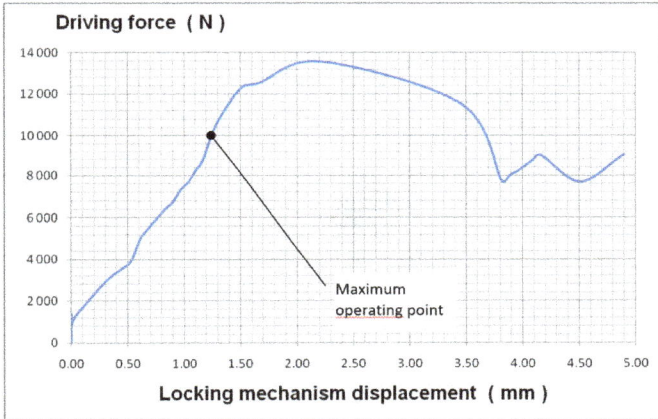

Fig. 7. Driving force in the function of the locking mechanism displacement.

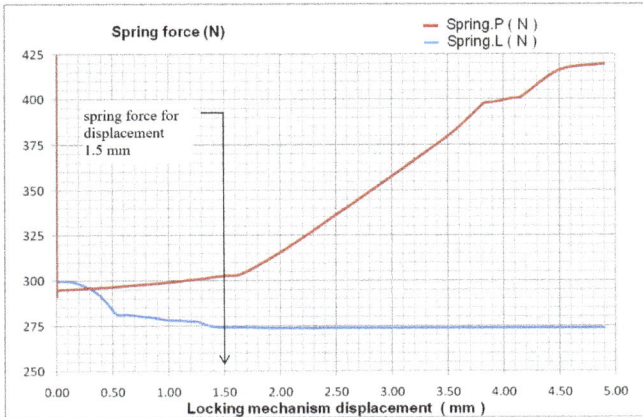

Fig. 8. Forces of springs as the function of time.

The mechanism will work properly, as for a given slider driving force of 10 kN, the slider displacement value is 1.23 mm (Fig. 7), i.e. under 1.5 mm. Thus, the mechanism design can be considered correct.

Conclusion

Dynamic simulations bring undeniable benefits to the design practice. This study shows that the designed mechanism will function properly. The investigation was performed in a multifaceted manner, without the need for costly experimental tests, which will be performed on the final version of the designed product only, as part of the device prototype testing stage.

It should be noted that as in the case of all technical calculations, the value of the results obtained depends on the accuracy of the boundary conditions set. As even the most refined software tools will miscalculate when fed incorrect data, the use of advanced computational methods, such as dynamic simulations, should be reserved for experienced engineers, practicing in the field.

What also needs to be emphasized is that for simple mechanisms, that is with a low degree of complexity, where it is easier to formulate the input data, the calculation results will be more trustworthy, and such is the case examined in this article.

References

[1] A. Arian, B. Danaei, H. Abdi, S. Nahavandi. Kinematic and dynamic analysis of the Gantry-Tau, a 3-DoF translational parallel manipulator, Applied Mathematical Modelling 51 (2017) 217–231. https://doi.org/10.1016/j.apm.2017.06.012

[2] J. Frączek, M. Wojtyra. Kinematyka układów wieloczłonowych – metody obliczeniowe. WNT, Warszawa, 2008.

[3] M. Gawrysiak. Mechatronika i projektowanie mechatroniczne. Białystok University of Technology, Białystok, 1997.

[4] A. Gronowicz. Podstawy analizy układów kinematycznych. Wroclaw University of Technology, Wrocław, 2003.

[5] P. Hejma, M. Svoboda, J. Kampo, J. Soukup. Analytic analysis of a cam mechanism. Procedia Engineering 177 (2017) 3-10. https://doi.org/10.1016/j.proeng.2017.02.175

[6] S. Miller. Teoria maszyn i mechanizmów – analiza systemów kinematycznych. Wroclaw University of Technology, Wrocław, 1996.

[7] J. Ormezowski. Analiza dynamiczna mechanizmu hamującego. Archiwum Motoryzacji 1 (2010) 27-34.

[8] A. Sapietova, J. Bukovan, M. Sapieta, L. Jakubovicova. Analysis and implementation of input load effects on an air compressor piston in MSC.ADAMS. Procedia Engineering 177 (2017) 554-561. https://doi.org/10.1016/j.proeng.2017.02.260

[9] P. Sperzyński, J. Szrek, A. Gronowicz. Synteza geometryczna mechanizmu realizującego trajektorię prostoliniową ocechowaną. Acta Mechanica et Automatica 4 (2010) 124-129.

[10] J. Vavro Jr., J. Vavro, P. Kováčiková, R. Bezdedová. Kinematic and dynamic analysis of planar mechanisms by means of the SolidWorks software. Procedia Engineering 177 (2017) 476-481. https://doi.org/10.1016/j.proeng.2017.02.248

Materials Research Forum LLC
https://doi.org/10.21741/9781644902059-40

Analysis of Neural Network Structure for Implementation of the Prescriptive Maintenance Strategy

FILO Grzegorz[1,a *] and LEMPA Paweł[1,b]

[1]Cracow University of Technology, Faculty of Mechanical Engineering,
Jana Pawła II Ave 37, 31-864 Cracow, Poland

[a]grzegorz.filo@pk.edu.pl, [b]pawel.lempa@pk.edu.pl

Keywords: Artificial Neural Network, Neuron Model, Layer Model, Prescriptive Maintenance, Input Signal Normalisation

Abstract. This paper provides an initial analysis of neural network implementation possibilities in practical implementations of the prescriptive maintenance strategy. The main issues covered are the preparation and processing of input data, the choice of artificial neural network architecture and the models of neurons used in each layer. The methods of categorisation and normalisation within each distinguished category were proposed in input data. Based on the normalisation results, it was suggested to use specific neuron activation functions. As part of the network structure, the applied solutions were analysed, including the number of neuron layers used and the number of neurons in each layer. In further work, the proposed structures of neural networks may undergo a process of supervised or partially supervised training to verify the accuracy and confidence level of the results they generate.

Introduction

Currently, the leading maintenance technology used practically in enterprises is predictive maintenance, which is one of the elements of the Industry 4.0 strategy. It allows for the optimal use of machines and devices, eliminating or reducing the adverse effects of failures downtimes and improving the organisation of maintenance works by carrying out regular inspections of the technical condition of machines. The prescriptive maintenance strategy is a further development of the predictive maintenance approach. According to this strategy, the system should eliminate or reduce the effects of undesirable situations in the industrial process and generate automatic recommendations and warnings for maintenance services regarding the possible occurrences of potential problems in the future.

A literature review by K. Lepenioti et al. [1] lists several methods used in prescriptive maintenance. The study provides the primary methods for implementing prescriptive analytics and identifies existing research challenges. The classification of methods for prescriptive maintenance includes the following categories:

• probabilistic models, including *Markov Decision Process*, *Hidden Markov Model* and *Markov Chain*, are usually used to calculate the likelihood of certain events occurring (like a malfunction). The usage of probabilistic models allows the avoidance of continuous monitoring of actual data looking for patterns of events that have happened before;

• mathematical programming is commonly used to solve complex decision-making problems, e.g. optimal or quasi-optimal allocation of scarce resources. It may use a variety of techniques and algorithms: *Mixed Integer Programming, Linear Programming, Binary Quadratic Programming, Non-Linear Programming, Robust and Adaptive Optimization, Dynamic Programming, Binary*

Linear Integer Programming, Stochastic Optimization, Conditional Stochastic Optimization, Constrained Bayesian Optimization, Fuzzy Linear Optimization, Optimal searcher path;

• evolutionary computation algorithms used for stochastic optimisation in a multi-agent environment, inspired by the process of biological evolution: *Genetic algorithms, Evolutionary Optimization, Greedy algorithm, Particle Swarm Optimization*;

• simulation is understood as the creation of simplified computer models and verification of their operation. It can be used in prescriptive analytics systems to increase the efficiency of the decision-making process regarding infrastructure security and product quality. Main simulation methods include: *Simulation over Random Forest, Risk Assessment* and *Stochastic simulation*;

• logic-based models are also used to improve the decision-making process. They usually store expert knowledge in the form of rules: *Association rules, Decision rules, Criteria-based rules, Fuzzy rules, Distributed rules, Benchmark rules*, etc.

• machine learning and data mining use the following methods: *K-means clustering, Reinforcement Learning, Privacy preservation, Boltzmann Machine, Nadaraya-Watson estimator*, and *Classical Artificial Neural Networks*. The mentioned methods are used to build network models using specific sets of input and output data (training samples), which are then can a prediction or recommendation for other, similar data sets. Combining neural networks with machine learning algorithms seems particularly useful in solving too complex problems for probabilistic models, mathematical programming, and logic-based models.

Implementations of Prescriptive Maintenance Strategy. The description of the conceptual model of the prescriptive maintenance strategy for the semiconductor production process was presented by Biebl et al. [2]. The core of the decision-making system is a 3-layer Bayesian network, trained to predict the causes of failures and generate recommendations regarding the possibility of potential problems. The study highlights the high requirements for ensuring the appropriate quality of training data and the need to classify service activities and generate documentation for the performance of these activities for individual components. The publication [3] presents the concept of building an integrated model covering both production planning and its maintenance using the prescriptive maintenance (PriMa) strategy. The proposed model covers maintenance, autonomous production control (APC), and production planning (PP). The decision-making process is based on the maintenance of knowledge databases and model-based learning algorithms. A proposal to build a learning-based reference model following the assumptions of the prescriptive maintenance strategy is included in the article [4]. The proposed reference model supports the implementation of the PM strategy, facilitates the integration of scientific data processing methods to predict potential future events, and identifies action fields to achieve higher prediction accuracy. The authors also postulate using a strategy of continuous improvement of the reference model during its operation by adding patterns of all newly occurring failures to the database. K. Matyas et al. in [5] proposed an innovative procedural approach to the recommended maintenance planning in manufacturing companies. Multi-dimensional data analysis and simulation tools were used to analyse historical data on product quality machine and production program failures. A rule-based reaction model predicts system failures and proposes recommended maintenance measures using identified data correlations and incoming real-time machine data. The results of actual industrial implementation in a 3D machining centre of an automotive company in Austria are presented. The system consists of four main modules: (1) overview, (2) machine health prediction, (3) planning support, and (4) key performance index (KPI). The authors declare that the system allows predicting about 43% of unplanned downtime of machines caused by mechanical failures. This effect leads to higher quality maintenance planning, including savings of up to 20% of the time spent on completing the entire maintenance process, including analysis, evaluation and manual activities. Additionally, a 30% reduction in maintenance costs was achieved by generating recommendations to avoid direct and indirect repair costs and an overall increase in equipment availability by 12%.

Neural Networks in the PM Models. In the available literature, neural networks are usually included as components of decision models involved in the prediction process. However, there are no scientific publications on the practical aspects of using neural networks, deep learning and machine learning methods in the prescriptive maintenance process concerning the company's maintenance. In particular, there is a lack of studies on structures and datasets containing actual industrial data that can be used in this process. Wang et al. [6] used a neural network to predict the sound attenuation properties of glass fibre mats. In this case, the authors used a classic 3-layer nonlinear network with one hidden layer. The training process used one hundred fifty independent data samples from experimental studies. In the input and hidden layers, neurons with tan-sigmoid activation function were used, while in the output layer - linear neurons. The Matlab system implemented the network model, and supervised training was carried out using six different algorithms. A network trained with the Bayesian Regularization (BR) algorithm obtained the best prediction results. A mean relative error below 5% was acquired by a network trained with the Bayesian Regularization (BR) algorithm. Lazakis et al. [7] proposed a hybrid methodology for identifying critical systems and components of a Panamax size container ship through reliability modelling and monitoring their physical parameters through neural networks. In particular, artificial neural networks were used to predict the upcoming values of all main engine cylinders exhaust gas temperatures. A three-layer network with hyperbolic tangent transfer function in the hidden layer and linear neurons in the output was designed and trained. The results predicted by the neural network were validated through comparison with actual observations recorded onboard the ship. Inspiring models of artificial neural networks that quantify membrane fouling causes in a reverse osmosis system were created by E. A. Roehl Jr et al. [8]. The models were developed and trained based on a six-year process database and comprised 59 hydraulic and water quality parameters, representing 190 runs between membrane cleanings. The runs were segmented into two phases which were modelled separately. The trained network allowed identification and prediction of four parameters within phase 1 and five parameters within phase 2 which became crucial for the correct operation of the system and avoiding failures. Furthermore, the obtained results were consistent with the previously known failure mechanisms. Unfortunately, the authors did not provide precise data on the architecture of the neural network and the types of activation functions used.

The carried out analysis shows that the use of neural networks and the related deep learning methods in the prescriptive maintenance process, especially about maintaining the company's operation, is currently in the phase of continuous growth. It is worth noting the growing interest in leading research centres. However, scientific publications on this subject are few. In particular, there are no studies on recommended network structures and real industrial data sets that can be used in the network training process. Hence, within this study, the tasks of formulating assumptions for constructing the neural network structure, selecting types of neurons, and normalising signals were undertaken.

The issues discussed in this article may be of interest to a wide audience using data-driven data analysis methods, including in materials sciences [9-11], machining [12-14], thermodynamics of internal combustion engines [15], energy[16, 17], mechanics of solids [18, 19], BIM [20, 21], road

Terotechnology XII

Materials Research Forum LLC

Materials Research Proceedings **24** (2022) 273-280

https://doi.org/10.21741/9781644902059-40

traffic engineering [22, 23], and even in military equipment [24, 25]. The developed identification (learning) techniques of neural networks can be used in analogous situations requiring the creation of appropriate predictive approximators [26-28] and classifiers [29].

Working Principle of a Neural Network

Using a model based on neural networks requires extensive research to develop the most effective network structure (number of layers, number and types of neurons) and the selection of a training algorithm because it may be of significant importance for the subsequent operation of the model in industrial conditions. The proposed neural network design strategy is shown in Fig. 1.

In the first phase, the inputs and outputs of the neural network are identified. The chosen input parameters are classified and normalised. Then, research is carried out to determine the structure and topology of the network, including the number of layers of neurons and the number and form of their activation functions in each of the layers. The final selection of individual parameters is assessed based on the obtained training results conducted using supervised learning techniques.

Input Parameter Identification. The first step in preparing input data is to develop models of behaviour of a selected device or a group of similar devices, their characteristics and maintenance procedures. The models include separate devices based on their technical specifications, mainly the type and ranges of operating parameters. This requires grouping (classification) of devices and then assigning individual groups to specific steps in business processes and determining the expected results of their use.

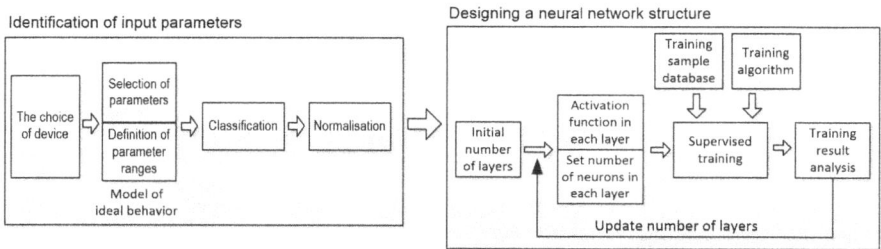

***Fig. 1.** Neural network design strategy for a selected device*

The conducted analysis allowed identification of the following tasks:

- selection of input parameters that have a key impact on the correct operation of the device;

- normalisation of input parameters. Typical input parameters u_i are assumed to be: voltage signal e.g. 0-10 V, current signal 4-20 mA, values from the continuous range $<a, b>$, values from the finite set $\{u_1, u_2, ..., u_k\}$ and logical parameters $\{false, true\}$. Values of logical parameters are mapped using a simple relation: $\{false, true\} \rightarrow \{0 \quad 1\}$. Parameters with continuous or quasi-continuous are normalised to the range $<0, 1>$ according to the dependency:

$$u_{norm} = f(u_i) = \frac{1}{b-a} \cdot u_i + \frac{a}{a-b}, \text{ where } u_i \in \langle a \quad b \rangle \text{ assuming } b > a . \quad (1)$$

For the parameter values from a finite set of size n, the following normalisation formula is proposed:

$u_{norm} = f(u_i) = \dfrac{ord(u_i) - 1}{k - 1}$, where *ord(i)* is the index of the *i*-th element and $u_i \in \{u_1 \quad \dots \quad u_k\}$; (2)

• creating the ideal reference output parameters: each identified input parameter is assigned a reference output value or range of values. This leads to designing a theoretical model of a device operating at its optimal conditions. The created model metrics also constitute a set of training and test samples for the neural network;

• creating the reference samples for the occurrence of emergency states: these samples combine values of the input parameters out of the allowable range with identifiers of the possible malfunctions or other undesirable exceptional events. Such a situation should exceed the threshold output value of a neuron or a group of neurons in the output layer.

Design of a Neural Network Structure. Based on previous publications on neural networks in the prescriptive maintenance strategy, a classic feed-forward network architecture was proposed, including the input layer, a certain number of hidden layers *m*, and the output layer. The scheme of the network is shown in Fig. 2.

The following requirements and recommendations were taken into account when creating the network structure:

• initial assumption of hidden layer number is *m*=3. However, there is a possibility of increasing the number in the case if the initial results of the effectiveness of the trained network are insufficient;

• number of input signals *u* is *k*;

• an initial number of neurons in hidden layers h_i, where $i = 1 \dots m$ depends on the layer number: $h_1, h_m = 2 \cdot k$, while $h_2, \dots, h_{m-1} = 3 \cdot k$;

• number of output parameters *y* represented by neurons in the last layer is *p*;

• activation functions are the *Swish* function *S(x)* in hidden layers and the *sigmoid* (*logistic*) function *σ(x)* in the output layer. The *Swish* function is mainly dedicated to the hidden layers of a multi-layer network. In contrast, the *sigmoid* function gives a convenient range of the output signal <0, 1>, allowing easy identification of the permitted and emergency operational states. The formulas and graphs of the activation functions and their derivatives are shown in Eq (3), Eq(4) and Fig. 3.

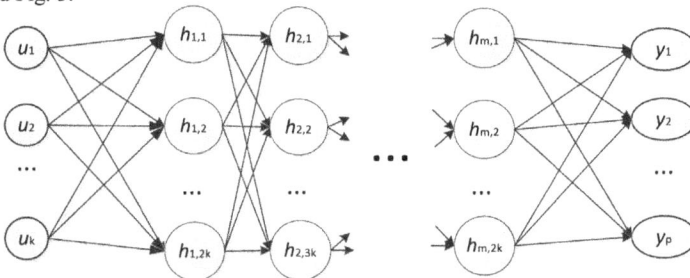

Fig. 2. Neural network architecture

$\sigma(x) = \dfrac{1}{1 + \exp(-x)}$, and the derivative can be calculated as: $\dfrac{d\sigma(x)}{dx} = \sigma(x) \cdot (1 - \sigma(x))$; (3)

$$S(x) = x \cdot \sigma(x) = \frac{x}{1 + \exp(-x)} \text{ , and the derivative is: } \frac{dS(x)}{dx} = S(x) + \sigma(x) \cdot (1 - S(x)). \tag{4}$$

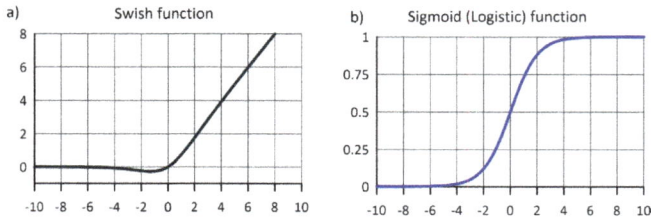

Fig. 3. *Activation functions: a) Swish, b) Sigmoid (Logistic)*

In the next step, the neural network should be subjected to a training process using sets of samples containing parameters and a device state corresponding to each sample (regular operation/occurrence of an emergency state). Verifying the correct operation of the synthesised neural network should take place by simulating the deviations of selected parameters from the standards and checking the related neural network outputs. It is expected that the acceptable efficiency understood as the percentage of correct messages for a given set of test samples should be about 75%. The verification is iterative, possibly changing the number of hidden layers and the number of neurons in each layer. The result of the work carried out within this study will be an object of a trained multilayer neural network. The object will be based on a selected device or process's behavioural patterns and performance characteristics.

Summary

The main contribution of this study to the development of the current state of the art is the proposal to develop an effective structure of the neural network concerning the Prescriptive Maintenance (RxM) strategy. This will fill the gap in research on the use of artificial intelligence (in particular neural networks) in this strategy and allow for its further automation process. The systematisation and analysis of data generated by enterprises to use them in the Prescriptive Maintenance strategy and the creation of behaviour patterns and corrective actions based on expert knowledge on devices, technologies, readout characteristics, etc., will also have a significant impact. The work proposes a methodology for identifying and normalising input and output signals to prepare training and test data samples. The architecture of a multi-layer neural network was also offered, which in the course of further work will be subjected to a training process using appropriate machine learning algorithms.

References

[1] K. Lepenioti, A. Bousdekis, D. Apostolou, G. Mentzas. Prescriptive analytics: Literature review and research challenges, International Journal of Information Management, 50 (2020), 57-70. https://doi.org/10.1016/j.ijinfomgt.2019.04.003

[2] F. Biebla, R. Glawara, A. Jalalic, F. Ansaria, B. Haslhoferc, P. de Boerd, W. Sihn. A conceptual model to enable prescriptive maintenance for etching equipment in semiconductor manufacturing, Procedia CIRP 88 (2020) 64-69. https://doi.org/10.1016/j.procir.2020.05.012

[3] R. Glawar, F. Ansari, C. Kardos, K. Matyas, W. Sihn. Conceptual Design of an Integrated Autonomous Production Control Model in association with a Prescriptive Maintenance Model (PriMa), Procedia CIRP 80 (2019) 482-487. https://doi.org/10.1016/j.procir.2019.01.047

[4] T. Nemeth, F. Ansari, W. Sihn, B. Haslhofer, A. Schindler. PriMa-X: A reference model for realizing prescriptive maintenance and assessing its maturity enhanced by machine learning, Procedia CIRP 72 (2018) 1039–1044. https://doi.org/10.1016/j.procir.2018.03.280

[5] K. Matyas, T. Nemeth, K. Kovacs, R. Glawar. A procedural approach for realizing prescriptive maintenance planning in manufacturing industries, CIRP Ann. Manuf. Technol 66 (2017) 461–464. http://dx.doi.org/10.1016/j.cirp.2017.04.007

[6] F. Wang, Z. Chen, C. Wu, Y. Yang. Prediction on sound insulation properties of ultrafine glass wool mats with artificial neural networks, Applied Acoustics 146 (2019) 164–171. https://doi.org/10.1016/j.apacoust.2018.11.018

[7] I. Lazakis, Y. Raptodimos, T. Verelas. Predicting ship machinery system condition through analytical reliability tools and artificial neural networks, Ocean Engineering 152 (2018) 404–415. https://doi.org/10.1016/j.oceaneng.2017.11.017

[8] E.A. Roehl Jr, D. A. Ladner, R. C. Daamen, J. B. Cook, J. Safarik, D. W. Philips Jr, P. Xie. Modeling fouling in a large RO system with artificial neural networks, Journal of Membrane Science 552 (2018) 95-106. https://doi.org/10.1016/j.memsci.2018.01.064

[9] J. Pietraszek, A. Gadek-Moszczak, N. Radek. The estimation of accuracy for the neural network approximation in the case of sintered metal properties. Studies in Computational Intelligence 513 (2014) 125-134. https://doi.org/10.1007/978-3-319-01787-7_12

[10] A. Szczotok, J. Pietraszek, N. Radek. Metallographic Study and Repeatability Analysis of γ' Phase Precipitates in Cored, Thin-Walled Castings Made from IN713C Superalloy. Archives of Metallurgy and Materials 62 (2017) 595-601. https://doi.org/10.1515/amm-2017-0088

[11] J. Pietraszek, A. Szczotok, N. Radek. The fixed-effects analysis of the relation between SDAS and carbides for the airfoil blade traces. Archives of Metallurgy and Materials 62 (2017) 235-239. https://doi.org/10.1515/amm-2017-0035

[12] N. Radek, J. Pietraszek, A. Goroshko. The impact of laser welding parameters on the mechanical properties of the weld, AIP Conf. Proc. 2017 (2018) art.20025. https://doi.org/10.1063/1.5056288

[13] N. Radek, J. Pietraszek, A. Gadek-Moszczak, Ł.J. Orman, A. Szczotok. The morphology and mechanical properties of ESD coatings before and after laser beam machining, Materials 13 (2020) art. 2331. https://doi.org/10.3390/ma13102331

[14] N. Radek, J. Konstanty, J. Pietraszek, Ł.J. Orman, M. Szczepaniak, D. Przestacki. The effect of laser beam processing on the properties of WC-Co coatings deposited on steel. Materials 14 (2021) art. 538. https://doi.org/10.3390/ma14030538

[15] M. Kekez, L. Radziszewski, A. Sapietova. Fuel type recognition by classifiers developed with computational intelligence methods using combustion pressure data and the crankshaft angle at which heat release reaches its maximum, Procedia Engineering 136 (2016) 353-358. https://doi.org/10.1016/j.proeng.2016.01.222

[16] Ł.J. Orman Ł.J., N. Radek, J. Pietraszek, M. Szczepaniak. Analysis of enhanced pool boiling heat transfer on laser-textured surfaces. Energies 13 (2020) art. 2700. https://doi.org/10.3390/en13112700

[17] R. Ulewicz, D. Siwiec, A. Pacana, M. Tutak, J. Brodny. Multi-criteria method for the selection of renewable energy sources in the polish industrial sector, Energies 14 (2021) art.2386. https://doi.org/10.3390/en14092386

[18] M. Zmindak, Ł. Radziszewski, Z. Pelagic, M. Falat. FEM/BEM techniques for modelling of local fields in contact mechanics, Communications - Scientific Letters of the University of Zilina 17 (2015) 37-46.

[19] A. Kubecki, C. Śliwiński, J. Śliwiński, I. Lubach, L. Bogdan, W. Maliszewski. Assessment of the technical condition of mines with mechanical fuses, Technical Transactions 118 (2021) art. e2021025. https://doi.org/10.37705/TechTrans/e2021025

[20] G. Majewski, M. Telejko, Ł.J. Orman. Preliminary results of thermal comfort analysis in selected buildings, E3S Web of Conf. 17 (2017) art. 56. https://doi.org/10.1051/e3sconf/20171700056

[21] M. Dobrzański. The influence of water price and the number of residents on the economic efficiency of water recovery from grey water, Technical Transactions 118 (2021) art. e2021001. https://doi.org/10.37705/TechTrans/e2021001

[22] A. Bakowski, V. Dekŷš, L. Radziszewski, Z. Skrobacki, P. Świetlik. Estimation of uncertainty and variability of urban traffic volume measurements in Kielce, 11th Int. Sci. Tech. Conf. Automotive Safety (2018) 1-8. https://doi.org/10.1109/AUTOSAFE.2018.8373314

[23] A. Bakowski, V. Dekŷš, L. Radziszewski, Z. Skrobacki. Validation of traffic noise models, AIP Conf. Proc. 2077 (2019) art.020005. https://doi.org/10.1063/1.5091866

[24] B. Szczodrowska, R. Mazurczuk. A review of modern materials used in military camouflage within the radar frequency range, Technical Transactions 118 (2021) art.e2021003. https://doi.org/10.37705/TechTrans/e2021003

[25] M. Morawski,T. Talarczyk, M. Malec. Depth control for biomimetic and hybrid unmanned underwater vehicles, Technical Transactions 118 (2021) art. e2021024. https://doi.org/10.37705/TechTrans/e2021024

[26] J. Pietraszek, R. Dwornicka, A. Szczotok. The bootstrap approach to the statistical significance of parameters in the fixed effects model. ECCOMAS 2016 – Proc. 7th European Congress on Computational Methods in Applied Sciences and Engineering 3, 6061-6068. https://doi.org/10.7712/100016.2240.9206

[27] J. Pietraszek, N. Radek, A.V. Goroshko. Challenges for the DOE methodology related to the introduction of Industry 4.0. Production Engineering Archives 26 (2020) 190-194. https://doi.org/10.30657/pea.2020.26.33

[28] H. Danielewski, A. Skrzypczyk, W. Zowczak, D. Gontarski, L. Płonecki, H. Wiśniewski, D. Soboń, A. Kalinowski, G. Bracha, K. Borkowski. Numerical analysis of laser-welded flange pipe joints in lap and fillet configurations, Technical Transactions 118 (2021) art. e2021030. https://doi.org/10.37705/TechTrans/e2021030

[29] J. Pietraszek, E. Skrzypczak-Pietraszek. The uncertainty and robustness of the principal component analysis as a tool for the dimensionality reduction. Solid State Phenom. 235 (2015) 1-8. https://doi.org/10.4028/www.scientific.net/SSP.235.1

Terotechnology XII
Materials Research Proceedings **24** (2022) 281-287

Materials Research Forum LLC
https://doi.org/10.21741/9781644902059-41

Analysis of Neural Network Training Algorithms for Implementation of the Prescriptive Maintenance Strategy

LEMPA Paweł[1,a] * and FILO Grzegorz[1,b]

[1]Cracow University of Technology, Faculty of Mechanical Engineering,
Jana Pawła II Ave 37, 31-864 Cracow, Poland

[a]pawel.lempa@pk.edu.pl, [b]grzegorz.filo@pk.edu.pl

Keywords: Neural Network Training, Multilayer Network Training, Supervised Training, Semi-Supervised Training, Prescriptive Maintenance

Abstract. This paper presents a proposal to combine supervised and semi-supervised training strategies to obtain a neural network for use in the prescriptive maintenance approach. It is required in this approach because of only partially labelled data for use in supervised learning, and additionally, this data is predicted to expand quickly. The main issue is the decision on which are suitable training methodologies for supervised learning, having in mind using this data and methods for semi-supervised learning. The proposed methods of training neural networks with supervised and semi-supervised training to receive the best results will be tested and compared in further work.

Introduction

One of the trends in modern applications of artificial intelligence methods is increasing productivity by implementing the prescriptive maintenance approach. This allows to speed up the response in a failure or downtime. In particular, the system can diagnose the system condition online, predict possible problems that may occur in the future, and inform the operator using a system of recommendations and warnings. Moreover, this approach is in line with the Industry 4.0 strategy. There are many approaches to building a prescriptive maintenance system [1] in the literature, including artificial neural networks with machine learning and data mining algorithms.

The practical use of artificial neural networks requires the design of the structure and the training process that can be performed using the machine learning method. Machine learning is characterised by high data processing and real-time computing potential, making it suitable for compensating for the shortcomings of traditional risk analysis methods in infrastructure maintenance management. He et al. developed a generative adversarial network (GAN)-based semi-supervised learning method to construct real-time risk-based early warning systems [2]. Fuzzy logic rules estimated the risk quantitatively, while a convolutional neural network (CNN) g warning models. Risk analysis can efficiently classify a dataset obtained from equipment or processes. However, contemporary publications indicate a high degree of complexity of this process, making it impossible to determine the risk for all possible situations. Therefore, the results of supervised training are insufficient and sometimes even impossible. At the same time, self-taught learning may be widely applicable to solve various practical classification problems, which was postulated by Raina et al. in [3].

Semi-supervised training combines conventional methods using standard samples, each containing a vector of inputs and a corresponding vector of expected outputs, with fully unsupervised training methods. In the second case, training samples only include an input model, and the whole process is self-organised. Thus, the network's responses are arranged through the spontaneous formation of local interactions between neurons. Some general principles of building an artificial network structure, data preparation and in-depth training using both approaches were presented by Goodfellow et al. in [4]. The research currently being carried out on semi-supervised artificial neural network training methods, algorithms, and practical applications cover many aspects, including:

• conducting the selection and classification processes on the available data to extract sample sets for combining the supervised and unsupervised training process. Data analysis is an indispensable part of the research process, especially when previous test results are unavailable. This is indicated, among others, by publications by Englund et al. [5] concerning meta-analyses of data selection criteria for stream predation experiments and by Hatush et al. [6], where the authors evaluate contractor prequalification data through the formulation of selection criteria and success factors;

• analysis of the available mechanisms and algorithms of semi-supervised training to solve the research problem posed, such as node label learning from unbalanced data [7]. The Hebb association rule, Kohonen's rule, and hybrid methods can be distinguished among the functional self-organisation algorithms. Bedekar et al. [8] used the Hebb rule to build a fault detection mechanism in power systems. B.S. Yang et al. [9] utilised the Kohonen network to diagnose faults of rotating machinery. In contrast, F.A. Souza et al. [10] used a Kohonen network with cascade perceptron to locate flaws in rural distribution feeders. Similarly, Z. Chen et al. [11] built a hybrid deep computation model for aeroplane engine fault detection. Verification of the results obtained using various types of self-organisation by comparing them with those generated by the neural network trained only in the fully supervised process indicates [8, 10] that the semi-supervised approach may positively affect the effectiveness of detecting emergencies;

• conducting the process of regularising the neural network to extend the possibility of generating correct answers for training patterns and new data. This kind of research was carried out by, e.g. Pretorius et al. concerning noise regularised deep neural networks [12] and Didcock et al. for neural network model averaging [13];

• analysis of the necessity and frequency of potential relearning the neural network model based on the obtained results of the preliminary training. An example of neural network dynamic relearning was provided by Li et al. [14]. This allows to acquire new knowledge on the requirements for the necessity and effects of cyclical re-training of neural networks in the context of the conditions changing with time during the operation of these networks;

• the issue of the performance of systems based on neural networks, including the impact of their speed on the confidence level of the generated results. Hence, sometimes there is a need to perform additional tests to determine the balance between the rate of action and the level of confidence in assessing the risk of emergencies. W. Jiang et al. dealt with optimising neural network operation for real-time multitasking applications [15].

This work will develop general guidelines for the neural network model training using different methodologies with full and partial supervision for the Prescriptive Maintenance strategy (RxM). It is advantageous when the created RxM system lacks the full expert knowledge necessary to determine the risk of all possible exceptions and emergencies. In particular, the results of analyses of the use of static and dynamic data samples and the possibility of independent adaptation of the neural network to the new realities of work (new devices, previously unknown types of failures), while maintaining an appropriate balance between speed and efficiency, may be helpful.

The issues discussed in this article may be of interest to a wide audience using data-driven data analysis methods, including in materials sciences [16-19], machining [20-22], construction of internal combustion engines [23], energy control [24, 25], mechanics of solids [27, 28], civil engineering [29, 30], and even in military equipment development [31, 32]. The provided analysis may be used in analogous situations requiring the creation of appropriate predictive approximators [33-35] and classifiers [36, 37].

Neural Network Training Approaches
The combination of the supervised and semi-supervised training strategies to obtain a neural network for use in the prescriptive maintenance approach can solve the problem of a big amount of data, which is only partially labelled, requires knowledge of experts, and will be continually expanded in time.

A proposition is to first use adequate training algorithms for supervised learning but with the possibility of based on its methods for semi-supervised learning. A combination of labelled and unlabelled data will be used for semi-supervised learning. Testing and comparing different training strategies for supervised learning will narrow the possible options for training methods for unsupervised learning.

The proposed strategy for neural network training algorithms to implement the prescriptive maintenance is shown in Fig. 1. The first stage describes preparing labelled data and experimenting with different types and methods of training neural networks using supervised learning. In the second stage, results and conclusions of the first stage with the unlabelled data are used to select proper training methods for semi-supervised learning.

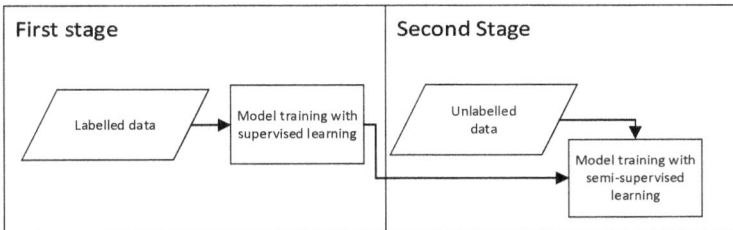

Fig. 1. Process of using supervised and semi-supervised learning for prescriptive maintenance

Supervised learning. There are two main types of learning problems. A classification involves predicting a class label, and Regression involves predicting a numerical label. The problem with Prescriptive Maintenance is that it can apply to both because maintaining the infrastructure in the enterprise in such a way that it retained all the desired functionalities and reflected the strategies and their compilations as faithfully as possible, at the same time omitting issues that are irrelevant to the strategy can be categorised in different ways. For input data (labelled and not), characteristics for maintenance processes such as device architecture (components of the device, device topology), age of devices, frequency of failure, frequency of maintenance, conditions in which they operate devices, measurement data from these devices, etc. are used. Another problem to be solved will be the development of metrics, which are connected with maintenance strategies, e.g. RBM (Risk Based Maintenance), CBM (Condition Based Maintenance) or TBM (Time Based Maintenance).

Multi-Task learning focuses on fitting a model one dataset for multiple related problems, and it improves when working on numerous tasks instead of one. It can be advantageous to solve problems with a large amount of labelled input data of one type that can be correlated with less labelled data. Thanks to that, each output can be predicted using different parts of the model. In active learning, the model can ask human operators during the learning process in case of vagueness. However, it is not an approach for solving semi-supervised learning problems.

In online learning, the available data and models have to be updated directly before prediction because observations are provided over time. Also, the probability distribution of observations is expected to change over time. In the case of prescriptive maintenance, where there is new data every chosen period, it is essential not to wait until the end of the process because it can not come.

Transfer learning. There is also the possibility to use Transfer learning, where learning knowledge of an already trained machine learning model is applied to a different but related problem. Which also described situations in prescriptive maintenance when new types of machines or processes are implemented. This new data can be similar to existing information that neural network recognises but still have some differences. It can work correctly when the amount of labelled data is much higher than unlabelled. This approach can be used for institutions already using trained models for labelled data.

Semi-supervised learning. Semi-supervised learning is a good alternative to transfer learning when the amount of unlabelled data significantly exceeds already defined data. Thanks to the information from supervised learning about labelled data, it will be possible to choose the correct assumptions for unlabelled data. It can be continuity assumptions, where the smoothness assumption also yields a bias for decision boundaries in regions with low-density, cluster assumption with the use of clustering algorithms, where discrete clusters of information are much more likely to share a label or manifold assumption, where labelled and unlabelled data help avoid the problem with dimensionality and distances and densities are used for learning.

The most popular options for training semi-supervised learning are generative models, graph-based models, support vector machines in semi-supervised versions (transductive support vector machine) and Multiview models.

Generative models can be used as an extension of classification from supervised learning. Still, if assumptions based on labelled data for unlabelled are wrong, the accuracy of the solution can be lower than the one based on. Essential for it is that unlabelled data have to have distinctive parameters that are recognisable. The possibility of using this approach for part of data for the prescriptive maintenance strategy is high.

For graph-based models, there is a node for labelled and unlabelled data. For graph construction, prescriptive maintenance expert knowledge has to be used. Data forms lower-dimensional manifolds in its embedding space, and the task of a classification algorithm is to separate these tangled manifolds. The important part is creating correct nodes for labelled data, which should be based on proper expertise. The transductive support vector machine model aims to identify unlabelled data for decision boundary to achieve maximal margin overall all information. The main problem is optimisation. Example solutions can be deterministic annealing, continuation method, semi-definite programming relaxation, or stochastic gradient descent.

Multiview models use at least three different learners trained on the same labelled data but with some differences in their bias. The final prediction is based on the votes of learners. Each learner can be a black box type and can be modified independently, which can be possibly used in prescriptive maintenance cases.

Summary

The main contribution of this study to the development of the current state of the art is the proposal to choose suitable training methodologies for the neural network concerning the Prescriptive

Terotechnology XII   Materials Research Forum LLC
Materials Research Proceedings **24** (2022) 281-287   https://doi.org/10.21741/9781644902059-41

Maintenance (RxM) strategy. This will help fill the gap in research on the use of machine learning for Prescriptive Maintenance and 1 allow for its further automation process. The analyses of existing research in this field with suggestions for the use of correct training algorithms for supervised and semi-supervised learning were proposed. Especially correlation between the selection of training for supervised neural networks and its influence on training for semi-supervised learning.

References

[1] K. Lepenioti, A. Bousdekis, D. Apostolou, G. Mentzas. Prescriptive analytics: Literature review and research challenges, International Journal of Information Management 50 (2020) 57-70. https://doi.org/10.1016/j.ijinfomgt.2019.04.003

[2] R. He, X. Li, G. Chen, G. Chen, Y. Liu. Generative adversarial network-based semi-supervised learning for real-time risk warning of process industries, Expert Systems with Applications 150 (2020) art. 113244. https://doi.org/10.1016/j.eswa.2020.113244

[3] R. Raina, A. Battle, H. Lee, B. Packer, A.Y. Ng. Self-taught learning: transfer learning from unlabeled data ACM International Conference Proceeding 227 (2007) 759-766. https://doi.org/10.1145/1273496.1273592

[4] I. Goodfellow, Y. Bengio, A. Courville, Deep Learning. MIT Press, November 2016, ISBN 9780262035613

[5] G. Englund, O. Sarnelle, S.D. Cooper. The importance of data-selection criteria: meta-analyses of stream predation experiments, Ecology 80 (1999) 1132-1141. https://doi.org/10.1890/0012-9658(1999)080[1132:TIODSC]2.0.CO;2

[6] Z. Hatush, M. Skitmore. Evaluating contractor prequalification data: selection criteria and project success factors, Construction Management and Economics 15 (1997) 129-147. https://doi.org/10.1080/014461997000000002

[7] M. Frasca, A. Bertoni, M. Re, G. Valentini. A neural network algorithm for semi-supervised node label learning from unbalanced data, Neural Networks 43 (2013) 84-98. https://doi.org/10.1016/j.neunet.2013.01.021

[8] P. P. Bedekar, S. R.Bhide, V. S. Kale. Fault section estimation in power system using Hebb's rule and continuous genetic algorithm, International Journal of Electrical Power & Energy Systems 33 (2011) 457-465. https://doi.org/10.1016/j.ijepes.2010.10.008

[9] B.S. Yang, T. Han, J.L. An. ART-KOHONEN neural network for fault diagnosis of rotating machinery, Mech. Sys. and Signal Proc. 18 (2004) 645-657. https://doi.org/10.1016/S0888-3270(03)00073-6

[10] F. A. Souza, M. F. Castoldi, A. Goedtel, M. da Silva. A cascade perceptron and Kohonen network approach to fault location in rural distribution feeders, Applied Soft Computing 96 (2020) art. 106627. https://doi.org/10.1016/j.asoc.2020.106627

[11] Z. Chen, X. Yuan, M. Sun, J. Gao, P. Li. A hybrid deep computation model for feature learning on aero-engine data: applications to fault detection, Applied Mathematical Modelling 83 (2020) 487-496. https://doi.org/10.1016/j.apm.2020.02.002

[12] A. Pretorius, H. Kamper, S. Kroon. On the expected behaviour of noise regularised deep neural networks as Gaussian processes. Pattern Rec. Letters 138 (2020) 75-81. https://doi.org/10.1016/j.patrec.2020.06.027

[13] N. Didcock, S. Jakubek, H.-M. Kögeler. Regularisation methods for neural network model averaging, Engineering Applications of Artificial Intelligence 41 (2015) 128-138. https://doi.org/10.1016/j.engappai.2015.02.005

[14] W. Li, W. Wang, X. Wang, S. Liu, L. Pei, F. Guo. A dynamic relearning neural network model for time series analysis of online marine data, Computers & Geosciences 73 (2014) 99-107. https://doi.org/10.1016/j.cageo.2014.09.006

[15] W. Jiang, Z. Song, J. Zhan, Z. He, X. Wen, K. Jiang. Optimized co-scheduling of mixed-precision neural network accelerator for real-time multitasking applications, Journal of Systems Architecture 110 (2020) art. 101775. https://doi.org/10.1016/j.sysarc.2020.101775

[16] J. Pietraszek, A. Gadek-Moszczak, N. Radek. The estimation of accuracy for the neural network approximation in the case of sintered metal properties. Studies in Computational Intelligence 513 (2014) 125-134. https://doi.org/10.1007/978-3-319-01787-7_12

[17] J. Pietraszek, R. Dwornicka, A. Szczotok. The bootstrap approach to the statistical significance of parameters in the fixed effects model. ECCOMAS 2016 – Proc. 7th European Congress on Computational Methods in Applied Sciences and Engineering 3, 6061-6068. https://doi.org/10.7712/100016.2240.9206

[18] A. Szczotok, J. Pietraszek, N. Radek. Metallographic Study and Repeatability Analysis of γ' Phase Precipitates in Cored, Thin-Walled Castings Made from IN713C Superalloy. Archives of Metallurgy and Materials 62 (2017) 595-601. https://doi.org/10.1515/amm-2017-0088

[19] J. Pietraszek, A. Szczotok, N. Radek. The fixed-effects analysis of the relation between SDAS and carbides for the airfoil blade traces. Archives of Metallurgy and Materials 62 (2017) 235-239. https://doi.org/10.1515/amm-2017-0035

[20] N. Radek, J. Pietraszek, A. Goroshko. The impact of laser welding parameters on the mechanical properties of the weld, AIP Conf. Proc. 2017 (2018) art.20025. https://doi.org/10.1063/1.5056288

[21] N. Radek, J. Pietraszek, A. Gadek-Moszczak, Ł.J. Orman, A. Szczotok. The morphology and mechanical properties of ESD coatings before and after laser beam machining, Materials 13 (2020) art. 2331. https://doi.org/10.3390/ma13102331

[22] N. Radek, J. Konstanty, J. Pietraszek, Ł.J. Orman, M. Szczepaniak, D. Przestacki. The effect of laser beam processing on the properties of WC-Co coatings deposited on steel. Materials 14 (2021) art. 538. https://doi.org/10.3390/ma14030538

[23] M. Kekez, L. Radziszewski, A. Sapietova. Fuel type recognition by classifiers developed with computational intelligence methods using combustion pressure data and the crankshaft angle at which heat release reaches its maximum, Procedia Engineering 136 (2016) 353-358. https://doi.org/10.1016/j.proeng.2016.01.222

[24] T. Lipiński. Corrosion resistance of 1.4362 steel in boiling 65% nitric acid, Manufacturing Technology 16 (2016) 1004-1009.

[25] Ł.J. Orman Ł.J., N. Radek, J. Pietraszek, M. Szczepaniak. Analysis of enhanced pool boiling heat transfer on laser-textured surfaces. Energies 13 (2020) art. 2700. https://doi.org/10.3390/en13112700

[26] M. Zmindak, L. Radziszewski, Z. Pelagic, M. Falat. FEM/BEM techniques for modelling of local fields in contact mechanics, Communications - Scientific Letters of the University of Zilina 17 (2015) 37-46.

[27] A. Kubecki, C. Śliwiński, J. Śliwiński, I. Lubach, L. Bogdan, W. Maliszewski. Assessment of the technical condition of mines with mechanical fuses, Technical Transactions 118 (2021) art. e2021025. https://doi.org/10.37705/TechTrans/e2021025

[28] S. Marković, D. Arsić, R.R. Nikolić, V. Lazić, B. Hadzima, V.P. Milovanović, R. Dwornicka, R. Ulewicz. Exploitation characteristics of teeth flanks of gears regenerated by three hard-facing procedures, Materials 14 (20210 art. 4203. https://doi.org/10.3390/ma14154203

[29] G. Majewski, M. Telejko, Ł.J. Orman. Preliminary results of thermal comfort analysis in selected buildings, E3S Web of Conf. 17 (2017) art. 56. https://doi.org/10.1051/e3sconf/20171700056

[30] M. Dobrzański. The influence of water price and the number of residents on the economic efficiency of water recovery from grey water, Technical Transactions 118 (2021) art. e2021001. https://doi.org/10.37705/TechTrans/e2021001

[31] B. Szczodrowska, R. Mazurczuk. A review of modern materials used in military camouflage within the radar frequency range, Technical Transactions 118 (2021) art.e2021003. https://doi.org/10.37705/TechTrans/e2021003

[32] M. Morawski,T. Talarczyk, M. Malec. Depth control for biomimetic and hybrid unmanned underwater vehicles, Technical Transactions 118 (2021) art. e2021024. https://doi.org/10.37705/TechTrans/e2021024

[33] T. Styrylska, J. Pietraszek. Numerical modeling of non-steady-state temperature-fields with supplementary data. Zeitschrift fur Angewandte Mathematik und Mechanik 72 (1992) T537-T539.

[34] J. Pietraszek, N. Radek, A.V. Goroshko. Challenges for the DOE methodology related to the introduction of Industry 4.0. Production Engineering Archives 26 (2020) 190-194. https://doi.org/10.30657/pea.2020.26.33

[35] H. Danielewski, A. Skrzypczyk, W. Zowczak, D. Gontarski, L. Płonecki, H. Wiśniewski, D. Soboń, A. Kalinowski, G. Bracha, K. Borkowski. Numerical analysis of laser-welded flange pipe joints in lap and fillet configurations, Technical Transactions 118 (2021) art. e2021030. https://doi.org/10.37705/TechTrans/e2021030

[36] J. Pietraszek. Response surface methodology at irregular grids based on Voronoi scheme with neural network approximator. 6th Int. Conf. on Neural Networks and Soft Computing, Jun 11-15, 2002, Springer, 250-255. https://doi.org/10.1007/978-3-7908-1902-1_35

[37] J. Pietraszek, E. Skrzypczak-Pietraszek. The uncertainty and robustness of the principal component analysis as a tool for the dimensionality reduction. Solid State Phenom. 235 (2015) 1-8. https://doi.org/10.4028/www.scientific.net/SSP.235.1

Terotechnology XII
Materials Research Proceedings 24 (2022) 288-293

Materials Research Forum LLC
https://doi.org/10.21741/9781644902059-42

Usage of Unmanned Aerial Vehicles in Medical Services: A Review

KOZIOŁ Anna[1,2,a] and SOBCZYK Andrzej[3,b] *

[1]Cracow University of Technology, Faculty of Material Engineering and Physics, Poland

[2]Norwegian University of Science and Technology, Norway

[3]Cracow University of Technology, Faculty of Mechanical Engineering, Poland

[a] anna.koziol@student.pk.edu.pl, [b] andrzej.sobczyk@pk.edu.pl

Keywords: Unmanned Aerial Vehicle (UAV), Drone, Medical Transport Delivery, Medical Emergencies

Abstract: The usage of drone technology has increased in a vast range of disciplines, including medical services. Drones can aerially deliver medical supplies and laboratory test samples during health emergencies such as the COVID-19 pandemic. It can also be used as a delivery device for an automated external defibrillator which might significantly increase the survival chances of out-of-hospital cardiac arrest victims. Significant cost savings compared with ground transportation and speed of delivery will probably drive drone implementation in various areas in the next few years.

Introduction

Unmanned aerial vehicle (UAV), also known as a drone is an aircraft operated without a human pilot on board [1]. This technology has already been deeply studied in the XX century. Their development is deeply associated with the military, where they had their initial use in World War II as targets for weapon accuracy practice [2]. Nowadays UAVs are revolutionizing civil society with technology that is advanced enough to be introduced to everyday life. Drones are used not only for public missions such as border surveillance or military training [3] but also in medicine and healthcare.

Through drones, the modern healthcare industry can improve the delivery of medical supplies to remote areas, provide access to automated external defibrillators (AED) for patients with cardiac arrest, provision disaster assessments, and deliver aid packages, vaccines, and medicines, blood [4]. In and after the era of COVID-19 UVAs were used in social distance inspection and Personal Protective Equipment (PPEs) delivery but also delivery of test kits, vaccines, and laboratory samples [5]. Because drones are generally resistant to dull, dirty, and dangerous missions [3], they are well suited for surveillance of disaster sites and areas with biological hazards [6]. The purpose of this review is to describe the most common healthcare applications of UVA technology and summarize current attempts to make pilotless aircrafts an inherent element of medical services and rescue. The main categories investigated in this article are: transporting and delivering medical supplies, samples, and biological material like blood or organs, delivering automated external defibrillators, searching for rescue operations, and drone usage during the COVID-19 pandemic.

Transportation

Biological Samples and Blood Products. Several reports have already demonstrated that using UVA technology for biological sample transportation does not affect the laboratory results if the temperature is ambient or cold and the time of flight is no longer than 40 minutes [7-9]. In the case of longer flights, lasting around 3 hours, particular attention needs to be paid to samples containing glucose and potassium, where 8% and 6.2% bias, respectively, were shown. Those changes were perceived as consistent with the magnitude of the temperature difference. Long drone flights of biological samples require stringent environmental controls to ensure consistent results [10]. However, there was no evidence of other significant effects on other blood components. Where the difference between the ambient and the unit temperature was approximately 20°C, there was no evidence of red blood cell hemolysis. Also, no significant changes in apheresis platelet count and pH were registered, which suggests that drone transportation is a viable option for the transportation of blood products [9]. Additionally, studies conducted on Borneo Island proved that drone transportation is more cost-effective than an ambulance. By estimating the Incremental Cost-Effectiveness Ratio we can conclude that the significantly shorter transport time of the drone offsets its cost per minute[11].

Organ and tissue transportation is also a field where UVA technology is promising. If made more efficient, the impact of transportation-related factors impacting transplantation outcomes could be substantially diminished. In April 2019, the company MissionGO cooperated with the University of Maryland Medical Center and for the first time delivered a kidney that was successfully transplanted.[12]. The model recently used for establishing parameters relevant for organ transportation was a six-rotor drone. Before and after the flight there was no drone-travel-related damage revealed[13].

Natural Disasters and Search and Rescue (SAR) Operation. Most natural disasters and mountain rescue victims are injured, lost or ill. According to the Wilderness Medical Society Practice Guidelines for Prevention and Management of Avalanche and Non-avalanche Snow Burial Accidents, it is crucial to reach the patient during the first hour of the rescue operation [14]. That is why the potential use of drones in searching for victims was investigated by the drone-snowmobile technique. This method is promising because a larger area can be searched faster compared to the classical line search technique [15]

Additionally, by using UVA technology it is possible to quickly respond to natural disasters by evaluating the damage and collecting real-time information. In Nepal after the earthquake in 2015 and in the Philippines after the typhoon in 2013 drones were used for mapping the most destroyed and needing help areas and determining road conditions [16].

Delivery of ADS. Out-of-hospital cardiac arrest (OOHCA) occurs often enough to alarm public health services and conduct additional research on the successful delivery of an automated external defibrillator(AED), especially since the use of AED significantly increases survival. Studies in Stockholm County showed that twenty locations were potential placements for a drone. The difference in response time of ambulances between urban and rural areas is substantial, which gives an opportunity to use UVA systems designed by using the Geographic Information System [17].

Another study was conducted in Salt Lake City County, where the current estimated travel time of an ambulance was compared to the estimated travel time of a network of AED-enabled medical drones. 96.4% of demands can be reached within 5 minutes using currently available emergency medical support vehicles and facility locations. Nevertheless, more factors need to be considered drone networks as a potential help in reducing the travel time of the AED [18].

In Wales the project Concept of Operation was developed, to identify the requirements associated with deploying the drone to deliver ADS beyond visual line-of-sight (BVLOS). Studies

Terotechnology XII
Materials Research Proceedings **24** (2022) 288-293

Materials Research Forum LLC
https://doi.org/10.21741/9781644902059-42

showed successful transportation of an ADS by parachute payload drops, final delivery of 4.5 km was completed in 2:50 minutes [19].

Nearby bystanders are their reactions play a considerable role in rescue and resuscitation. The simulation study evaluated the efficiency of a drone in providing the early location of a possible drowning victim in comparison with standard procedure. The median time from start to contact with the manikin was 4:34 min, which gives a shorter time than for the search party of surf lifeguards. A drone transmitting live video on a tablet may be used for providing an earlier location of submerged victims and is time-saving in comparison to traditional search parties [20].

Studies over bystander experience showed three main categories: technique and preparedness, support through conversation with the dispatcher, and aid and decision-making. None of the participants hesitated to retrieve the AED and all of them found the interaction with the AED drone less difficult than performing CPR or handling their own mobile phone during T-CPR, which makes good sense to continue studies on this topic in the future [21]. Using UAV technology was also proved as promising by studies piloted in the community of Caledon, Ontario. Drone-delivered AEDs may be feasible and effective, but successful uptake in smaller communities will require work on education about cardiac arrest literacy levels [22].

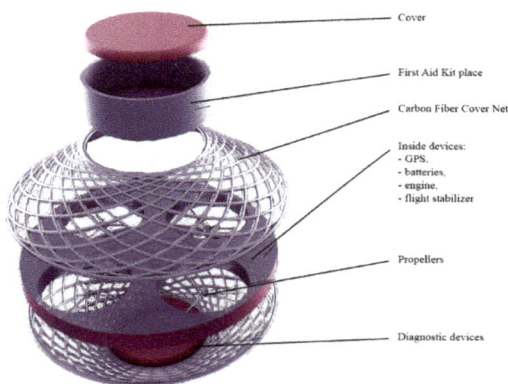

Fig.1. *Conceptual design of diagnostic-rescuing drone [23]*

Drones During COVID-19 Era
The pandemic period offered a real opportunity to use UAVs for providing healthcare support to the COVID-19 victims. Quick, effective, and contactless transport of vaccination and medication can help in reducing the number of infections. In case of a strict lockdown, drones can also be utilized for providing food and other supplies in areas the most affected by COVID-19 [24, 25]. Mini-drone systems can also provide real-time activities that can be used to pose a severe outbreak to community security [26]. Additionally, by using a thermal camera and high precision infrared, drone systems are capable of detecting victims of SARS-COV-2, squaring their temperature, or conducting the test, which also helps in detecting infected patients on a street. During the pick of the pandemic, China used drones to disinfect streets and distribute medications [27].

In September 2019 researchers conducted the first successful BVLOS mission to deliver diabetes medication. That showed the drone's capability to carry medical supplies reliably. In this state of crisis, this function could be used for the rapid delivery of medical supplies, as well as groceries, as witnessed in some parts of China, the USA, and Australia [23]

Conclusions

Until nowadays, drones were considered more as an essential component of militaries. This perception has changed what led to the explosion of the drone industry. As shown in this review, the potential for drone use in clinical microbiology, epidemiology, transportation of medical supplies and even mapping the virus spread is enormous.

UAVs may significantly increase access to health care for the communities lacking good infrastructure. However, challenges like legal medical issues or national airspace legalization need to be taken into consideration, which could also help in developing new transportation in healthcare and life-saving technology ideas.

Further work on the development of service flying drones will be an impulse for development both for works related to the construction of light internal combustion engines [28] and methods for the analysis of experimental data [29, 30] and scenarios of possible failures [31].

References

[1] Unmanned Aircraft System (UAS): regulatory framework and challenges, NAM/CAR/SAM Civil – Military Cooperation Havana, Cuba, 13 – 17 April 2015

[2] J.F. Keane, S.S. Carr. A Brief History of Unmanned Aircraft. Johns Hopkins APL Tech Dig. 32 (2013) 558-571.

[3] S.G. Gupta, M.M. Ghonge, P.M. Jawandhiya. Review of Unmanned Aerial System (UAS). Int. J. Adv. Res. Comp. Eng. Technol. 2 (2013) 1646-1658. https://doi.org/10.2139/ssrn.3451039

[4] M. Balasingam. Drones in medicine—The rise of the machines. Int. J. Clin. Pract. 71 (20170 art. e12989. https://doi.org/10.1111/ijcp.12989

[5] L. Ramadass, S. Arunachalam, Z. Sagayasree. Applying deep learning algorithm to maintain social distance in public place through drone technology. Int. J. Pervasive Comput. Commun. 16 (2020) 223-234. https://doi.org/10.1108/IJPCC-05-2020-0046

[6] J.C. Rosser, V. Vignesh, B.A. Terwilliger, B.C. Parker. Surgical and medical applications of drones: A comprehensive review, Journal of the Society of Laparoendoscopic Surgeons 22 (2018) art. e2018.00018. https://doi.org/10.4293/JSLS.2018.00018

[7] T.K. Amukele, L.J. Sokoll, D. Pepper, D.P. Howard, J. Street. Can unmanned aerial systems (Drones) be used for the routine transport of chemistry, hematology, and coagulation laboratory specimens? PLoS ONE 10 (2015) art. e0134020. https://doi.org/10.1371/journal.pone.0134020

[8] T.K. Amukele, J. Street, K. Carroll, H. Miller, S.X. Zhang. Drone transport of microbes in blood and sputum laboratory specimens. J. Clin. Microbiol. 54 (2016) 2622-2625. https://doi.org/10.1128/JCM.01204-16

[9] T.K. Amukele, P.M. Ness, A.A.R. Tobian, J. Boyd, J. Street. Drone transportation of blood products. Transfusion 57 (2017) 582-588. https://doi.org/10.1111/trf.13900

10. T.K. Amukele, J. Hernandez, C.L. Snozek, R.G. Wyatt, M. Douglas, R. Amini, J. Street. Drone Transport of Chemistry and Hematology Samples over Long Distances. Am. J. Clin. Pathol. 148 (2017) 427-435. https://doi.org/10.1093/ajcp/aqx090

[11] M.A. Zailani, R.Z. Azma, I. Aniza, A.R. Rahana, M.S. Ismail, I.S. Shahnaz, K.S. Chan, M. Jamaludin, Z.A. Mahdy. Drone versus ambulance for blood products transportation: an economic evaluation study. BMC Health Serv. Res. 21 (2021) art. 1308. https://doi.org/10.1186/s12913-021-07321-3

[12] D. Freeman. A drone just flew a kidney to a transplant patient for the first time ever. It won't be the last. MACH (2019). https://www.nbcnews.com/mach/science/drone-just-flew-kidney-transplant-patient-firsttime-ever-it-ncna1001396. [viewed 10.03.2022]

[13] J.R. Scalea, S. Restaino, M. Scassero, S.T. Bartlett, N. Wereley. The final frontier Exploring organ transportation by drone. Am. J. Transplant. 19 (2019) 962-964. https://doi.org/10.1111/ajt.15113

[14] C. Van Tilburg, C.K. Grissom, K. Zafren, S. McIntosh, M.I. Radwin, P. Paal, P. Haegeli, W.W.R. Smith, A.R. Wheeler, D. Weber, B. Tremper, H. Brugger. Wilderness Medical Society Practice Guidelines for Prevention and Management of Avalanche and Nonavalanche Snow Burial Accidents. Wilderness Environ. Med. 28 (2017) 23–42. https://doi.org/10.1016/j.wem.2016.10.004

[15] Y. Karaca, M. Cicek, O. Tatli, A. Sahin, S. Pasli, M.F. Beser, S. Turedi. The potential use of unmanned aircraft systems (drones) in mountain search and rescue operations. Am. J. Emerg. Med. 36 (2018) 583-588. https://doi.org/10.1016/j.ajem.2017.09.025

[16] Z.B. Htet. Disaster drones: great potential, few challenges? RSIS Commentary (2016) CO16253.

[17] A. Claesson, D. Fredman, L. Svensson, M. Ringh, J. Hollenberg, P. Nordberg, M. Rosenqvist, T. Djarv, S. Österberg, J. Lennartsson, Y. Ban. Unmanned aerial vehicles (drones) in out-of-hospital-cardiac-arrest. Scand. J. Trauma, Resusc. Emerg. Med. 24 (2016) art. 124. https://doi.org/10.1186/s13049-016-0313-5

[18] A. Pulver, R Wei, C. Mann. Locating AED Enabled Medical Drones to Enhance Cardiac Arrest Response Times, Prehospital Emergency Care 20 (2016) 378-389. https://doi.org/10.3109/10903127.2015.1115932

[19] N. Rees, J. Howitt, N. Breyley, P. Geoghegan, C. Powel. A simulation study of drone delivery of Automated External Defibrillator (AED) in out of Hospital Cardiac Arrest (OHCA) in the UK. PLoS ONE 16 (2021) art. E0259555. https://doi.org/10.1371/journal.pone.0259555

[20] A. Claesson, L. Svensson, P. Nordberg, M. Ringh, M. Rosenqvist, T. Djarv, J. Samuelsson, O. Hernborg, P. Dahlbom, A. Jansson, J. Hollenberg. Drones may be used to save lives in out of hospital cardiac arrest due to drowning. Resuscitation 114 (2017) 152-156. https://doi.org/10.1016/j.resuscitation.2017.01.003

[21] J. Sanfridsson, J. Sparrevik, J. Hollenberg, P. Nordberg, T. Djärv, M. Ringh, L. Svensson, S. Forsberg, A. Nord, M. Andersson-Hagiwara, A. Claesson. Drone delivery of an automated

external defibrillator - A mixed method simulation study of bystander experience. Scand. J. Trauma, Resusc. Emerg. Med. 27 (2019) art. 40. https://doi.org/10.1186/s13049-019-0622-6

[22] K. Sedig , M.B. Seaton , I.R. Drennan, S. Cheskes, K.N. Dainty. "Drones are a great idea! What is an AED?" novel insights from a qualitative study on public perception of using drones to deliver automatic external defibrillators, Resuscitation Plus 4 (2020) art. 100033. https://doi.org/10.1016/j.resplu.2020.100033

[23] A.Kozioł. The concept of rescuing drone. B.Tech. thesis, Politechnika Krakowska, Kraków, 2020.

[24] M. Sharma. Drone Technology for Assisting COVID-19 Victims in Remote Areas: Opportunity and Challenges, J. Med. Syst. 45 (2021) art. 85. https://doi.org/10.1007/s10916-021-01759-y

[25] D.S. Jat, C. Singh (eds.). Artificial Intelligence-Enabled Robotic Drones for COVID-19 Outbreak; Intelligent Systems and Methods to Combat Covid-19, Springer, 2020.

[26] J. Euchi. Do drones have a realistic place in a pandemic fight for delivering medical supplies in healthcare systems problems?, Chinese J. Aeronaut. 34 (2021) 281-190. https://doi.org/10.1016/j.cja.2020.06.006

[27] M. Sharma, How drones are being used to combat COVID-19,' Geospatial World, 2020. https://www.geospatialworld.net/blogs/how-drones-are-being-used-to-combat-covid-19/ [viewed 10.03.2022]

[28] M. Kekez, L. Radziszewski, A. Sapietova. Fuel type recognition by classifiers developed with computational intelligence methods using combustion pressure data and the crankshaft angle at which heat release reaches its maximum, Procedia Engineering 136 (2016) 353-358. https://doi.org/10.1016/j.proeng.2016.01.222

[29] J. Pietraszek, A. Szczotok, N. Radek. The fixed-effects analysis of the relation between SDAS and carbides for the airfoil blade traces. Archives of Metallurgy and Materials 62 (2017) 235-239. https://doi.org/10.1515/amm-2017-0035

[30] J. Pietraszek, N. Radek, A.V. Goroshko. Challenges for the DOE methodology related to the introduction of Industry 4.0. Production Engineering Archives 26 (2020) 190-194. https://doi.org/10.30657/pea.2020.26.33

[31] G. Filo, J. Fabiś-Domagała, M. Domagała, E. Lisowski, H. Momeni. The idea of fuzzy logic usage in a sheet-based FMEA analysis of mechanical systems, MATEC Web of Conf. 183 (2018) art.3009. https://doi.org/10.1051/matecconf/201818303009

Terotechnology XII

Materials Research Forum LLC

Materials Research Proceedings 24 (2022) 294-300

https://doi.org/10.21741/9781644902059-43

Analysis of Urban Traffic Noise at Weekends – Case Study

RADZISZEWSKI Leszek[1, a], BĄKOWSKI Andrzej[2,b*], BURDZIK Rafał[3,c] and WARCZEK Jan[4,d]

[1,2]Kielce University of Technology, Faculty of Mechatronics and Mechanical Engineering, Al.Tysiąclecia Państwa Polskiego 7, 25-314 Kielce, Poland

[3,4] Silesian University of Technology, Faculty of Transport and Aviation Engineering, ul. Akademicka 2A, 44-100 Gliwice, Poland

[a]lradzisz@tu.kielce.pl, [*b]abakowski@tu.kielce.pl, [c]rafal.burdzik@polsl.pl, [d]jan.warczek@polsl.pl

Keywords: Variability of Urban Traffic Noise at Weekend, Cnossos-EU Model Validation

Abstract. The study carried out an analysis of the urban traffic noise parameters on Sundays and Saturdays. The results of noise simulations according to the Cnossos-EU model were compared with the sound level calculated by a permanent automatic sound and traffic volume monitoring station. The variations in results were evaluated. Analyzes carried out showed that the traffic of passenger vehicles is the main source of road noise. A very good agreement of the noise values determined according to the Cnossos-EU model and the measured ones was obtained. The maximum noise values on Sundays are only slightly smaller than on Saturdays. The shape of the noise diagram and the noise values at individual hours of the day on Saturdays are different than on Sundays. An experimental model of noise variability at weekends has been proposed. The equations describing the variability of the equivalent sound level were validated. Fit factor R^2 of the proposed equations to the experimental data ranges from 0.85 to 0.94.

Introduction

Studies of the harmful impact of urban transport means on the environment are currently presented in the literature in numerous publications [1-4] but mainly for weekdays. Some of the effects are an annoyance and sleep disturbance. Also, long-term health effects such as cardiovascular disease have also been related to traffic noise [5]. To assess urban noise, cities regularly produce noise maps. The paper [6] presents seasonal and weekday influences on noise indicators based on a noise map. For the noise mapping realized by computational methods, important are reliability and uncertainty [7]. The acoustic climate assessment for noise mapping needs the selection of acoustic hazards in the analyzed areas. Also, input noise level data are burdened with certain uncertainties. The idea of applying interval arithmetic for the assessment of acoustic models' uncertainty is formulated in the paper [8]. The problem of estimation of the long-term environmental noise hazard indicators and their uncertainty is also presented in the paper [9]. Vehicle traffic parameters such as flow, speed, and structure have a significant impact on the air and ground pollution with exhaust emissions, noise, vibrations, and other phenomena that create environmental hazards. The paper [10] presents an interesting study on the impact of tire pressure on noise generated by vehicles. An additional factor influencing the acoustic climate is the modifications to the road infrastructure, the aim of which may also be to reduce the emission of traffic noise to the environment [11]. That is while it is necessary to permanently monitor over a long period of time vehicle traffic and analyze the recorded data [12]. The variations in traffic volume and noise are of interest e.g. in dynamic traffic management systems and navigation services, assessing the environmental effects of traffic. Stationary monitoring stations or low-cost wireless sensor systems and participative citizenship initiatives measure urban noise [13]. Determining the models

Terotechnology XII
Materials Research Proceedings **24** (2022) 294-300

Materials Research Forum LLC
https://doi.org/10.21741/9781644902059-43

describing the acoustic field caused by road noise forces the solution of many practical engineering problems. Much attention is paid to the microphone positions near intersections. In these places, conventional noise mapping methods suffer from numerical errors that often occur in large-scale city noise mapping calculations using automatically generated geographic inputs. Problems of this type are avoided, for example, by applying modifications to noise modeling methods. One of the noise mapping optimization methods is presented in [14]. Increasingly, in noise modeling, attention is paid to the variability of the frequency structure. Octave noise models are sometimes better at capturing the actual hearing experience of people in a given area. In other words, the same level does not mean the same acoustic nuisance. An example of such an approach to noise modeling is the publication [15]. Due to the complexity of the problem of noise modeling, its solution is sometimes sought with the use of artificial neural networks or by the use of genetic algorithms [16,17]. Analysis of the distribution of the traffic volume showed that in the case of interrupted traffic flows, e.g. in urban areas, it is usually non-normal. Traffic in urbanized areas can be analyzed depending on the adopted time interval and location of the road in the communication system [18,19]. Within one week, the traffic volume on weekdays differs significantly from the traffic on weekends and holidays. The period of one day can be divided into 24 hours. In each subsequent hour of the day, vehicle traffic parameters may be different. Besides the general shape of the daily flow profile, the shape of the pick periods is important for traffic management, as well.

The problem of traffic noise at weekends has been analyzed much less frequently in the literature, which is mainly due to the lower traffic intensity, especially for heavy goods vehicles. However, as the authors' research has shown, it does not cause significant changes in the sound pressure level. The goal of this research was to obtain an insight into the traffic volume profiles and to detect traffic noise patterns according to the day of the weekend and Cnossos-EU model validation. To evaluate the results, a traffic noise experimental curve model was proposed.

Measurements and Calculations Results
The traffic volumes and noise analyzed in this study were measured [12] by the permanent station recording traffic volume, velocity, and sound pressure levels located in Popieluszka Av. in Kielce. The analysis was determined in accordance with the procedure presented in [13], with the difference that the noise was recorded every 1 hour. The box plots of vehicle speed on Saturdays and Sundays in 2016, are shown in Figure 1.

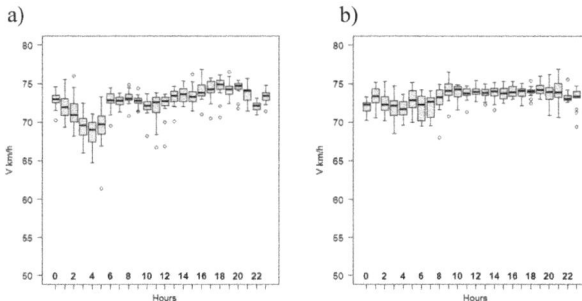

Fig. 1. *Box plots for weekends of the relation between the hourly traffic speed and time for 24 h period a) on Saturdays, b) on Sundays.*

It is worth noting that the speed charts on Saturdays and Sundays have different courses. On Saturdays from midnight, the median speed decreases to its minimum value at 4 AM and then increases to 7 AM. Then, in the subsequent hours, its value changes slightly and the highest occurs at 6 PM. Compared to Saturdays on Sundays, both the shape of the speed chart and the speed values at individual hours of the day are different. The analysis of the share of individual groups of vehicles in the traffic structure [12] showed that the dominant group is passenger vehicles, for which the share in the traffic flow was on Sundays 84%. The next group was medium-heavy vehicles, for which the share in the analyzed period amounted on Sundays to 11%. The share of heavy vehicles mounted on Sundays to 2%. For this reason, in the further part of the work, the traffic of passenger vehicles will be analyzed. Figure 2 presents traffic volume box plots for individual hours of the day for passenger vehicles on weekends [12]. On Saturdays (Fig. 2 a), it can be seen that the number of vehicles increases from 5 AM and stabilizes between noon at 12 AM and 14 PM to around 1000 per hour. From 15 PM the number of vehicles is gradually decreasing. On Sundays, the number of vehicles only increases from 6 AM to 3 PM and gradually declines from 4 PM. One can also notice a certain symmetry in the graph of changes in the number of vehicles in the range from 8 AM to 11 PM (Fig. 2 b).

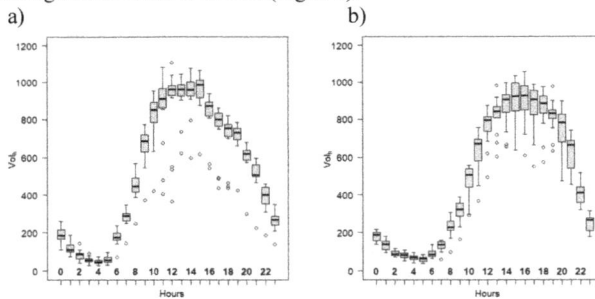

Fig. 2. *Volume of passenger vehicle traffic for individual hours of the day: a) on Saturdays, b) on Sundays.*

The differences in the number and speed of vehicles on Saturdays and Sundays have an impact on the chart shape and time distribution of noise values. Smaller numbers of vehicles on Sundays contribute to reducing the noise level, especially in the evening and at night, which can be seen in Fig. 3. On the other hand, during the day, on Sundays, passenger vehicles and medium-heavy vehicles drive at higher speeds and more dynamically than on Saturdays, which means that the maximum noise values are only slightly reduced. These conclusions are confirmed by the results of simulations made in accordance with the Cnossos-EU model.

Terotechnology XII

Materials Research Proceedings **24** (2022) 294-300

Materials Research Forum LLC

https://doi.org/10.21741/9781644902059-43

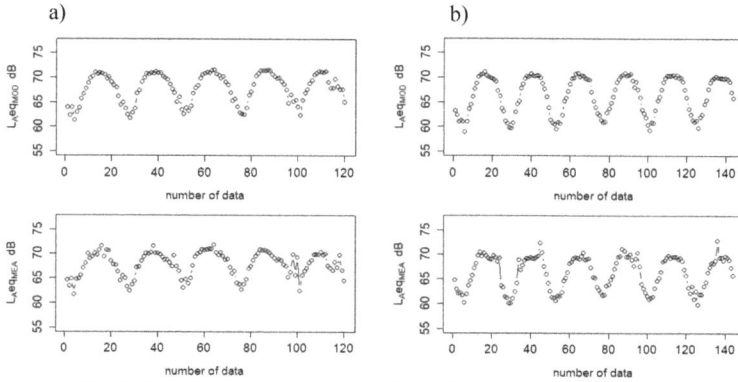

Fig. 3. *Simulated and measured values of equivalent sound level*
for all vehicles in 2016 for a) 5 Saturdays and b) 6 Sundays.

The comparison of the median results of the simulations carried out according to the Cnossos-EU [13] model and the experimental ones are shown in Fig. 4. The relationships, shown in Figures 4a and 4b, indicate both some similarities and differences in traffic noise on weekends. Noise values on Saturdays and Sundays are similar, but their time distributions are varied. One can also notice a certain symmetry in the graph of changes in noise in the range: for Saturdays from 5 AM to 11 PM (Fig. 4a) and for Sundays from 9 AM to 11 PM (Fig. 4b).

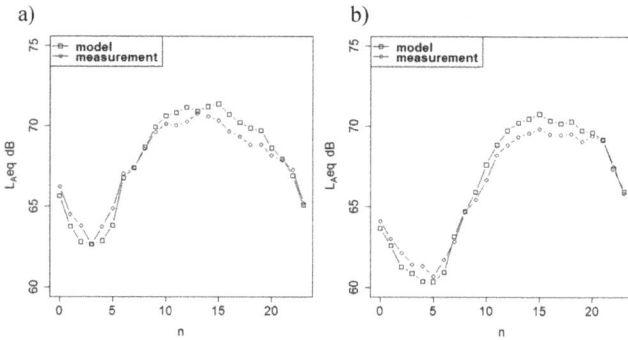

Fig. 4. *Median of equivalent sound level for all vehicles*
in 2016 for a) Saturdays, b) Sundays.

One can notice a very good agreement of the obtained results in these figures. This compliance is confirmed by the value of the RMSE parameter, which for Saturdays is 0.68 dB(A) and for Sundays, it is 0.66 dB(A).

Materials Research Forum LLC

https://doi.org/10.21741/9781644902059-43

Traffic Noise Forecasting on Saturdays and Sundays

The division of the day into three-time intervals, i.e. day, evening, and night, is justified for physiological and normative reasons related to the harmful influence of transport on humans. On the other hand, it is complicated by the mathematical description of the variability of the values of vehicle motion parameters as a function of time, which was presented in [12].

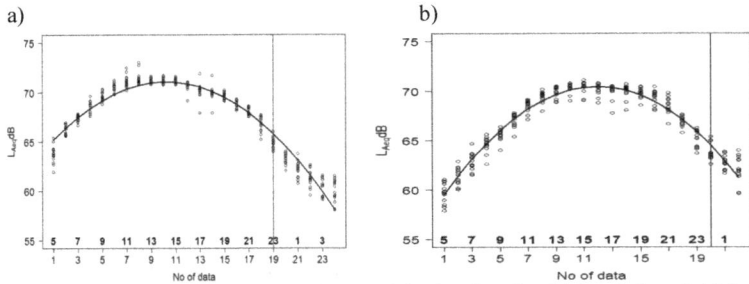

Fig. 5. Dependence of L_{Aeq} on hours of the day for all vehicles a) from 5 AM on Saturdays until 4 AM on Sundays b) from 5 AM on Sundays until 2 AM in Mondays.

For this reason, the authors decided to develop an experimental model of noise variability on weekends in 2016 and conduct an analysis in the time interval from 5 AM on Saturdays to 4 AM on Sundays and from 5 AM on Sundays until 2 AM in Mondays. Such a transformation in the time domain does not affect the value of the sound intensity level but facilitates its mathematical description of the variability.

Changes in the sound intensity level can be described by the equation:
– for the time interval from 5 AM on Saturdays to 4 AM on Sunday

$$L_{Aeq} = 63.84 + 1.40 \cdot h - 0.07 \cdot h^2$$
(1)

– for the time interval from 5 AM on Sundays until 2 AM on Mondays

$$L_{Aeq} = 57.41 + 2.18 \cdot h - 0.09 \cdot h^2$$
(2)

where h – number of data.

The coefficients of fitting model curves to the experimental data are approximately $R^2 = 0.94$. The validation of the experimental model was carried out on the set of measurement data from 2011 (noise was calculated according to the Cnossos-EU model). The determined values of the RMSE parameter for Saturdays are 1.39 dB(A) and for Sundays 1.26 dB(A), which confirms the possibility of practical use of the proposed model for noise forecasting. The analysis of the measured sound level values on Sundays showed that from 11 AM to 9 PM the value of 68 dB(A) was exceeded. For this reason, simulations of the impact of limiting vehicle speed to 50 km/h on noise values were carried out. Calculations have shown that the noise values are reduced by about 3 dB(A) and that the level of 68 dB(A) is not exceeded.

Conclusions

1. The noise level exceeds the permissible values both on Saturdays and Sundays, which is mainly due to the number and speed of passenger vehicles.
2. Compared to Saturdays on Sundays, both the shape of the noise diagram and the noise values at individual hours of the day are different
3. Time distributions of noise values at the weekend can be described by a second-order polynomial and the RMSE parameter values are in the range from 0.70 dB(A) to 1.40 dB(A)
4. The simulations of the impact of the vehicle speed limit to 50 km/h on the median noise values (according to the Cnossos-EU model) showed that the noise values were reduced by about 3 dB(A) and that the 68 dB(A) level was not exceeded.

References

[1] S. Domazetovska, M. Anachkova, V. Gavriloski, Z. Petreski. Influence of the traffic flow in urban noise pollution, Proc. 2020 Int. Congress on Noise Control Engineering, INTER-NOISE 2020, Seoul, Korean Society of Noise and Vibration Engineering, art. 166585.

[2] R. Benocci, A. Molteni, M. Cambiaghi, F. Angelini, H.E. Roman, G. Zambon. Reliability of DYNAMAP traffic noise prediction. Applied Acoustics 156 (2019) 142-150. https://doi.org/10.1016/j.apacoust.2019.07.004

[3] T. Figlus, J. Gnap, T. Skrucany, P. Szafraniec. Analysis of the influence of different means of transport on the level of traffic noise. Zeszyty Naukowe Politechniki Śląskiej, series Transport, 97 (2017) 27-38. https://doi.org/10.20858/sjsutst.2017.97.3

[4] R. Sanchez-Sanchez, J.C. Fortes-Garrido, J.P. Bolivar. Noise monitoring networks as tools for smart city decision-making. Archives of Acoustics 43 (2018) 103-112. https://doi.org/10.24425/118085

[5] E. M. Salomons, M. Berghauser Pont. Urban traffic noise and the relation to urban density, form, and traffic elasticity. Landscape and Urban Planning 108 (2012) 2-16. https://doi.org/10.1016/j.landurbplan.2012.06.017

[6] P. Mioduszewski, J.A. Ejsmont, J. Grabowski, D. Karpiński. Noise map validation by continuous noise monitoring. Applied Acoustics 72 (2011) 582-589. https://doi.org/10.1016/j.apacoust.2011.01.012

[7] J. Wierzbicki, W. Batko. Uncertainty of noise mapping software. Proc. 7th European Conf. Noise Control 2008, EURONOISE 2008 (2008): 1955-1958.

[8] W. Batko, P. Pawlik, New approach to the uncertainty assessment of acoustic effects in the environment. Archives of Acoustics 37.1 (2012): 57-61.

[9] W.M. Batko, B. Stępień. Type A standard uncertainty of long-term noise indicators. Archives of Acoustics 39 (2014): 25-36. https://doi.org/10.2478/aoa-2014-0004

[10] J. Warczek, R. Burdzik, Ł. Konieczny, G. Siwiec. Frequency analysis of noise generated by pneumatic wheels. Archives of Acoustics 42 (2017) 459-467. https://doi.org/10.1515/aoa-2017-0048

[11] M. Smiraglia, R. Benocci, G. Zambon, H.E. Roman. Predicting Hourly Traffic Noise from Traffic Flow Rate Model: Underlying Concepts for the DYNAMAP Project, Noise Mapp. 3 (2016) 130–139. https://doi.org/10.1515/noise-2016-0010

[12] A. Bąkowski, L. Radziszewski. Analysis of the Traffic Parameters on a Section in the City of the National Road during Several Years of Operation. Communications – Scientific Letters of the University of Zilina 24 (2022) A12-A25. https://doi.org/10.26552/com.C.2022.1.A12-A25

[13] A. Bąkowski, V. Dekys, L. Radziszewski, Z. Skrobacki. Validation of Traffic Noise Models, AIP Conf. Proc. 2077 (2019) art. 020005. https://doi.org/10.1063/1.5091866

[14] E.M. Salomons, H. Zhou, W.J.A. Lohman. Efficient numerical modeling of traffic noise. Journal of the Acoustical Society of America 127 (2010) 796–803. https://doi.org/10.1121/1.3273890

[15] A. Can, L. Leclercq, J. Lelong, D. Botteldooren. Traffic noise spectrum analysis: Dynamic modeling vs. experimental observations. Applied Acoustics 71 (2010) 764–770 https://doi.org/0.1016/j.apacoust.2010.04.002

[16] P. Kumar, S.P. Nigam, N. Kumar. Vehicular traffic noise modeling using artificial neural network approach. Transportation Research Part C:Emerging Technologies 40 (2014) 111-122. https://doi.org/10.1016/j.trc.2014.01.006

[17] X. Zhang, H. Kuehnelt, W. De Roeck. Traffic Noise Prediction Applying Multivariate Bi-Directional Recurrent Neural Network. Applied Sciences 11 (2021) art. 2714. https://doi.org/10.3390/app11062714

[18] X. Lu, J. Kang, P. Zhu, J. Cai, F. Guo, Y. Zhang. Influence of urban road characteristics on traffic noise, Transportation Research Part D:Transport and Environment 75 (2019) 136-155. https://doi.org/10.1016/j.trd.2019.08.026

[19] J. Pietraszek, N. Radek, K. Bartkowiak. Advanced Statistical Refinement of Surface Layer's Discretization in the Case of Electro-Spark Deposited Carbide-Ceramic Coatings Modified by a Laser Beam. Solid State Phenomena 197 (2013) 198-202. https://doi.org/10.4028/www.scientific.net/SSP.197.198

Terotechnology XII
Materials Research Proceedings **24** (2022) 301-308

Materials Research Forum LLC
https://doi.org/10.21741/9781644902059-44

The Assessment of the Environmental Impact of Textile Cleaning Processes in the Aquatic Environment on Human Health

PARASKA Olga[1,a *], IVANISHENA Tetiana[2,b], ZAKHARKEVICH Oksana[3,c], HES Lubos[4,d] and RADEK Norbert[5,e]

[1,2]Khmelnytskyi National University, Department of Chemistry and Chemical Engineering, st Instytutska 11, 29016, Khmelnickiy, Ukraine

[3]Khmelnytskyi National University, Department of Technology and Design of Sewing Products, st. Instytutska 11, 29016, Khmelnickiy, Ukraine

[4]Technical University of Liberec, Department of Textile Evaluation, st. Studentska 1402/2, 46117 Liberec, Czech Republic

[5]Kielce University of Technology, Faculty of Mechatronics and Mechanical Engineering, Al. 1000-lecia Państwa Polskiego 7, 25-314 Kielce, Poland

[a]olgaparaska@gmail.com,[b]itso77@ukr.net, [c]zbir_vukladach@ukr.net, [d]lubos.hes@gmail.com, [e]norrad@tu.kielce.pl

Keywords: Textile Cleaning Processes, Wet Cleaning, Biosurfactants, Life Cycle Assessment, Environmental Impact

Abstract. This article analyzes the technological process of textile product cleaning in the aquatic environment from the point of view of identified sources of hazard. It is established at what particular stages of cleaning of the products, the negative affects on the environment, workers and consumers occur. According to Life Cycle Assessment (LCA), detailed information was received at an enterprise about the technological processes and the impact of their individual factors on the environment and human health. Creating closed water cycles will reduce the impact on the environment and human health. The use of new types of biosurfactants in detergent compositions will allow reducing the duration of technological operations, and the number of cycles of textile product processing, to increase their exploitation life and improve the quality of textile product cleaning.

Introduction

In modern conditions, the activity of the garment care industry involves finding opportunities to reduce costs, minimize risks, and increase business stability and competitive advantages [1 – 3]. In order to improve the efficiency of textile cleaning technologies in the aquatic environment, the quality of services provided to consumers, and the environmental safety of the processes, it is necessary to take into account the social, economic, and environmental impacts of technologies throughout the life cycle of textile products [2, 4, 5]. The implementation of the Life Cycle Assessment (LCA) procedure allows us to determine the interaction between energy, material costs, and technology over the life cycle of textiles, as well as in the long run – to determine the impact of these interactions on the environment and society (from raw materials production through processing, manufacturing, distribution, use, disposal or recycling).

According to [1, 2, 5], the implementation of the LCA method is one of the leading tools for sustainable business development, which allows for increasing the efficiency of the enterprise activity by studying the environmental impact while the exportation of textile products/services.

The use of the LCA method is driven by the desire to increase the market value of products/services by disclosing information about their safety. In addition, the LCA allows you to set and optimize the most energy and resource-intensive stages of technological processes, as well as to create the preconditions for cost reduction through the rational use of raw materials, resources, and energy. Information obtained from LCA is important in making strategic business development decisions and innovating when cleaning textiles at an enterprise.

The implementation of LCA in the process of textile products cleaning allows us to create new, environmentally friendly auxiliaries and technologies; introduce eco-labeling and certification; an increase production efficiency; the identification of cost reduction opportunities; make decisions based on sustainable development goals; compare alternative opportunities for sustainable development of the enterprise; improving the quality of service provision, extending the life of textile products.

Methodology
The LCA method is a part of the integrated Life Cycle Sustainability Assessment method in accordance with ISO 14000 series standards, in particular ISO 14040: 2006 Environmental management – Life Cycle Assessment – Principles and framework, ISO 14044: 2006 Environmental management – Life Cycle Assessment – Requirements and guidelines. This method also includes Life Cycle Costing and Social Life Cycle Assessment [6, 7]. Other similar tools include Environmental Impact Assessment (EIA), Ecological Risk Assessment (ERA), Material Flows Analysis (MFA), and Cost-Benefit Analysis (CBA). The LCA is a multilateral and multifactorial environmental management tool aimed at assessing environmental impacts.

The sequence of LCA conducting [6, 7]:
– Preparatory stage – determining the purpose and scope: determining the purpose of the study, the circle of persons to whom the results of the study will be communicated; description of the technology life cycle system under study, and determination of boundaries and functions.
– The main stage:
- life cycle inventory analysis: data collection, data calculation, flow, and emission distribution.
- life cycle impact assessment: selection of impact categories and their indicators, distribution of inventory results, calculation of results of category indicators.
– Final stage:
- interpretation of the results of the analysis: consideration of the results of the inventory analysis and impact assessment, preparation of the report on the results of the conducted assessment, formation of conclusions and recommendations.
- environmental indicators of the study: waste generation during operation, consumption of water resources, wastewater discharge, emissions into the atmosphere, greenhouse gas emissions, physical factors (noise, vibration, radiation, thermal pollution), and impact on soils.

In the LCA study of the technological process of textile products cleaning in the aquatic environment at the company LLC TPP "Universal" (Ukraine) [8] estimation techniques were used. The analysis of the use and utilization of the main solvents (water) and means, without production and extraction of raw materials for their production. Thus, cleaning technologies for environmental and human health impacts and the effectiveness of the removal of contamination from textiles were analyzed.

Results and Discussion

In the technological processes of textile products cleaning in the aquatic environment, they use detergents and auxiliaries, which can remain on the products, affect the health of workers and consumers, as well as the environment [3, 9-11]. Modern washing machines and water purifiers [3, 8] using the latest technologies consume less electricity when cleaning clothes, but energy production can cause air pollution after a while. Therefore, when choosing professional equipment, first of all, pay attention to the technical and economic indicators of the equipment. It should be borne in mind that washing machines and water purifiers consume a large amount of water as a cleaning medium.

In the process of textile products cleaning in the aquatic environment at the enterprise LLC TPP "Universal" they use professional tools of different producers [8]: Alberti Angelo, Colortex (Italy), Bufa, Kreussler, Seitz (Germany), LLC "Sphere-93" (Ukraine). The volumes of use of professional means in some technological operations at the enterprise are given in Table 1.

Table 1. The volume of use of professional means in the enterprise

Technological operation	Consumption of preparations per year, l (kg)
stain removal	117.29
wet cleaning, laundering	149.84
finishing	113.28

According to Table 1, we can conclude that the processes of product cleaning and the preparations they use have some impact on the environment and human health. On the basis of the received information, we will analyze the activity of the textile cleaning company in accordance with the LCA methodology.

The first stage defines the scope and purpose of the research, for this purpose is established:

1. Functions of the production system – efficiency of removal of pollution from textile products.
2. Functional unit – consumption of water, energy, chemicals.

In determining the functional unit and reference flows distinguish the following steps: identification of functions; choice of functions and definition of the functional unit; identification of product performance and definition of reference flow.

The purpose of the functional unit is to quantify the function provided by the production system. Thus, the first step is to identify the goal that the production system provides, that is, to identify its function.

After determining the functional unit, it is necessary to determine the number of products required to perform the function, expressed in terms of the functional unit. Reference flow is related to product performance and is usually defined as the result of a standard measurement method. The nature of this measurement and calculation depends on the object of study. Reference flow – the number of required output streams from processes in a given production system that is required to perform a function expressed by a functional unit (ISO 14041 [12]).

In the laundering process or wet cleaning, water is used as the primary solvent for wetting the products, and in the subsequent stages, the primary solvent is used in combination with more effective means. The advantage of this technology is the easy removal of dirt from the surface of textiles.

To narrow the boundaries of the system we use the rule of "trimming" – the exclusion of flows that do not have a significant impact on the environment, as well as the corresponding individual processes. This simplifies the model of the production system.

An operational analysis of a textile cleaning service provider contains the following input and output indicators. Inputs: materials (chemicals, detergents, auxiliaries), water (drinking and technological); the amount of energy used; types of services. Outputs: clean textile products; detergent remains (waste); emission (evaporation) of washing solutions. Indicators of operational characteristics of the technological process of textile products cleaning in an aqueous environment are shown in Table 2.

To determine the boundaries of a single process, a link can be established with the sites within the given list to determine the smallest components of the production system for which data is acceptable. Due to the variability of certain technological processes carried out on a particular site, the boundaries of a single process are set in order to minimize the need for distribution procedures.

Types of effect, methods used and interpretation of results: vapors of chemicals emitted into the atmosphere, hydrosphere, water (Tables 2, 3) and affecting the environment, health of staff who remain in working space for a long time.

Table 2. Indicators of operational characteristics of the technological process

Operation Performance Indicators	Group	The grounds for selecting an indicator, its purpose
Materials		
Chemicals consumption per product, (l, kg, m³)	E	Set the amount of chemicals used to reduce consumption
Water consumption per product, general use, m³	E	Set water consumption to reduce consumption
Quality of textiles cleaning,%	E	Setting the amount of re-cleaning to reduce it
Electricity		
Calculation of electricity costs per 1 kg of textile products, total energy use, kW, UAH	E	Set electricity costs to reduce them
Calculation of electricity consumption for technological purposes, kW	E	Establishment of electricity bill
Textiles (products)		
Number of textile products that have lost their consumption properties, item	E	Establishment of the service life of textile products of different assortment, with the aim of reducing the environmental impact during the life cycle
Waste		
Waste (emissions), m³	E	Set the amount to minimize waste
Quantity of waste per 1 kg of textile products	F	Set waste minimization characteristics in accordance with future regulations in order to reduce them
Drainage into water and soil		
Amount of wastewater in the process of cleaning 1 kg of textile products	D	Set the amount of wastewater in the process of cleaning 1 kg of textile products

Terotechnology XII
Materials Research Proceedings 24 (2022) 301-308

Materials Research Forum LLC
https://doi.org/10.21741/9781644902059-44

Table 3. *Environmental pollution of the textile cleaning process in the aquatic environment*

Impact on the environment	Emissions
Air	- detergent, auxiliaries evaporation
Water	- wastewater after the process of water treatment - wastewater after the washing process

Inventory analysis is the basis and indispensable element of any LCA. During this phase, all material, energy, and source streams associated with the production system are collected and systematized for LCA in the environment. After the inventory analysis, the following conclusions were obtained: during the process of textile products cleaning, there are vapors of chemicals, and wastewater contaminates the remains of detergents.

Carbon dioxide emissions that affect climate change, as well as sulfur and nitrogen oxides, which affect oxidation processes, i.e. CO_2 and SO_2 equivalents, are potential impacts of the activity of the garment care industry.

The first stage of the impact assessment is to distribute the data collected from the inventory analysis of emissions, use of natural resources, and land by impact classes. The impact class includes climate change, oxidation, ecotoxicity, and carcinogen substances.

The characterization process involves the transformation of environmental variables into a single cost system. In practice, this is done by multiplying the variable, which is obtained in the inventory analysis by the coefficient of performance.

Normalization occurs after characterization. When normalized, the impact class indicators correlate with relevant data in a particular area; also calculate the relative share of the influence of the class of action in the ratio of the reference value. Normalization is required before the final consolidation of the data into a common impact indicator. At the accenting stage, the normalized values of the impact class indicators are combined into integrated impact indicators if the action of the system under study is to be expressed by means of integrated indicators. An integrated impact indicator can be obtained by multiplying the normalized impact class indicators by the weight factor of each impact class and compiling the results obtained. In practice, only mandatory structural parts that comply with the standard are performed at the LCA stage. The results are represented as the equivalents of CO_2 and SO_2. The calculations made at this stage for different impact classes are given in Table 4.

Integrated indicators of the impact of the technological process of textile cleaning in the aquatic environment are given in Table 5.

According to the integration of the general results, it is proved that the share introduced by the processes of textile products cleaning in the aquatic environment is insignificant in comparison with other technologies. Creating closed water supply cycles, reducing the duration of technological operations, and the use of new types of biosurfactants, will reduce the contamination burden on the environment and human health, the number of product treatment cycles in operation, and improve the quality of cleaning.

Table 4. *Performance evaluation indicators for different impact classes*

Impact class	Units	Indicator	Normalization factor	Normalization
Energy	kW	2.7	201.5 kW · eq / h	0.013
Water used	l	13.3	417580 l · eq	0.000032
Materials:	kg	0.006	27.13 kg · eq	0.00022
Preparations, Auxiliaries	l	0.1		0.0037
Climate change	kg CO_2 · eq	0,25	12300 kg CO_2 · eq	0.00002
Oxidation	kg SO_2 · eq	-	58.9 kg SO_2 · eq	-
Formation photochemical oxidants	kg C_xH_y · eq	-	32.20 kg C_xH_y · eq	-
Eutrophication (wastewater)	kg PO_4 · eq	0.005	8.01 kg PO_4 · eq	0.00062
Liquid waste	kg	-	0.00719 kg · eq	-
Total	-	-	-	0.0176

Table 5. *Integrated indicators of the impact of the process of textile products cleaning*

Weight coefficient			Integrated Impact Indicators		
Local	Regional	Global	Local	Regional	Global
10.57	11.77	12.76	0.137	0.153	0.106
10.39	8.61	9.12	$3.3 \cdot 10^{-4}$	$2.7 \cdot 10^{-4}$	$2.9 \cdot 10^{-4}$
8.34	9.23	10.56	0.033	0.036	0.041
8.02	9.66	12.71	$1.6 \cdot 10^{-4}$	$1.9 \cdot 10^{-4}$	$2.5 \cdot 10^{-4}$
7.23	9.41	10.09	-	-	-
13.68	11.68	7.56	-	-	-
2.57	3.52	3.8	0.0016	0.002	0.002
14.91	11.23	8.92	-	-	-
Total					
75.71	75.11	75.52	0.172	0.19	0.149

The third step is the method of evaluating the action. In various countries in the LCA process, a great deal of effort is being made to unify and systematize the environmental factors and methods of impact on the environment [6, 7, 12]. The Eco-indicator 95 method considers energy costs for raw materials, the final disposal of waste, and related environmental impacts and harms.

The fourth stage is the interpretation of the results. At this stage, the results of inventory analysis and impact assessment according to the tasks and technologies used are brought together.

Based on the results of the LCA, detergents, and technologies [13, 14] were developed for textile products cleaning in the aquatic environment. According to [13, 14] the cost of reagents, the time of operations, the cycle of processing products, energy costs are reduced, the environmental safety of the process is improved, high quality of cleaning products is achieved, and their service life is extended.

Materials Research Forum LLC

https://doi.org/10.21741/9781644902059-44

Summary
Thus, in the process of inventory analysis, it was found that during the textile products cleaning in the aquatic environment, emissions into the air are vapors of chemicals, and the main components of wastewater are surfactants that are part of detergents.

The technological process of textile product cleaning from the point of view of the identified sources of environmental danger was analyzed, and it was determined which stages of the technological process of the products cleaning have a negative impact on the environment and employees of the enterprise as well as the most harmful preparations.

According to the integration of the results of the study, the share of textile cleaning technologies in the aquatic environment is insignificant, which makes it possible to extend the scope of their use in the process of removing contaminants from the products, restoring their consumption properties, by creating closed water supply cycles and using new types of detergents compositions with low environmental impact.

The developed methodology for conducting LCA technologies for textile product cleaning in the aquatic environment can be applied to individual technological operations of product cleaning, as well as implemented at various enterprises of the garment care industry, which will minimize the negative impact on the environment and human health, reduce the cost of water, detergents, and auxiliaries, improve the quality of removal of contamination from textile products.

References
[1] S. Muthu. Environmental impacts of the use phase of the clothing life cycle. Handbook of Life Cycle Assessment (LCA) of Textiles and Clothing, 2015, pp. 93-102.

[2] Greening the economy through Life Cycle Thinkin. UNEP DTIE Sustainable Consumption and Production Branch. Paris France, 2012.

[3] Professional Textile Care Industry CINET [online]. 2020. [viewed 2022-01-20]. Available from: https://www.cinet-online.com.

[4] B. Shen, Q. Li, C. Dong, P. Perry. Sustainability issues in textile and apparel supply chains, *Sustainability:* Special Issue *9* (2017) 1592. https://doi.org/10.3390/su9091592

[5] R. Lomonaco-Benzing, J. Ha-Brookshire. Sustainability as social contract: textile and apparel professionals' value conflicts within the corporate moral responsibility spectrum, *Sustainability* *8*(12) (2016) 1278. https:// doi.org/10.3390/su8121278

[6] Environmental management. ISO 14040: 2006. Environmental management – Life cycle assessment – Principles and framework. [online]. 2015. [viewed 2022-01-20]. Available from:https://www.iso.org/standard.

[7] Life cycle assessment. ISO 14044: 2006. Environmental management – Life cycle assessment – Requirements and guidelines [online]. 2015. [viewed 2022-01-20]. Available from: https://www.iso.org/standard.

[8] List.in.ua. Garment care industry. [online]. 2014. [viewed 2022-01-20]. Available from: http://www.universal-ua.com.

[9] K. L. Mittal, T. Bahners. Textile finishing: Recent developments and future trends, John Wiley & Sons, 2017.

[10] V.G. Gerasimova, N.Ye. Dyshinevich, G.V. Golovaschenko. Modern features of regulation of the safe use of detergents in the European Union, Customs Union and Ukraine, Ukrainian Journal of Modern Problems of Toxicology 3(62) (2013) 5-11. https:// doi.org/10.33273/2663-4570.

[11] S. Khan, A. Malik. Environmental and health effects of textile industry wastewater, in environmental deterioration and human health, Springer (2014) 55-71. https://doi.org/10.1007/978-94-007-7890-0_4

[12] Inventory analysis. ISO 14041:1998 Environmental management – Life cycle assessment – Goal and scope definition and inventory analysis [online]. 2014. [viewed 2022-01-10]. Available from: https://www.iso.org/standard.

[13] O. Paraska, T. Rak, D. Chervonyuk. Use of ecologically friendly surface active substances in the finishing of textiles. In: I. Frydrych, G. Bartkowiak, M. Pavlowa (eds.), Innovations in clothing design, materials, technology and measurement methods – monograph. Lodz, 2015, 218-225.

[14] O. Paraska. Development of scientific basis of resource-saving technologies for cleaning the textile garments in the aquatic environment. Kherson, 2021. Dissertation. Kherson National Technical University.

Terotechnology XII
Materials Research Proceedings **24** (2022) 309-315

Materials Research Forum LLC
https://doi.org/10.21741/9781644902059-45

Development of the Mobile Application to Calculate Parameters of Underwear Patterns

ZAKHARKEVICH Oksana[1,a], KOSHEVKO Julia[2,b] *, SHVETS Galyna[3,c], KULESHOVA Svetlana[4,d], BAZYLIUK Elvira[5,e], PARASKA Olga[6,f] and KAZLACHEVA Zlatina[7,g]

[1-4]Khmelnytskyi National University, Department of Technology and Design of Sewing Products, st Instytutska 11, 29016, Khmelnickiy, Ukraine

[5]Khmelnytskyi National University, Department of Design, st. Instytutska 11, 29016, Khmelnickiy, Ukraine

[6]Khmelnytskyi National University, Department of Chemistry and Chemical Engineering, st. Instytutska 11, 29016, Khmelnickiy, Ukraine

[7]Trakia University, Faculty of Technics and Technologies, st. Graf Ignatiev 38, 8600 Yambol, Bulgaria

[a]zbir_vukladach@ukr.net, [b]juliakoshevko@gmail.com, [c]galyamet@ukr.net, [d]kuleshova_lana@ukr.net, [e]e.bazyliuk@gmail.com, [f]olgaparaska@gmail.com [g]zlatinka.kazlacheva@trakia-uni.bg

Keywords: Mobile Applications, Design Calculation, Underwear Parts, Psychographic Profile

Abstract. The study aims to develop a mobile application to support the process of designing underwear. The application allows calculating parameters of underwear patterns. The method of semantic differential has been used to evaluate the developed mobile application. The resulting psychographic profiles of the application indicate a positive assessment by experts. The developed app allows achieving the same precision of the calculation as other calculation methods while the speed is much preferable. Besides that, the risk of accidental mistakes due to the human factor is excluded from the process of designing underwear patterns.

Introduction

COVID19 pandemic situation resulted in the global use of mobile technologies in the fields, which usually were not influenced by the new era of high-tech innovations. The range of papers is dedicated to using mobile devices for learning [1-5, 13-15]. While most of the researchers are concerned with the overall influence and/or impact of mobile learning on students in different areas [2-5, 13-15], several papers present possibilities for incorporating such technologies in the field of design including clothing design [1]. Many professionals in the field of garment design and apparel manufacturing are more than unambiguous positive on the question of using smartphones as a means to integrate innovations into garment design. Moreover, if such means are integrated into the education process it is even more impactful for the industry as it carries on the high quality of the future performance of clothing designers and patternmakers. The most obvious way to develop and improve communication channels of the apparel companies is to develop specific mobile applications that will serve as an interpreter between a customer and a company itself. The main hypothesis of the current study is that the mobile application is the way to improve not only communication channels but the design process itself.

Materials Research Forum LLC

https://doi.org/10.21741/9781644902059-45

Based on the monitoring of the existing mobile applications in the field of apparel design and manufacturing that was described in [9-11] we found out that there are no applications that purposed for the calculation of underwear patterns. Though, there are some applications for the calculation of other garment types patterns such as "CloStyler" [11], "SHOES Step-by-Step", "RDMK Step-by-Step" [10], etc. These apps allow seeing the results of calculations instantly. A user can see the scheme by which he is instructed when constructing a pattern. The calculation itself is done in the same order as the pattern drafting method instructs. The main purpose of such an app is to reduce the time wasted while constructing patterns and at the same time to increase their accuracy.

The mobile applications that were found on the market and in the scientific papers are mostly concerned with top wear garments such as coats, jackets, dresses, skirts, and trousers. Therefore, the focus of the current work is underwear, the design process of which is not supported by mobile technologies till now.

Literature analysis of the scientific papers did not allow to find papers about mobile apps to support the underwear design process. Though, it was discovered that breast volume parameters and design of the bra are discussed from more than one perspective [6-8]. Besides that, it was determined that pattern design systems (PDS) that are usually used for garment design are not so useful in a case when underwear is in question. Due to the specifics of the manufacturing that is often referred to as small enterprises and bespoke clothing, it is much more convenient to use a simple calculation app that would be cheaper and faster than traditional PDS.

Therefore, the main goal of the paper is to develop a mobile application to support the process of designing the patterns of underwear garments.

Methodology

We used the method of prototyping to explore different aspects of the intended design of the app. On the preliminary study of the current research, we developed a prototype of the mobile application named "N_Underwear" that was described in the work [16]. The prototype was ranked by students studying apparel design as well as several clothing designers and researchers in the field. Results showed the necessity to improve the app as it allows to calculate only the parameters of the draft of the basic thongs. On the other hand, we discovered that the app allowed improving the accuracy of patterns' construction. Besides that, all of the respondents indicated that it was convenient to use an app instead of the classic calculator or PDS especially if there is no one.

To develop the app we choose the MIT App Inventor (USA) that allowed creating the application for the Android operating system.

The semantic differential method described in [12] was used to evaluate the developed application. At the first stage of using this method, pairs of words with opposite meanings are formed, which form a semantic differential. Each pair of Kansei words is a bipolar pair for a separate attribute of the developed application: speed, accuracy, complexity, convenience, relevance (needs). The scales of the semantic differential for each attribute of the mobile application were represented as bipolar pairs expressed by adjectives or adverbs. The scales are presented in the form of horizontal rulers in the questionnaire. Each scale has seven gradations of values, which are expressed in numerical form (-3, -2, -1, 0, +1, +2, +3). For ease of representation of Kansei words, all bipolar pairs were encoded with the first letters of words, which is common practice: SQ (Slow-Quick); CS (Complicated-Simple); AI (Accurate-Inaccurate); FU (User-Friendly interface – User-Unfriendly interface); NU (Necessary-Unnecessary). At the next stage of the study, the application was tested by experts and evaluated using the questionnaire. The

Terotechnology XII
Materials Research Proceedings **24** (2022) 309-315

Materials Research Forum LLC
https://doi.org/10.21741/9781644902059-45

expert group consisted of 22 students and 14 faculty members (teachers etc.), clothing designers, and patternmakers.

The results of the preliminary survey were used to improve the prototype. The main features that were improved are as follows: the icon of the app was designed anew to be more recognizable (Fig. 1); the calculation of the bra patterns was added; the interface design was changed; the language of the app prototype was changed to English with some instructions in Ukrainian given simultaneously by using slash mark (/).

The input data for calculation are body measurements and the amount of eases. The body measurements and amount of eases must be entered by a user. The app gives users the option of completing the text fields by a shorthand method based on what has been typed before. Otherwise, the fields will be autocompleted by zeros. Flowchart displaying the work of the app "N_Underwear" is given in Fig. 2. The order of calculation is performed according to the pattern drafting method. The names of the constructive segments correspond to the points in the given figures of patterns blocks (Fig. 3).

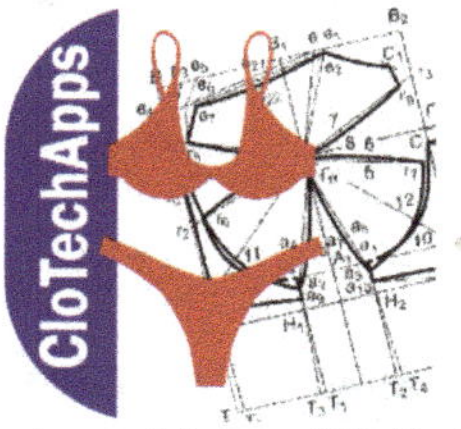

Fig. 1. The icon of the app "N_Underwear".

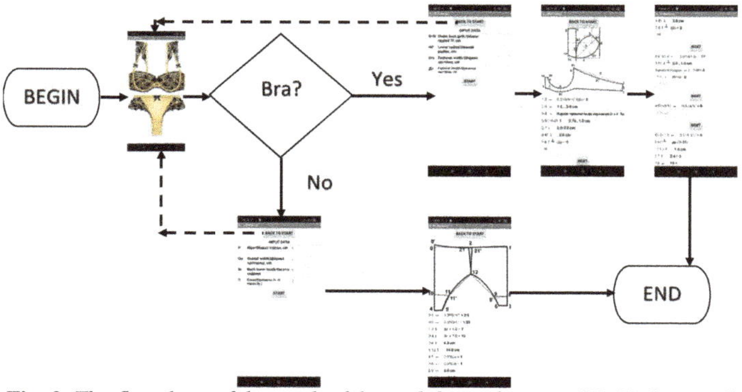

Fig. 2. The flowchart of the work of the mobile application "N_Underwear".

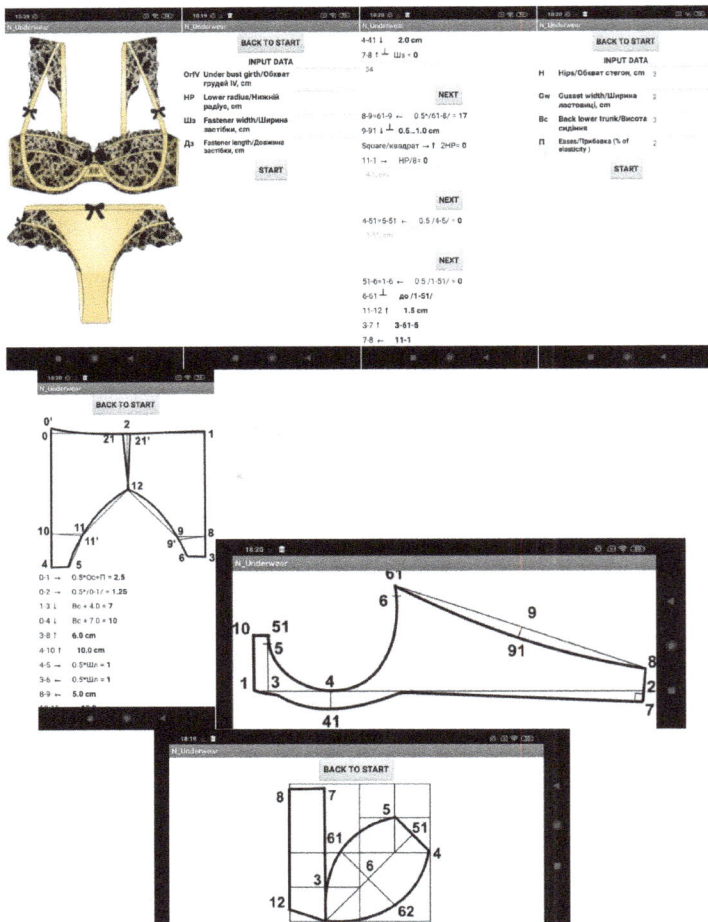

Fig. 3. *Dialog boxes for working with a mobile application "N_Underwear".*

Materials Research Forum LLC
https://doi.org/10.21741/9781644902059-45

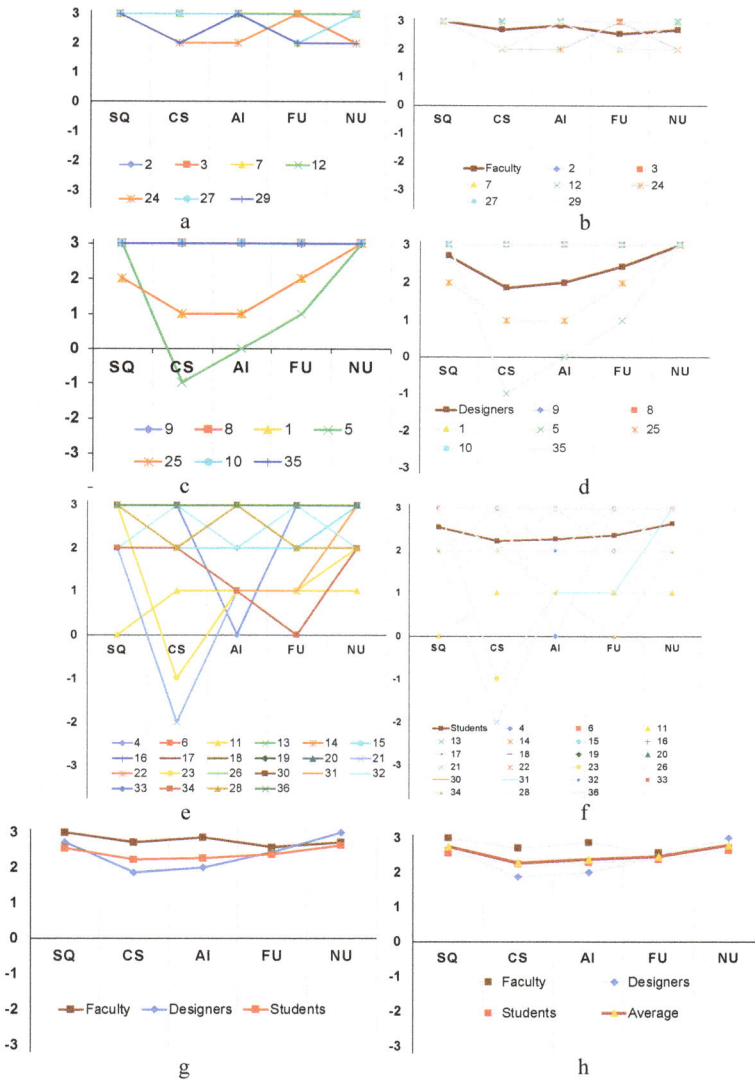

Fig. 4. *Psychographic profiles of the mobile application "N_Underwear": a, b – expert group of faculty members (group K1); c, d – an expert group of clothing designers (group K2); e, f – an expert group of students (group K3); g, h – the average values of the evaluation coefficients of the three expert groups (K).*

Results and Discussion

As a result of the survey, psychographic profiles of the developed mobile application were obtained (Fig. 4, a – h). The profiles display the average values of the evaluation coefficients for each pair of Kansei words. As the coefficients correspond to the positive values of Kansei words, the results of the survey indicate that the experts approved the mobile application "N_Underwear".

As one can see from Fig. 4, the application is assessed mostly with marks related to the positive meaning of Kansei Words. While the average values of the evaluation coefficients of the three expert groups form psychographic profiles that have identical forms experts of group K3 showed much more contraversions points while assessing the app. The most controversial point is the CS (Complicated-Simple) when assessed by students. It might be explained by their luck of the experience. This attribute evaluated by professional designers as well as teachers got the same level of positive response as all other attributes.

Summary

As a result of the evaluation of the developed mobile application "N_Underwear" by the methods of Kansei Engineering, the level of its competitiveness was confirmed by the positive values of the average evaluation coefficients in the psychographic profiles: 2.75 (quick); 2.27 (simple); 2.38 (accurate); 2.46 (convenient); 2.78 (necessary).

The developed app allows achieving the same precision of the calculation as other calculation methods while the speed is much preferable. Besides that, the risk of accidental mistakes due to the human factor is excluded from the process of designing underwear patterns.

The application is the only one on the market of mobile applications in the field of garment design that allows calculating parameters of underwear patterns.

Acknowledgment

The authors would like to thank Nataliya Dik (the director and owner of private garment company "Novodvorskaya Underwear") and Oksana Korinyovska for their support and technical assistance in this study.

References

[1] D. Borisenko. The use of mobile applications in the development of a design product in the training of future design professionals, Information Technologies and Learning Tools 68 (2018) 47-63. https://doi.org/10.33407/itlt.v68i6.2224

[2] M. Abner, F. Baytar. Apps to increase student engagement: a case of textiles and apparel sustainability education, International Journal of Fashion Design, Technology and Education 12 (2019) 56–64. https://doi.org/10.1080/17543266.2018.1477996

[3] I.S.H. Wai, S.S.Y. Ng, D.K. Chiu, K.K. Ho, P. Lo. Exploring undergraduate students' usage pattern of mobile apps for education, Journal of Librarianship and Information Science 50 (2018) 34-47. https://doi.org/10.1177/0961000616662699

[4] J. Abdurrahman, M. Beer, P. Crowther. Pedagogical requirements for mobile learning : a review on MOBIlearn Task Model, Journal of Interactive Media in Education 1 (2015) art. 12. http://doi.org/10.5334/jime.ap

[5] T. Page. Application-based mobile devices in design education, International Journal of Mobile Learning and Organisation 8 (2014) 96-111. https://doi.org/10.1504/IJMLO.2014.062347

[6] K. Shin. Patternmaking for underwear design 2nd Ed. Createspace Independent Publishing Platform, 2015.

[7] A. Peterson. Exploratory study on breast volume and bra cup design, Journal of Textile and Apparel, Technology and Management 11(1) (2019) 1-13.

[8] Z. Li. Bra in the New Era: A Study from the Perspective of Feminism. Proc. 2021 Int. Conf. Social Development and Media Communication (SDMC 2021) 367-371. https://doi.org/10.2991/assehr.k.220105.069

[9] T. Zhylenko, A. Kudryavtsev, O. Zakharkevich. Mobile Application to Calculate the Parameters of Top Wear Basic Design, Science and Innovation 15 (2019) 24-34. https://doi.org/10.15407/scin15.03.024

[10] O. Zakharkevich, J. Koshevko, S. Kuleshova, S. Tkachuk, A. Dombrovskyi. Development of the mobile applications for using in apparel and shoes design, Vlakna a Textil 28 (2021) 105-122.

[11] O. Zakharkevich, I. Poluchovich, S. Kuleshova, J. Koshevko, G. Shvets, A. Shvets. "cloStyler" – Mobile application to calculate the parameters of clothing blocks. IOP Conference Series: Materials Science and Engineering 1031 (2021) art. 012031. https://doi.org/10.1088/1757-899X/1031/1/012031

[12] S. Kuleshova, O. Zakharkevich, J. Koshevko, O. Ditkovska. Development of expert system based on Kansei Engineering to support clothing design process, Vlakna a Textil 3 (2017) 30-41.

[13] F. Rosell-Aguilar. State of the app: A taxonomy and framework for evaluating language learning mobile applications. CALICO Journal 34 (2017) 243–258. https://doi.org/10.1558/cj.27623

[14] C.H. Pereira, R. Terra. A mobile app for teaching formal languages and automata. Computer Applications in Engineering Education 26 (2018) 1742-1752. https://doi.org/10.1002/cae.21944

[15] R. Thomas, M. Fellowes. Effectiveness of mobile apps in teaching field-based identification skills. Journal of Biological Education, 51 (2017) 136-143. https://doi.org/10.1080/00219266.2016.1177573

[16] N. Dik. Udoskonalennya protsesiv proektuvannya zhinochoyi bilyzny v khudozhniy systemi «Ansambl'» z urakhuvannyam stratehiyi rozvytku asortymentu. M.Sc. thesis, Khmelnitskyi National University, 2020.

Keyword Index

About the Editor

Norbert Radek

Norbert Radek is an associate professor at the Kielce University of Technology as a Head of the Department of Exploitation Engineering and Industrial Laser Systems (Faculty of Mechatronics and Mechanical Engineering) and Director of Centre for Laser Technology of Metals. In 2000, he obtained the M.Sc. degree in specialization laser and plasma technologies in the mechanical engineering field at TU Kielce. He defended his Ph.D. degree in the 2006 specialization in surface engineering under the supervision of prof. Bogdan Antoszewski TU Kielce. He obtained his habilitation in welding technologies in 2013 at the University of Žilina. He specializes in surface engineering, especially in laser modifications of the surface layer, and in special varnish coatings, including railway and military ones.

www.ingramcontent.com/pod-product-compliance
Lightning Source LLC
Chambersburg PA
CBHW071325210326
41597CB00015B/1353